New Developments in Mode-Water Research

Atsushi Kubokawa • Shang-Ping Xie
Fumiaki Kobashi • Humio Mitsudera
Editors

New Developments in Mode-Water Research

Dynamic and Climatic Effects

Previously Published in Journal of Oceanography
Volume 68, Number 1, 2012

Springer

Editors
Atsushi Kubokawa
Faculty of Environmental Earth Science
Hokkaido University
Sapporo
Japan

Shang-Ping Xie
International Pacific Research Center
and Department of Meteorology
SOEST, University of Hawaii at Manoa
Honolulu, HI
USA

Fumiaki Kobashi
Faculty of Marine Technology
Tokyo University of Marine
Science and Technology
Tokyo
Japan

Humio Mitsudera
Institute of Low Temperature Science
Hokkaido University
Sapporo
Japan

ISBN 978-4-431-56330-3 ISBN 978-4-431-54162-2 (eBook)
DOI 10.1007/978-4-431-54162-2
Springer Tokyo Heidelberg New York Dordrecht London

Printed on acid-free paper

Springer is part of Springer Science+Business Media (www.springer.com)

J Oceanogr (2012) 68:1–3
DOI 10.1007/s10872-011-0090-8

New developments in mode-water research: an introduction

Shang-Ping Xie · Atsushi Kubokawa · Fumiaki Kobashi · Humio Mitsudera

Published online: 4 December 2011

Two landmark papers appeared in 1969, laying the foundation for this special section in *Journal of Oceanography*. Masuzawa (1969) coined the term "subtropical mode water" (STMW) in the North Pacific, characterizing it as a thick, vertically uniform layer of σ_θ 25.2–25.6 that forms a distinct mode in water volume censuses as a function of temperature, salinity or density. By carefully studying historical hydrographic observations, Uda and Hasunuma (1969) discovered an eastward subtropical countercurrent (STCC) in the southern subtropical gyre where the Sverdrup theory predicts a broad westward current. This countercurrent is peculiar in its direction against both the prevailing northeast trade winds and the westward current underneath. Uda and Hasunuma noted in passing that the STMW resides north of the STCC, splitting the thermocline into a northward-shoaling upper branch and a downward-sloping lower thermocline. Since then, studies of mode water and STCC went their separate ways. The STMW is a favorite subject for water-mass analysts, with other mode waters found in the 1990s in the central and eastern North Pacific. The strong deviation of STCC from the barotropic Sverdrup streamfunction, on the other hand, captured the imagination of dynamists, who proposed a flurry of theories that attempt to explain this peculiar current.

A self-consistent theory of STCC had to wait for 30 years, identifying none other than mode waters as the culprit. As often happens in history, seemingly distinct phenomena—mode water and STCC in this case—turn out to be closely related from a fresh perspective under the light of new theories. Based on a ventilated thermocline model, Kubokawa (1999) showed that mode waters of different densities, formed in the deep winter mixed layer around the Kuroshio Extension east of Japan, will cross their paths on their way south in the subtropical gyre, stack up in the vertical, and shoal the upper thermocline, creating a surface-intensified STCC on the south flank of the mode-water pool. Subsequent observational and modeling studies lend support for this theory of STCC. Thus, mode waters are far more than just thick water masses moving around passively in the subtropical gyre but assume the role of a dynamical agent that causes surface currents to deviate substantially from the Sverdrup dynamics. In retrospect, the dynamical effect of mode water may not be too surprising given that mode waters are characterized by their low potential vorticity, and that potential vorticity is a dynamical tracer conservative following a parcel's trajectory on an isopycnal.

S.-P. Xie (✉)
International Pacific Research Center and Department of Meteorology, SOEST, University of Hawaii at Manoa, Honolulu, HI 96822, USA
e-mail: xie@hawaii.edu

S.-P. Xie
Physical Oceanography Laboratory, Ocean University of China, Qingdao 266100, China

A. Kubokawa
Faculty of Environmental Earth Science, Hokkaido University, Sapporo 060-0810, Japan

F. Kobashi
Faculty of Marine Technology, Tokyo University of Marine Science and Technology, Tokyo 135-8533, Japan

F. Kobashi
Research Institute for Global Change, Japan Agency for Marine-Earth Science and Technology, Yokosuka 237-0061, Japan

H. Mitsudera
Institute of Low Temperature Science, Hokkaido University, Sapporo 060-0819, Japan

Important technological advances for the past decade enabled impressive progress in the description and understanding of mode-water formation, variability, and its dynamical and climatic effects. Argo profiling floats, in synergy with satellite altimetry, revealed the important role played by meso-scale eddies in mode-water formation, with STMW preferentially formed in anticyclonic eddies/meanders of the Kuroshio Extension. High-resolution ocean general circulation models (GCMs) that resolve meso-scale eddies offer a useful tool to study variability and interaction of jets, eddies, and mode waters. Results from satellite observations and coupled ocean–atmosphere GCMs showed that STCC, its variability and change leave visible imprints on, and induce feedback from, the atmosphere.

A symposium entitled "New Developments of Mode-Water Research: Dynamic and Climatic Effects" was convened on the Shinagawa campus of the Tokyo University of Marine Science and Technology on 26 March 2010 as part of the Spring Meeting of the Oceanographic Society of Japan. Twelve talks were presented to an enthusiastic audience of about 100 people. A consensus was reached to showcase recent progress in mode-water research in a compendium of papers in a scientific journal. This special section of 10 papers in *Journal of Oceanography* is the result. It includes both papers presented at the symposium and those submitted in response to a call for special-section contributions. They cover a wide range of topics from the formation, circulation and variability of mode waters, their dynamic effect on surface current, and climate impact.

The special section features three review papers. Oka and Qiu (2012) showcase progress in studying the formation and circulation of mode waters based on recent observations. Kobashi and Kubokawa (2012) give a historical account of the developments in dynamical understanding of STCC, and review recent results on dynamic and climatic effects of mode-water ventilation. Masuzawa (1969) drew an analogy between the North Pacific STMW and the 18°C water of the North Atlantic. Joyce (2012) highlights new perspectives of the North Atlantic 18°C water gained from the field experiment Clivar mode water dynamics experiment" (CLIMODE).

Seven original research articles in the special section present the latest research into mode water and its dynamic and climatic effects. A pair examines mode water formation processes: in a non-eddy resolving ocean GCM, Nishikawa and Kubokawa (2012) study the formation of a mixed layer depth front in winter over the northwest subtropical gyre and its role in mode water subduction; the analysis of Argo observations by Kouketsu et al. (2012) reveals a strong modulation of the winter mixed layer by eddies, with large depths observed inside anti-cyclonic eddies. Toyama and Suga (2012) examines the influence of mode waters on the central water, a water mass in the main pycnocline characterized by a constant density ratio on T–S diagrams.

Two papers examine mode-water effects on interannual variability in eastward surface jets based on an eddy-resolving ocean GCM: Sasaki et al. (2012) suggest that the anomalous southward intrusion of low potential-vorticity water strengthens the Hawaii Lee Countercurrent on interannual timescales, a result they support with Argo observations; Nonaka et al. (2012) show that mode-water ventilation is the dominant mechanism for STCC variability on interdecadal timescales and remains in a significant role on interannual timescales. The observational analysis by Kobashi and Xie (2012) confirms that local wind is important for STCC variability on interannual timescales, and suggests an interaction between local wind and STCC in late spring. This revives the wind-driven idea of Yoshida and Kidokoro (1967), not for the mean STCC but for seasonal and interannual variations. Based on a global warming simulation, Xu et al. (2012) show that mode waters weaken their low potential-vorticity characteristic as ocean stratification increases and the winter mixed layer shallows along the Kuroshio Extension, reducing the Kuroshio Extension's influence on the STCC.

Papers in the special section point to new directions for mode-water research. How do mode waters vary on decadal and longer timescales, and how will they change in response to global warming? What causes mode water variability, and how does it affect the surface circulation and climate of the central and southern subtropical gyre? What are the roles of meso-scale eddies in the formation and dissipation of mode waters, and in their variability and change? We hope that this special section will serve as a sign post in our endeavor to answer these challenging questions.

Acknowledgments We, the editorial team for this special section (A.K., the guest editor in chief), wish to thank all the reviewers for their time and effort essential to maintain the scientific rigor of this special section, T. Saino and T. Hibiya, Chief Editors of the journal, for their editorial guidance. This work is supported by the Japan Agency for Marine-Earth Science and Technology, the National Basic Research Program of China (3012CB955600), the US National Science Foundation (0854365), and Grants-in-Aid for Scientific Research of the Ministry of Education, Culture, Sports, Science and Technology, Japan (22106007, 23340139).

References

Joyce TM (2012) New perspectives on eighteen-degree water formation in the North Atlantic. J Oceanogr 68. doi:10.1007/s10872-011-0029-0

Kobashi F, Kubokawa A (2012) Review on North Pacific Subtropical Countercurrents and Subtropical Fronts: role of mode waters in ocean circulation and climate. J Oceanogr 68. doi:10.1007/s10872-011-0083-7

Kobashi F, Xie SP (2012) Interannual variability of the North Pacific Subtropical Countercurrent: role of local ocean-atmosphere interaction. J Oceanogr 68. doi:10.1007/s10872-011-0048-x

Kouketsu S, Tomita H, Oka E, Hosoda S, Kobayashi T, Sato K (2012) The role of meso-scale eddies in mixed layer deepening and mode water formation in the western North Pacific. J Oceanogr 68. doi:10.1007/s10872-011-0049-9

Kubokawa A (1999) Ventilated thermocline strongly affected by a deep mixed layer: a theory for subtropical countercurrent. J Phys Oceanogr 29:1314–1333

Masuzawa J (1969) Subtropical mode water. Deep Sea Res 16:463–472

Nishikawa S, Kubokawa A (2012) Mixed layer depth front and subduction of low potential vorticity water under seasonal forcings in an idealized OGCM. J Oceanogr 68. doi:10.1007/s10872-011-0086-4

Nonaka M, Xie SP, Sasaki H (2012) Interannual variations in low potential vorticity water and the subtropical countercurrent in an eddy-resolving OGCM. J Oceanogr 68. doi:10.1007/s10872-011-0042-3

Oka E, Qiu B (2012) Progress of North Pacific mode water research in the past decade. J Oceanogr 68. doi:10.1007/s10872-011-0032-5

Sasaki H, Xie SP, Taguchi B, Nonaka M, Hosoda S, Masumoto Y (2012) Interannual variations of the Hawaiian Lee Countercurrent induced by low potential vorticity water ventilation in the subsurface. J Oceanogr 68. doi:10.1007/s10872-011-0074-8

Toyama K, Suga T (2012) Roles of mode waters in formation and maintenance of Central Water in the North Pacific. J Oceanogr 68. doi:10.1007/s10872-011-0040-5

Uda M, Hasunuma K (1969) The eastward subtropical countercurrent in the western North Pacific Ocean. J Oceanogr Soc Jpn 25:201–210

Xu LX, Xie SP, Liu Q, Kobashi F (2012) Response of the North Pacific Subtropical Countercurrent and its variability to global warming. J Oceanogr 68. doi:10.1007/s10872-011-0031-6

Yoshida K, Kidokoro T (1967) A subtropical countercurrent in the North Pacific—an eastward flow near the Subtropical Convergence. J Oceanogr Soc Jpn 23:88–91

Contents

J Oceanogr (2012) 68:5–20
DOI 10.1007/s10872-011-0032-5

SPECIAL SECTION: REVIEW

New developments in mode-water research: Dynamic and climatic effects

Progress of North Pacific mode water research in the past decade

Eitarou Oka · Bo Qiu

Received: 19 February 2011 / Revised: 27 April 2011 / Accepted: 28 April 2011 / Published online: 4 June 2011

Abstract This article reviews the progress in research on North Pacific mode waters of the past decade from the physical oceanographic perspective. The accumulation of satellite altimeter sea surface height data, the rapid growth of the Argo profiling float array, and the advancement in eddy-resolving ocean general circulation models have greatly improved the traditional views on the mode waters that were formed prior to the 1990s based on the analyses of historical temperature/salinity data. Areas where significant progress was made include: (1) descriptions of the mode waters' distributions and properties with fine spatial scales, particularly in their formation regions in winter where observational data had been insufficient; (2) clarifications of the mode waters' formation and subduction processes relating to the large-scale mean circulation, as well as to the time-varying mesoscale eddy field; (3) impacts of the mode waters' circulation and dissipation processes on the climate and biogeochemical processes; and (4) dynamic versus thermodynamic causes underlying the mode waters' decadal changes. In addition to the review, future directions for mode water research are also presented.

Keywords Mode waters · North Pacific · Physical oceanography · Seasonal to decadal variability · Eddy modifications · Climate and biogeochemical impacts

E. Oka (✉)
Atmosphere and Ocean Research Institute,
The University of Tokyo, Kashiwa, Chiba 277-8564, Japan
e-mail: eoka@aori.u-tokyo.ac.jp

B. Qiu
Department of Oceanography,
University of Hawaii at Manoa, Honolulu, HI 96822, USA

1 Introduction

In the subtropical gyres and part of the subpolar gyres of the world oceans, there are distinct water masses characterized by a nearly vertically homogeneous layer or low potential vorticity (PV) lying just above or within the permanent pycnocline. These water masses, called "mode waters," are formed as a deep mixed layer on the warm side of a strong current or front in late winter because of convective mixing, and are left in the subsurface as a pycnostad after being capped by the seasonal pycnocline in spring. Some part of the formed mode waters is entrained into the mixed layer in the following winter, modulating the sea surface temperature, while the remaining part enters the permanent pycnocline (this process is called subduction) and is then advected to spread over a much wider area than its formation regions, carrying temperature, salinity, and PV anomalies. The formation, circulation, and dissipation of mode waters and their variability are related to ocean-atmosphere interaction and various upper-ocean dynamic and thermodynamic processes on a wide range of time scales. Moreover, they are believed to play an important role in biogeochemical processes such as the oceanic uptake of atmospheric CO_2 (e.g., Bates et al. 2002) and the nutrient cycling in the oligotrophic subtropical gyres (e.g., Palter et al. 2005; Krémeur et al. 2009; Sukigara et al. 2011).

The subtropical gyre in each ocean basin contains three types of mode waters formed in the western, eastern, and poleward parts of the gyre, except that the South Hemisphere gyres share the Subantarctic Mode Water formed just north of their common southern boundary, the subantarctic front (Hanawa and Talley 2001). In the North Pacific, a thermostad of 16–18°C lying in the upper permanent thermocline in the northwestern part of the subtropical gyre was identified and named the subtropical mode water (STMW) by Masuzawa (1969) because it produces a mode in the volume distribution on the temperature-salinity diagram for the subtropical gyre.[1] Since then STMW was little explored for nearly 2 decades, but drew attention again in the late 1980s when the Ocean Mixed Layer Experiment (Toba et al. 1991) was implemented as one of the Japanese activities in the World Climate Research Program. Possible roles of STMW as a heat reservoir memorizing the wintertime ocean-atmosphere interaction have inspired a number of studies based on shipboard observations and data analyses that helped clarify the formation, circulation, and long-term variation of STMW.

STMW is formed as a deep winter mixed layer just south of the Kuroshio and the Kuroshio Extension (KE) between ~132°E and near the dateline, where warm surface water supply from the Kuroshio/KE underlying the strong East Asian winter monsoon results in large oceanic buoyancy loss in winter (Hanawa 1987; Hanawa and Hoshino 1988; Suga and Hanawa 1990; Bingham 1992). Due to the downstream cooling of the Kuroshio/KE, colder and denser STMW tends to be formed progressively in the farther eastern part of the formation region (Suga and Hanawa 1990; Bingham 1992). After spring, the main body of STMW, capped by the seasonal pycnocline and isolated from atmospheric contact, is advected southwestward by the "Kuroshio Countercurrent" and is subducted into the permanent pycnocline (Hanawa 1987; Suga et al. 1989; Suga and Hanawa 1990, 1995a; Bingham 1992).

The volume and properties of STMW formed every winter exhibit significant year-to-year variations (Hanawa 1987; Suga et al. 1989). On interannual time scales, thicker and colder STMW tends to be formed in years with stronger winter monsoons because of larger oceanic heat loss and increased southward Ekman transport (Suga and Hanawa 1995b; Yasuda and Hanawa 1997, 1999; Taneda et al. 2000; Hanawa and Kamada 2001; Hanawa and Yoritaka 2001). On decadal to interdecadal time scales, on the other hand, intensification of the westerlies leads to the formation of warmer STMW through gyre spin-up and increased warm water advection by the Kuroshio (Hanawa and Kamada 2001). The local existence of STMW south of Japan is also influenced by path variations of the Kuroshio (Yoshida 1964; Kawabe 1995). The westward advection of STMW from the region southeast of Japan is blocked when the Kuroshio takes a large meander path (Bingham et al. 1992; Suga and Hanawa 1995b, c).

In the late 1990s, the other two mode waters in the North Pacific were identified in succession, as expected from isopycnal PV maps presented in an earlier study (Talley 1988). Nakamura (1996) and Suga et al. (1997) separately described a thermostad of 8.5–11.5 and 10–13°C lying in the lower permanent pycnocline in the central part of the subtropical gyre using different climatological data and termed it the central mode water (CMW). They inferred that CMW is formed between KE and its northern bifurcation (Suga et al. 1997) and between the northern bifurcation and the subarctic front[2] (Nakamura 1996), which renders CMW much colder, fresher, and denser than STMW. The eastern subtropical mode water (ESTMW) was identified by Hautala and Roemmich (1998) as a weak thermostad of 16–22°C lying in the upper permanent pycnocline in the eastern part of the subtropical gyre. It is formed as a relatively deep winter mixed layer southwest of the subarctic front, which extends in the northwest-southeast direction in this region (Yuan and Talley 1996; Suga et al. 2004). It is often stratified in terms of temperature and salinity in a compensating way within its pycnostad, possibly because of cross-frontal intrusions (Hautala and Roemmich 1998; Sprintall and Roemmich 1999).

The two new mode waters formed and advected anticyclonically in the central to eastern part of the gyre contrast with the classical STMW that is largely confined to the Kuroshio recirculation region in the northwestern part of the gyre. They are interesting from two points of view (Hautala and Roemmich 1998; Hanawa and Talley 2001). First, their low PV signature is a good tracer of the ventilation process in the subtropical gyre. Second, if their volume and properties formed every winter change from year to year, their advection in the permanent pycnocline might transport temperature and salinity anomalies from the mid-latitude ocean surface to the equatorial upwelling region and cause interdecadal climate variability, as hypothesized by Hanawa (1996) and Gu and Philander

[1] The term "mode water," introduced by Masuzawa (1969), was subsequently applied to any thick, broadly distributed, near-surface layer characterized by low PV (Hanawa and Talley 2001). Therefore, mode waters nowadays are not necessarily characterized by uniformity in terms of temperature and salinity. In other words, temperature and salinity might be stratified in a compensating way within a pycnostad of mode waters.

[2] In this review, the term "subarctic front" is used for a density-compensating front between the warmer, saltier water in the subtropics and the colder, fresher water in the subarctics, characterized by the outcrop of the 33.0–33.8 isohalines (Roden 1970, 1972; Zhang and Hanawa 1993; Yuan and Talley 1996), as in the literature of physical oceanography. This front often separates into two or more fronts, particularly in the eastern North Pacific, and is also called the subarctic frontal zone. In the literature of fisheries oceanography (e.g., Favorite et al. 1976; Yasuda 2003), the term "subarctic front" is referred to the front at the southern boundary of subsurface temperature inversions characterizing the subarctics, represented by the 4°C isotherm standing almost vertically below the 100-m depth (Uda 1963; Favorite et al. 1976). This front is called the "polar front" in this review, as in the literature of physical oceanography.

(1997). An initial attempt using historical temperature data (Yasuda and Hanawa 1997) demonstrated that CMW becomes colder during the decade after the 1976/1977 regime shift (e.g., Nitta and Yamada 1989; Trenberth 1990) than the previous decade, probably because of larger oceanic heat loss and increased southward Ekman transport associated with the intensification of the westerlies. On the other hand, Hautala and Roemmich (1998) did not find a substantial change in the ESTMW volume and temperature between 1970–1979 and 1991–1997, although the lack of data during the intervening period might hinder the detection of ESTMW changes associated with the 1976/1977 regime shift.

The aforementioned studies on the North Pacific mode waters prior to around 2000 were comprehensively reviewed by Hanawa and Talley (2001). In the past decade following their review, mode water research has been advanced greatly because of the rapid development of research tools. One of the biggest factors is the accumulation of satellite-based data such as sea surface height and the development of subsurface observing systems represented by Argo (Roemmich et al. 2001). Compared to the historical data until the 1990s that had insufficient coverage away from coasts and were spatially limited to a number of repeat sections, the unprecedented Argo array of 3,000 profiling floats provides us now with temperature and salinity data down to 2,000-dbar depth at 10-day intervals with a horizontal resolution of 3° in latitude and longitude (Roemmich et al. 2009; Freeland et al. 2010). This makes it possible to take snapshots of the entire structure of mode waters at various times of year, providing new insights on their formation, circulation, dissipation, and temporal variations. In addition, the long-term satellite-based data have revealed the variability of currents/fronts and eddy fields that are closely related to mode water variability, particularly in the energetic Kuroshio-Oyashio Extension region (e.g., Qiu 2002) where STMW and CMW are formed (Qu et al. 2002). Another important factor is the introduction and progress of modeling studies of the North Pacific mode waters. As the model resolution improved from non-eddy-permitting to eddy-permitting and further to eddy-resolving (e.g., Hasumi et al. 2010; Masumoto 2010), the structures of the currents/fronts and the associated mode waters became reproduced more realistically, helping to understand the underlying dynamics, particularly on mesoscales that are not fully resolved by observations. It is also worth noting that while most of the studies until the 1990s were performed by a single research group at Tohoku University, Japan, their activities have stimulated many scientists to participate in the mode water research, generating a variety of views and approaches.

This paper reviews the progress of physical oceanographic research on the North Pacific mode waters in the past decade following the review of Hanawa and Talley (2001). Readers may refer to the review of Joyce (2011) in this issue, and Kelly et al. (2010), for the mode water research progress in the North Atlantic and the comparison with the North Pacific. Research progress on the formation and subduction of the North Pacific mode waters is described in Sect. 2, and that on the other processes such as their circulation, modification, and dissipation is summarized in Sect. 3. After reviewing the studies on their decadal variability in Sect. 4, we present future directions for the mode water research in Sect. 5.

2 Formation and subduction

In the classical ventilated thermocline theory (Luyten et al. 1983), water in the subtropical gyre is pushed down (or subducted) from the base of the Ekman layer into the permanent pycnocline because of Ekman pumping and is subsequently transported by the anticyclonic Sverdrup flow along isopycnal surfaces, conserving its PV. Practically, water is subducted not from the base of the Ekman layer but from that of the late winter mixed layer (Stommel 1979), which generally shoals equatorward within the subtropical gyre. Consequently, lateral induction through the sloping mixed layer base increases the subduction rate as compared to the estimation from Ekman pumping alone (Williams 1989, 1991; Marshall et al. 1993; Huang and Qiu 1994; Qiu and Huang 1995).

The PV (Q) of water subducted at the base of the winter mixed layer is expressed as,

$$Q = \frac{f}{\rho_0} \frac{\mathbf{u}_b \cdot \nabla \rho_m}{w_b + \mathbf{u}_b \cdot \nabla h} \quad (1)$$

where f is the planetary vorticity, ρ_0 the reference density, ρ_m the mixed layer density, h the mixed layer depth (MLD), $\mathbf{u}_b$ and w_b the horizontal and vertical velocities at the base of the winter mixed layer, respectively, and ∇ is the horizontal differential operator (Williams 1989, 1991). The distribution of $-w_b$ (called the vertical pumping term) is relatively uniform over the North Pacific subtropical gyre (Huang and Qiu 1994), and the water acquires low Q if $-\mathbf{u}_b \cdot \nabla h$ is large or $-\mathbf{u}_b \cdot \nabla \rho_m$ is small. In other words, mode waters are formed in the case of large lateral induction associated with an MLD front (Kubokawa and Inui 1999; Kubokawa 1999; Nishikawa and Kubokawa 2007) or small density advection associated with a small downstream gradient of the mixed layer density (Fig. 1).

In the early 2000s, the mechanisms of large-scale formation and subduction of the three mode waters in the North Pacific were examined, based mainly on the subduction theory including Eq. 1 and by non-eddy-permitting

Fig. 1 Schematic illustrating the two mechanisms of mode water subduction: **a** large lateral induction and **b** small density advection. *Solid* and *dashed lines* denote isopycnals and the base of the winter mixed layer, respectively

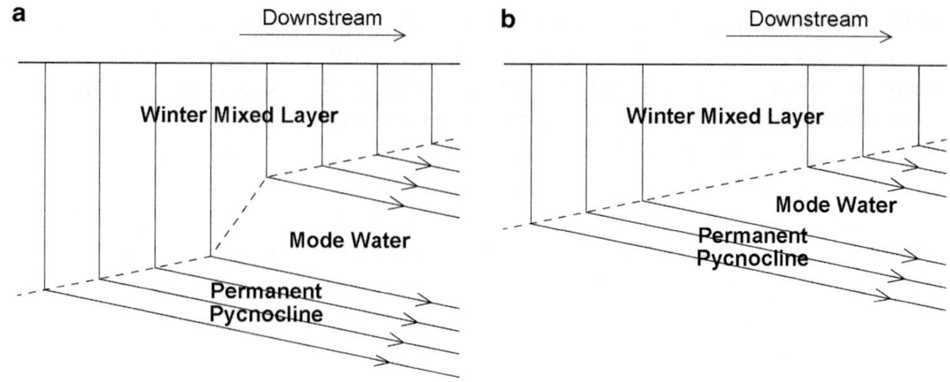

models with a horizontal resolution of 1° or 2°. The STMW formation region is characterized by the deep winter mixed layer south of the Kuroshio and KE accompanied by a sharp MLD front at its southern end (Suga and Hanawa 1990; Oka and Suga 2003), and STMW is subducted into the permanent pycnocline by crossing this front to the south because of large lateral induction (Kubokawa and Inui 1999; Tsujino and Yasuda 2004). Similarly, CMW enters the permanent pycnocline by crossing an MLD front at the eastern end of the formation region, namely the deep winter mixed layer north of KE (Xie et al. 2000; Qu et al. 2002; Tsujino and Yasuda 2004; Hosoda et al. 2004). In contrast, the ESTMW formation region is characterized by a weak winter MLD maximum centered at 30°N, 140°W that is much shallower than those in the STMW and CMW formation regions. This maximum is formed in spite of relatively uniform, weak wintertime cooling over the eastern North Pacific because stratus clouds (Ladd and Thompson 2000, 2001) and the saline Ekman transport from the south (Toyoda et al. 2004) preclude the development of the seasonal pycnocline in summer and precondition the deepening of winter mixed layer. The ESTMW formation region is also characterized by a small horizontal gradient of the winter mixed layer density because the horizontal changes of the mixed layer temperature and salinity, between warmer, saltier water to the southwest and colder, fresher water to the northeast, compensate each other. Consequently, low PV of ESTMW is generated by the small density advection mechanism (Xie et al. 2000; Hosoda et al. 2001; Ladd and Thompson 2001). These subduction mechanisms proposed by the modeling studies are supported by a winter mixed layer climatology constructed by Suga et al. (2004) for STMW and ESTMW, but not for CMW. In their climatology, the winter MLD gradually decreases eastward in the eastern part of the CMW formation region, which implies that the water is subducted mainly because of small density advection (Suga et al. 2004, 2008). This discrepancy between the models and the observations is probably because the past models tend to produce too deep winter mixed layers in the CMW formation region, particularly in its eastern part (Ladd and Thompson 2001; Tsujino and Yasuda 2004).

As the horizontal resolution of ocean general circulation models improved from non-eddy-permitting to eddy-permitting, they better reproduced the frontal structures in the Kuroshio-Oyashio Extension region that are essential for the formation of STMW and CMW. Qu et al.'s (2002) model with a resolution of 1/4° (longitude) × 1/4° (latitude) reproduced the KE and subarctic fronts, and the formation of STMW and CMW south of the respective fronts. Furthermore, Tsujino and Yasuda's (2004) model with a resolution of 1/4° × 1/6° reproduced the northern bifurcation of KE (Kuroshio bifurcation) as well and captured the formation of two varieties of CMW, the lighter variety (L-CMW) south of the Kuroshio bifurcation front and the denser variety (D-CMW) south of the subarctic front. Although the Kuroshio bifurcation front in the model seems somewhat too strong (Tsujino and Yasuda 2004), the formation of two varieties of CMW is actually observed in a synoptic hydrographic section along 179°E (Mecking and Warner 2001) and a repeat section along 165°E (Oka and Suga 2005).

In the latter 2000s, accumulation of in situ observation data including those from Argo profiling floats enables us to capture more detailed, unsmoothed structures of mode waters at various times of year, particularly at higher latitudes in winter where an insufficient number of historical data had existed because of unfavorable weather conditions for shipboard measurements. An intense hydrographic survey carried out northeast of Japan in July 2002 revealed a new type of pycnostad with $\theta = 5$–7°C, $S = 33.5$–33.9, and $\sigma_\theta = 26.5$–26.6 kg m^{-3} (θ and σ_θ are potential temperature and density; S is salinity) around 43°N, 160°E south of the subarctic front and the polar front[3] (Saito et al. 2007), as suggested by earlier studies[4] (Yasuda 2003; Oka

[3] An intense hydrographic survey conducted in spring 2003 indicates that the subarctic front and the polar front coincide with each other in this longitude range (Eitarou Oka, personal communication 2011).

[4] In Yasuda (2003), TRMW is referred to as "Dense Central Mode Water."

and Suga 2005). This pycnostad is colder, fresher, and a little denser than D-CMW, and was named the transition region mode water (TRMW) by Saito et al. (2007) because its formation region corresponds to the transition region described in the literature of fisheries oceanography (e.g., Favorite et al. 1976; Yasuda 2003). TRMW is often stratified in terms of temperature and salinity within its pycnostad as in the case of ESTMW, probably because of isopycnal intrusion that characterizes the transition region (Saito et al. 2007).

The TRMW distribution region presented by Saito et al. (2007) corresponds to the region where the deepest winter mixed layer in the North Pacific appears in previous climatologies (Huang and Qiu 1994; Ladd and Thompson 2000; Suga et al. 2004), although MLD there might be overestimated because the smoothing process in the climatologies mixes different water types across the subarctic front, causing artificial cabbeling near the sea surface (Ohno et al. 2004; Suga et al. 2004). This MLD maximum is possibly generated for the following reasons. First, it lies in the "stability gap" described by Roden (1970) and Yuan and Talley (1996), a zonal band of lateral minimum in the vertical stability located south of the subarctic front (Ladd and Thompson 2000; Suga et al. 2004; Saito et al. 2007). Second, the quasi-stationary jet associated with the polar front (Isoguchi et al. 2006) brings warmer and saltier water from the south, enhancing the oceanic heat loss and making the surface water denser in winter (Saito et al. 2007). Actually, recently developed high-resolution surface heat flux data and Argo float data revealed locally enhanced oceanic heat loss and deep mixed layer formation just southeast of the polar front in winter (Hiroyuki Tomita, personal communication, 2011).

The entire structure of the CMW and TRMW formation regions and the subduction mechanism of these waters were presented by a recent study using Argo float data during 2003–2008 (Oka et al. 2011a). To the north of KE, two zonally elongated regions of deep winter mixed layer extend along 33°–39°N and 39°–43°N from the east coast of Japan ($\sim$142°E) to $\sim$160°W. The southern region corresponds to the formation region of L-CMW, while the northern region to that of D-CMW and TRMW, in the western (eastern) part of which TRMW (D-CMW) is predominantly formed (Fig. 2). Both regions extend beyond the longitude range of the Kuroshio bifurcation front to the west and east, suggesting that the two regions are not separated by the bifurcation front, but rather enhance it, because these two adjacent low PV waters should accompany sharply inclined isopycnals in between. From the eastern part of both regions east of 170°E where the winter mixed layer becomes gradually shallower, warmer, and lighter to the east/downstream, L-CMW with $\theta = 11$–15°C, $S = 34.3$–34.6, and $\sigma_\theta = 25.7$–26.2 kg m^{-3} and D-CMW and TRMW (mostly the former) with $\theta = 7$–11°C, $S = 33.6$–34.2, and $\sigma_\theta = 26.1$–26.4 kg m^{-3} are subducted into the permanent pycnocline because of small density advection as inferred by the previous climatological studies (Suga et al. 2004, 2008). On the other hand, D-CMW and TRMW (mostly the latter) with $\sigma_\theta = 26.5$–26.6 kg m^{-3} formed only in the western part of the formation region

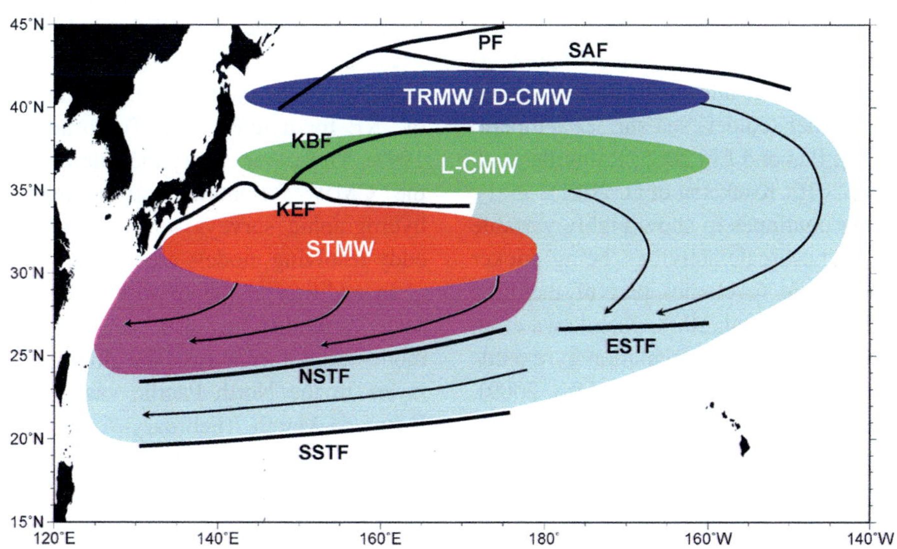

Fig. 2 Schematic illustrating the relationship between the mode waters and frontal structures in the western to central North Pacific. *PF* polar front, *SAF* subarctic front, *KBF* Kuroshio bifurcation front, *KEF* Kuroshio Extension front. NSTF, SSTF, and ESTF are the northern, southern, and eastern subtropical fronts identified in Kobashi et al. (2006), respectively. *Red*, *green*, and *blue ovals* denote the formation regions of STMW, L-CMW, and TRMW/D-CMW, respectively. *Pink* (*light blue*) *shadings* with *arrows* indicate the spreading of STMW (L-CMW and D-CMW) after subduction

where winter MLD gradually increases eastward/downstream are entrained into the mixed layer in the following winter and cannot be subducted into the permanent pycnocline, which likely explains the large discrepancy in the temperature-salinity relations between the winter mixed layer and the permanent pycnocline in this σ_θ range, argued by Suga et al. (2008). The downstream thickening of CMWs in the western part of their formation regions preconditions their permanent subduction in the eastern part, as speculated by earlier studies (Ladd and Thompson 2000; Mecking and Warner 2001).

In addition to the large-scale mode water formation and subduction mentioned above, effects of mesoscale eddies (Marshall 1997; Hazeleger and Drijfhout 2000) and the anticyclonic recirculation gyre, both of which characterize the highly variable and nonlinear KE system (e.g., Qiu 1999, 2002; Ebuchi and Hanawa 2001), on the STMW formation and subduction have been clarified using Argo float data combined with altimetric sea surface height data (Uehara et al. 2003; Pan and Liu 2005; Qiu et al. 2006; Oka 2009; Oka et al. 2011b; Kouketsu et al. 2011) and state-of-the-art, eddy-resolving models with a horizontal resolution of 1°/10° or better (Rainville et al. 2007; Nishikawa et al. 2010). The recirculation gyre consists of several anticyclonic circulations associated with crests of the Kuroshio/KE meander (e.g., Kawai 1972), in each of which thick STMW with a characteristic temperature is formed in winter, possibly because water is effectively cooled and homogenized there (Qiu et al. 2006; Rainville et al. 2007; Oka 2009; Nishikawa et al. 2010). The formation process is also modulated by prevailing mesoscale eddies; thicker (thinner) STMW tends to be formed in anticyclonic (cyclonic) eddies, in which background stratification is weaker (stronger) in association with the deeper (shallower) permanent pycnocline and oceanic heat loss is larger (smaller) because of warmer (colder) sea surface temperature (Uehara et al. 2003; Pan and Liu 2005; Rainville et al. 2007; Nishikawa et al. 2010; Kouketsu et al. 2011).

After spring, STMW continues to show highly variable distributions created by mesoscale activities, being thicker (thinner) in the anticyclonic (cyclonic) part of the flow (Uehara et al. 2003; Rainville et al. 2007; Nishikawa et al. 2010; Oka et al. 2011b). While the recirculation gyre tends to confine STMW in its formation region (Oka 2009), anticyclonic eddies transport part of it southward through the southern boundary of the region across the westward mean flow because of southward migration of eddies that trap STMW, as observed by Takikawa et al. (2005), and to eddy mixing (Rainville et al. 2007; Nishikawa et al. 2010). This also causes STMW to cross the winter MLD front and to be thereby subducted into the permanent pycnocline. Such eddy subduction contributes to approximately half of the total STMW subduction rate (Nishikawa et al. 2010). Thus, the STMW subduction revealed by Argo float data and high-resolution models is quite different from the traditional climatology-based picture that the thick body of STMW is advected by the southwestward mean flow as the season progresses (Sect. 1).

Subduction across the mean flow has also been reported for CMW. In synoptic meridional sections in the formation region, L-CMW (D-CMW) occasionally intrudes southward into the KE (Kuroshio bifurcation) front or exists as isolated patches south of the front (Mecking and Warner 2001; Oka and Suga 2005). As an extreme case, a high-resolution hydrographic survey southeast of Japan found a D-CMW patch trapped in a subsurface mesoscale eddy approximately 1,000 km south of KE in fall, which was estimated to be subducted only a half year ago in association with the instability of the KE front (Oka et al. 2009). Such cross-frontal subduction could modulate the CMW distribution determined by the large-scale subduction by the mean flow, particularly in the density range of $\sigma_\theta = 26.5$–26.6 kg m^{-3}, which is not ventilated through the large-scale CMW subduction (Oka et al. 2011a).

Furthermore, the CMW formation can also be dominated by mesoscale activities, as inferred from intermittent CMW distributions based on Argo float data from a single year (Bingham and Suga 2006) and from eddy-resolving numerical models (Nishikawa et al. 2010). In fact, a recent composite analysis of Argo float data and altimetric sea surface height data revealed that roughly half of winter mixed layers deeper than 150 dbar in the CMW formation region are formed in anticyclonic eddies (Kouketsu et al. 2011). Past shipboard observations also demonstrated that inside anticyclonic eddies (or warm-core rings) pinched off from the KE, thick STMW is modified to CMW or TRMW because of winter cooling and interaction with the ambient water while the eddies are migrating in the region north of the KE for a few years (Tomosada 1986; Yasuda et al. 1992). To further clarify the role of mesoscale activities in the CMW formation and subduction, high-resolution hydrographic surveys north of KE in conjunction with eddy-resolving models are strongly desired.

In addition to the mesoscale variability, shorter timescale variability has also been demonstrated to affect the mode water formation. The timing of the deepest mixed layers in the North Pacific varies among regions, being February–March (February–April) in the STMW and L-CMW (D-CMW and TRMW) formation regions (Oka et al. 2007; Ohno et al. 2009). The mixed layer deepening prior to these "late winter" months is dominated by episodic events on time scales of several days that are related to the passages of synoptic low-pressure disturbances, as revealed by in situ buoy observations and high-resolution numerical model simulations (Qiu et al. 2004; Rainville et al. 2007; Cronin et al. 2008; Jensen et al. 2011). The importance of synoptic-scale atmospheric forcing is further supported by short-term MLD variations observed by profiling floats (Yoshida and Hoshimoto 2006; Oka et al. 2007), with the use of a relatively small value of temperature and density as a threshold to determine MLDs (e.g., de Boyer Montégut et al. 2004).

3 Circulation and dissipation

The development of observations and models in the past decade enabled us to investigate the circulation, modification, and dissipation of the mode waters that had been inadequately studied until the 1990s. By using data from 20 profiling floats that were concentratively deployed in the anticyclonic recirculation gyre south of the first quasi-stationary meander of the KE under the Kuroshio Extension System Study project (KESS; http://www.uskess.org), Qiu et al. (2006) explored the seasonal evolution of the vertical structure of STMW in its formation region in 2004. As the seasonal pycnocline develops and gradually thickens after April, the underlying STMW is eroded from its top, becoming thinner, whereas its bottom remains largely at a constant depth of $\sim$500 dbar. From a PV budget analysis for this STMW under the assumptions that horizontal advection is negligible (because the KE is in a stable state in 2004, as explained in Sect. 4) and that the vertical eddy diffusivity (K_v) at the lower boundary of STMW is 0.1×10^{-4} m^2 s^{-1} as previously observed in the permanent pycnocline (e.g., Gregg and Sanford 1980; Ledwell et al. 1993), K_v at the upper boundary of STMW is estimated to be 2–5 $\times$ 10^{-4} m^2 s^{-1}, which is an order of magnitude larger than that in the permanent pycnocline. A recent dissolved oxygen budget analysis using data from a profiling float equipped with a fluorometer and an oxygen sensor estimated a similar K_v value of 1.7×10^{-4} m^2 s^{-1} near the upper boundary of STMW (Sukigara et al. 2011). Such large vertical diffusivity possibly stems from the sharp drop in stratification from the seasonal pycnocline to the STMW pycnostad, which acts as a barrier for the downward transmission of internal gravity waves generated in the surface mixed layer and contributes to the enhanced eddy mixing as these waves are reflected at the upper boundary of STMW (Qiu et al. 2006). The large vertical diffusivity has important implications not only for the modulation of sea surface temperature in the STMW formation region (e.g., Hanawa and Sugimoto 2004), but also for the nutrient supply from the subsurface to the oligotrophic surface layer in the subtropical gyre (e.g., Mori et al. 2008; Sukigara et al. 2011).

Meanwhile, shipboard observations using a microstructure profiler indicated that K_v is relatively uniform between the top and bottom of STMW, and is much smaller than the above estimates, being 10^{-6} to 10^{-5} m^2 s^{-1} in summer 2006 and 10^{-7} to 10^{-5} m^2 s^{-1} in winter 2007 (Mori et al. 2008). Why is there such a large discrepancy between the budget-based and observed values of K_v? The most plausible explanation is that like the deepening of the mixed layer in the fall and early winter seasons (Sect. 2), erosion of STMW is dominated by episodic atmospheric disturbances on time scales of several days, during which K_v becomes a few orders of magnitude larger but is missed by shipboard microstructure profiler measurements due to rough sea conditions. In fact, an analysis of data from an acoustic Doppler current profiler moored in the STMW formation region under the KESS project revealed that following each wind storm event during July–December 2004, the upper ocean inertial velocity shear tends to increase in amplitude and to propagate downward to the top layer of STMW (Luc Rainville, personal communication, 2010). Another possible cause for the discrepancy is interannual variation of K_v in STMW resulting from the changing density structure of STMW. Since thinner STMW is formed and more strongly dissipated in 2006 than in 2004 (Sect. 4), STMW is less vertically homogeneous during 2006, which is likely to hinder the intensive breaking of downward-propagating inertial waves at its upper boundary. The third possible cause is the importance of horizontal processes that were assumed to be negligible in Qiu et al. (2006). After spring, STMW patches originating from different regions and having different densities are stirred by mesoscale eddies and circulations and interleave with each other, and this helps destroy the vertically uniform structure of STMW, particularly when the KE is in an unstable state (Oka et al. 2011b; Sect. 4).

As noted in Sect. 2, part of STMW is transported southward from its formation region, being subducted into the permanent pycnocline. Subsequently, it is advected southwestward, carrying temperature and salinity anomalies, and reaches the western boundary one to several years later (Oka 2009). A portion of this "old" STMW is then transported back to the formation region by the Kuroshio/KE and is obducted onto the winter mixed layer, modulating the sea surface temperature there (Qiu and Huang 1995; Qiu 2002; Endoh et al. 2006; Liu and Hu 2007). Besides this anticyclonic circulation of STMW, eastward advection of STMW by the KE from the region southeast of Japan to the east of the dateline over a period of 1 year was suggested from a lag correlation analysis of winter sea surface temperature anomalies (Sugimoto and Hanawa 2005a).

The Kuroshio and KE influence not only STMW, but also CMW and TRMW to the north. Saito et al. (2011) analyzed quasi-Lagrangian observations by an isopycnal Argo float to indicate that TRMW becomes warmer and saltier after its formation in late winter, and is modified to

D-CMW within 1 year. They attributed this change partly to double-diffusive salt-finger convection between TRMW and the warmer, saltier, and equally dense surface water of the Kuroshio origin that is advected from the south. This scenario is consistent with the observed facts that the upper portion of TRMW is characterized by a high Turner angle implying active salt-finger convection and by inhomogeneity in terms of temperature and salinity relative to the lower portion (Toyama and Suga 2010, 2011) and that the proportion of TRMW (D-CMW) decreases (increases) eastward/downstream in their formation region (Oka et al. 2011a; Saito et al. 2011). Thus, TRMW is likely a transient feature in the TRMW/D-CMW formation region and can be regarded as the most upstream portion of D-CMW.

After L-CMW and D-CMW are subducted from the eastern part of the formation regions (Oka et al. 2011a), they are advected anticyclonically in the subtropical gyre toward the western boundary (Suga et al. 2004; Tsujino and Yasuda 2004; Oka et al. 2011a). As they circulate southwestward in the southern part of the gyre, D-CMW takes a more zonal path than L-CMW because of the beta spiral effect (Stommel and Schott 1977; Kubokawa 1999), which makes the two waters overlap each other (Fig. 2; Oka and Suga 2005; Kobashi et al. 2006; Xie et al. 2011). This stack of CMWs in the western to central part of the gyre, as well as STMW in the western part, has been shown to generate the Subtropical Countercurrent (Uda and Hasunuma 1969) and the associated subtropical fronts (Kubokawa 1997, 1999; Kubokawa and Inui 1999; Aoki et al. 2002; Kobashi et al. 2006; Yamanaka et al. 2008; Xie et al. 2011), whose thermal effects on the atmosphere locally enhance precipitation and generate positive wind stress curl anomalies in winter to spring (Kobashi et al. 2008; Xie et al. 2011). This is a new dynamic role of the mode waters in climate, which contrasts with the traditional role as a heat reservoir memorizing the wintertime ocean-atmosphere interaction (Xie et al. 2011; see also the review of Kobashi and Kubokawa 2011 in this issue for details). Another interesting feature is that as L-CMW and D-CMW circulate anticyclonically, their pathways shift to the inner side of the subtropical gyre represented by geostrophic streamlines (Oka and Suga 2005). This is consistent with the modeling study of Nonaka and Xie (2000) that demonstrated that temperature anomalies subducted from the northern part of the gyre in association with the subduction of low PV water circulate more zonally in the southern part of the gyre as high baroclinic mode Rossby waves than a passive tracer subducted from the same location, reaching the western boundary at a latitude higher than that of the North Equatorial Current bifurcation (e.g., Qiu and Chen 2010a), and are therefore unable to flow into the equatorial region.

ESTMW is also advected southwestward from the formation region along the outer path in the subtropical gyre (Hautala and Roemmich 1998), but its low PV signal is rapidly dissipated to the downstream (Ladd and Thompson 2001; Hosoda et al. 2001; Suga et al. 2004; Sugimoto and Hanawa 2005b). This is because (1) ESTMW is much thinner than the other mode waters in the North Pacific and is characterized by a weak PV minimum (Suga et al. 2004; Sugimoto and Hanawa 2007); (2) owing to the subsurface temperature/salinity stratification in the eastern part of the subtropical gyre, salt-finger convection occurs vigorously between ESTMW and the underlying permanent pycnocline water, which erodes ESTMW from its bottom after spring (Sugimoto and Hanawa 2005b, 2007; Shimada et al. 2007; Toyama and Suga 2010, 2011); and (3) owing to stratus clouds over the formation region, ESTMW is not capped by the strong seasonal pycnocline after spring, which allows atmospheric disturbances to continually modify ESTMW from its top (Sugimoto and Hanawa 2007). Nevertheless, subducted temperature/salinity anomalies in association with ESTMW might be transported to the downstream even after the low PV signal of ESTMW is lost. A recent analysis of Argo float data showed that isopycnal θ–S (or spiciness) anomalies propagate in the range of $\sigma_\theta = 25.0$–25.5 kg m^{-3} from the eastern subtropical Pacific to the western tropical Pacific, along a path and with a speed corresponding to the mean geostrophic flows (Sasaki et al. 2010). This propagation might be associated with the subduction and circulation of ESTMW, and its influence on the equatorial region, which has been hypothesized to be important for interdecadal climate variability of the North Pacific (Hanawa 1996; Gu and Philander 1997), should be examined quantitatively in future studies.

4 Decadal variability

The ocean-atmosphere system in the mid-latitude North Pacific exhibits significant decadal to interdecadal variability (e.g., Nitta and Yamada 1989; Tanimoto et al. 1993; Graham 1994; Trenberth and Hurrell 1994). Its state is often described using the Pacific decadal oscillation (PDO) index, which is defined as the leading principal component of North Pacific sea surface temperature variability poleward of 20°N (Hare 1996; Zhang 1996). A positive (negative) PDO index indicates a stronger (weaker) Aleutian low, stronger (weaker) westerlies, and a lower (higher) sea surface temperature except near the North American coast. An abrupt change of sign in the PDO index and the subsequent persistence for a few decades, called a regime shift, occurred in 1925, 1947, and 1977, exerting widespread impacts on the climate and marine ecosystems in the North Pacific (Mantua et al. 1997; Mantua and Hare 2002).

This observed long-term ocean-atmosphere variability is widely believed to be controlled by the ocean, which has much larger heat content and longer time scales than the atmosphere. As an important step for clarifying its mechanism, the response of the ocean interior to the decadal/interdecadal surface forcing has been explored. Pioneering works of Deser et al. (1996) and Schneider et al. (1999) analyzed historical temperature data to demonstrate that following the 1976/1977 regime shift when the PDO index turned positive, cold anomalies were subducted from the sea surface in the central North Pacific and then propagated downward in the permanent thermocline. Since the subduction area corresponded roughly to the central location of decadal sea surface temperature variability (Tanimoto et al. 1993) and also to the CMW subduction region, CMW was expected to play a key role in transmitting the decadal atmospheric changes to the upper ocean in the mid-latitude North Pacific (Suga et al. 1997). Following the data analysis of Yasuda and Hanawa (1997), numerical simulations using various surface forcing were performed in the early 2000s to investigate decadal variability of CMW, with particular attention to the 1976/1977 regime shift (Inui et al. 1999; Xie et al. 2000; Kubokawa and Xie 2002; Ladd and Thompson 2002; Hosoda et al. 2004). Their results, with some differences among models, showed that changes in the thermodynamic forcing (increases of upward surface heat flux and southward Ekman transport) and the dynamic forcing (increase of Ekman pumping) associated with the regime shift resulted in the formation of thicker and colder CMW and the eastward shift of its circulation path in the subtropical gyre, respectively, and their combined effects altered the subsurface thermal structure of the gyre. An analysis of the 180° repeat hydrographic section across the CMW formation region consistently indicated that the CMW temperature abruptly increased by 1°C from 1988 to 1989 when the PDO index transiently turned negative (Suga et al. 2003). Ladd and Thompson (2002) further generalized that when the PDO index is positive, thicker and colder CMW tends to be formed and the gyre circulation transporting it tends to be strengthened, and vice versa. Such a relation between the PDO index and the mode waters with no time lag was also demonstrated by recent modeling studies for CMW (Qu and Chen 2009) and STMW (Davis et al. 2011).

On the other hand, recent observational studies proposed a new mechanism of decadal STMW variability, which involves delayed oceanic response to the PDO-related surface wind forcing. Specifically, Qiu and Chen (2005) identified a decadal oscillation of the KE current system between two dynamic states based on the satellite altimeter sea surface height data during 1993–2004. This oscillation originates from the large-scale wind stress curl forcing in the central North Pacific around 160°W where the forcing has its largest amplitude (Qiu 2003). When the PDO index is positive (negative), negative (positive) sea surface height and permanent thermocline depth anomalies are generated there and propagate westward to the southeast of Japan over a period of ~3 years at the speed of first-mode baroclinic Rossby waves. As a result, the upstream KE jet becomes unstable and weak (stable and strong), and is accompanied by a weak (strong) southern recirculation gyre and high (low) regional eddy activity, which is against our expectation that a stronger mean flow favors baroclinic instability and leads to higher eddy activity (Qiu and Chen 2005, 2010b, 2011; Kelly et al. 2007; Taguchi et al. 2007). In the past 18 years, the KE was in the stable state in 1993–1994, 2002–2005, and 2010 and in the unstable state in 1995–2001 and 2006–2009.

The decadal KE variability is closely related to the STMW formation in the recirculation gyre. An analysis of temperature profiles in the KE region during 1993–2004 revealed that the STMW thickness also exhibited significant decadal variability and was large (small) when the KE was in the stable (unstable) state (Qiu and Chen 2006), which has been confirmed by several studies based on observational data (Pan and Liu 2005; Qiu et al. 2007; Sugimoto and Hanawa 2010; Oka et al. 2011b) and reanalysis product (Miyazawa et al. 2009). This occurs because during an unstable KE period, active southward transport of high PV water from the Mixed Water region across the KE due to the stronger eddy activity (Qiu and Chen 2006; Qiu et al. 2007), as well as the higher background stratification associated with the shallower permanent pycnocline in the weaker recirculation gyre (Sugimoto and Hanawa 2010), is unfavorable for the development of deep winter mixed layer, and the reverse is true during the stable KE period. For the period of 1993–2004, Qiu and Chen (2006) also demonstrated that the STMW thickness was highly correlated with the pre-existing upper-ocean stratification and had little correlation with the wintertime surface heat flux forcing. In other words, the thickness is determined by the dynamic forcing rather than the thermodynamic forcing, which contrasts with our traditional view that the latter forcing is dominant. A subsequent analysis of long-term temperature data during 1971–2007 showed that the contributions from the pre-existing stratification and the wintertime surface heat flux are comparable, with the former (latter) being more important after (before) around 1990 (Iwamaru et al. 2010). It has also been pointed out that the pre-existing stratification is determined not only by the dynamic state of the KE current system, but also by the summer surface heating, whose intensity depends on the activity of tropical cyclones (Kako and Kubota 2007; Tomita et al. 2010).

The decadal KE variability is important not only for the formation of STMW, but also for its subsequent evolution.

Recent analyses of Argo float and high-resolution shipboard observation data (Oka 2009; Oka et al. 2011b) showed that the STMW circulation in 2008 was much more turbulent than that in 2006. In 2006, the STMW formed in several anticyclonic circulations in the recirculation gyre tended to be continually trapped in its respective circulations after spring and to remain in the formation region until late fall (Oka 2009). In 2008, in contrast, it tended to gradually migrate southward. Simultaneously, STMWs with different temperatures formed in different longitude ranges were gradually stirred, then interleaved with each other, and were finally vertically mixed to form a STMW with an intermediate temperature (Oka et al. 2011b). The difference in the STMW circulation between the 2 years is attributable to the dynamic state of the KE current system. The KE was in the unstable state in both 2006 and 2008, but its path was more variable in 2008, which was the peak year of the unstable period of 2006–2009. The associated stronger eddy activity in the STMW formation region in 2008 was hypothesized to enhance the eddy transport of STMW in both the meridional and zonal directions (Oka et al. 2011b).

Although the seasonal evolution of the STMW distributions in the stable KE period has not been investigated in detail (because of the lack of Argo floats in 2002–2005), it is possible to put forth the following hypothesis about the decadal variation in the STMW subduction and dissipation. In the unstable KE period, the volume of STMW formed in winter is smaller than in the stable KE period because the formation region is partially masked by cyclonic eddies (Rainville et al. 2007) and the STMW thickness is smaller (as mentioned above), but the volume of STMW subducted into the south of the formation region is larger because of the enhanced eddy transport. Also, STMW in the formation region is dissipated after spring, mainly by vertical processes in the stable KE period and by horizontal processes in the unstable period. This is because the thicker and more uniform STMW in the stable period is more susceptible to the intensive breaking of downward-propagating inertial waves at its upper boundary (Sect. 3), while the stronger eddy activity in the unstable period helps destroy the vertically uniform structure of STMW (Oka et al. 2011b). Such hypothesized STMW variability would possibly affect the climate and primary production through the impact on the Subtropical Countercurrent (Yamanaka et al. 2008; Xie et al. 2011) and the nutrient supply from the subsurface to the surface layer (Sukigara et al. 2011), respectively.

The decadal variability of STMW temperature has also been explored. Modeling studies have demonstrated that on decadal time scales, stronger westerlies in the central North Pacific lead to the formation of warmer STMW several years later through an increase of the Kuroshio heat transport (Ladd and Thompson 2002; Yasuda and Kitamura 2003; Lee 2009), as previously shown by observational studies (Yasuda and Hanawa 1997; Hanawa and Kamada 2001). In addition, the Argo float data analysis (Oka 2009) suggested that the zonal temperature change in the STMW formed east of the Izu-Ogasawara Ridge ($\sim$140°E) is small in the stable KE period and large in the unstable KE period. This might be explained as follows. In the stable period, the KE takes a relatively straight path and is accompanied by a single, contiguous southern recirculation gyre, within which STMW having a zonally uniform temperature is formed. In the unstable period, on the other hand, the southern recirculation gyre is broken into several anticyclonic circulations, within each of which STMW with a different temperature is formed. In this latter case, the longer KE path results in a greater oceanic heat loss along the path, leading to a gradual decrease of the STMW temperature to the downstream. The float data analysis also revealed a significant interannual change in the STMW salinity (Oka 2009), which may be related to the variability of the North Pacific Tropical Water (Cannon 1966; Suga et al. 2000). With the accumulation of additional Argo float data in the coming years, we can expect the decadal variability in the STMW properties and its mechanisms to be clarified further.

The decadal variability of ESTMW has been least explored among the three mode waters. A hindcast run for 1965–1993 (Ladd and Thompson 2002) showed that the ESTMW density fluctuated on decadal time scales due primarily to the buoyancy flux change, while a recent data assimilation experiment for the 1990s (Toyoda et al. 2011) demonstrated that the ESTMW formation rate varied interannually owing to changes not only in the surface cooling in winter and the pre-existing stratification controlled by the low-level cloud coverage in summer, but also in the Ekman convergence of salt. These studies, however, pertain to the pre-Argo period during which we had inadequate salinity data. Given the important roles played by salinity in the ESTMW formation and dissipation (Sects. 2 and 3), future analyses of Argo float data and the reanalysis product assimilating them are expected to reveal new features and mechanisms of the ESTMW decadal variability.

5 Future directions

We have reviewed the progress in the observational and modeling studies of the North Pacific mode waters from the past decade. It is instructive to end this review by offering some directions for the mode water studies of the next decade. The first target is the role of mesoscale eddies, which prevail in the regions surrounding the KE jet (Mizuno and White 1983; Itoh and Yasuda 2010a, b; Itoh et al. 2011), in the formation and subduction of L-CMW, D-CMW, and TRMW. "Eddy-resolving" synoptic shipboard observations, such as the one conducted by Oka et al. (2011b), in conjunction with eddy-resolving simulations will contribute greatly to the refined depiction of the large-scale CMW formation and subduction revealed by Argo float data (Oka et al. 2011a).

Nearly 5 years have passed since the Argo float array in the North Pacific was completed, and the collected data have improved the climatology-based, classical views on STMW and CMW. However, the formation, subduction, and circulation processes of ESTMW have not been fully investigated using the float data. Future studies are encouraged to verify the various processes of ESTMW put forth by the modeling studies in the early 2000s.

During the past decade, accumulation of satellite altimeter sea surface height data has significantly enhanced our understanding of the decadal variability of the KE current system (Qiu and Chen 2005, 2010b, 2011). In a similar manner, accumulation of Argo float data will undoubtedly unravel the decadal variability of the mode waters in the coming decade. In Sect. 4, we have presented our hypothesis on the decadal variability of STMW associated with that of the KE current system. CMW is also expected to exhibit decadal variability owing to both the thermodynamic (surface heat flux and meridional Ekman transport) and dynamic forcings. We anticipate two types of dynamic forcing on CMW that originate from the PDO-related surface wind forcing in the central North Pacific. One is the decadal variability of eddy activity north of the KE, which would control the number and distribution of anticyclonic eddies that are the probable formation sites of thick CMW (Kouketsu et al. 2011). A recent study by Qiu and Chen (2011) found that the decadal mesoscale eddy variability had an opposite phase between the upstream (140°–152°E) and downstream (152°–165°E) KE regions. These downstream decadally varying eddy signals can potentially induce the long-term CMW variability. The other is the observed (Qiu and Chen 2005) and simulated (Nonaka et al. 2006; Taguchi et al. 2007) decadal shifts of the thermohaline fronts bounding the CMW formation regions. If these shifts are accompanied by changes in the winter mixed layer properties in the interfrontal regions, the CMW density can fluctuate without a significant change in the thermodynamic forcing (Suga et al. 2011). It is of interest to examine the degree to which the large-scale CMW subduction scheme for 2003–2008 (Oka et al. 2011a) applies to other periods. The maximum σ_θ of CMW that is subducted into the permanent pycnocline (26.4 kg m^{-3} for 2003–2008) can fluctuate on decadal time scales, and this may cause the decadal apparent oxygen utilization changes observed in the subsurface centered at $\sigma_\theta = 26.6$ kg m^{-3} in the subtropical and subarctic North Pacific (e.g., Andreev and Kusakabe 2001; Ono et al. 2001; Watanabe et al. 2001; Emerson et al. 2004; Mecking et al. 2008). Investigation into the decadal mode water variability and its mechanisms will also be facilitated by further development of numerical models in the coming decade. Models that are able to reproduce the recently identified structures and variability of currents, eddies, and mode waters are called for.

The decadal variability of STMW and CMW can influence the sea surface temperature in their formation regions through re-emergence (Hanawa and Sugimoto 2004; Sugimoto and Hanawa 2005a) or obduction after circulating in the subtropical gyre (Qiu and Huang 1995; Qiu 2002; Endoh et al. 2006). It can also potentially affect the sea surface temperature in the southern region through its impact on the Subtropical Countercurrent (Yamanaka et al. 2008; Xie et al. 2011; Kobashi and Xie 2011; Nonaka et al. 2011) and the Hawaiian Lee Countercurrent (Sasaki et al. 2011). On the other hand, the decadal variability of ESTMW can possibly modulate the sea surface temperature in the equatorial upwelling region. Quantification of these influences and clarification of the underlying mechanisms will lead to a better understanding of the ocean's role in the decadal climate variability.

Multidecadal variability or longer-term changes of the mode waters have been examined only recently. The STMW temperature observed in the winter formation region at 141°–150°E and in summer at 137°E exhibited an increasing trend of 3–5 $\times$ 10^{-2}°C/year^{-1} for the period of 1971–2007 (Iwamaru et al. 2010). Coupled climate model simulations predicted that in response to global warming, the densities of the mode waters tended to become lower (Lee 2009; Luo et al. 2009), although their subduction rates increased in Lee's model and decreased in Luo et al.'s model. It has also been predicted that banded structures of the sea surface temperature warming that were slanted in the northeast-southwest direction appear in response to global warming because the shallower winter mixed layer in the Kuroshio-Oyashio Extension region results in weaker mode water subduction, weaker Subtropical Countercurrent, and smaller thermal advection from the southwest (Xie et al. 2010, 2011; Xu et al. 2011). Such long-term mode water changes and their consequences are expected to become increasingly important not only from the physical, but also the biogeochemical, perspectives in the coming decades.

Acknowledgments The authors thank Atsushi Kubokawa for the opportunity to write this review and two anonymous reviewers for helpful comments on the manuscript. This review is based on discussions by the authors in summer 2010 when EO visited the University of Hawaii at Manoa for 1 month under the Overseas Internship Program for Outstanding Young Earth and Planetary Researchers provided by the Department of Earth and Planetary Science, the University of Tokyo. EO is also supported by the Japan Society for

Promotion of Science [KAKENHI, Grant-in-Aid for Scientific Research (B), no. 21340133] and the Ministry of Education, Culture, Sports, Science and Technology, Japan (MEXT; Grant-in-Aid for Scientific Research on Innovative Areas under grant no. 22106007). BQ is supported by NSF through grant OCE-0926594 and NASA through contract 1207881 as part of NASA's Ocean Surface Topography Mission.

References

Andreev AG, Kusakabe M (2001) Interdecadal variability in dissolved oxygen in the intermediate water layer of the western subarctic gyre and Kuril Basin (Okhotsk Sea). Geophys Res Lett 28:2453–2456

Aoki Y, Suga T, Hanawa K (2002) Subsurface subtropical fronts of the North Pacific as inherent boundaries in the ventilated thermocline. J Phys Oceanogr 32:2299–2311

Bates NR, Pequignet AC, Johnson RJ, Gruber N (2002) A variable sink for atmospheric CO_2 in subtropical mode water of the North Atlantic Ocean. Nature 420:489–493

Bingham FM (1992) Formation and spreading of subtropical mode water in the North Pacific. J Geophys Res 97:11177–11189

Bingham FM, Suga T (2006) Distributions of mixed layer properties in North Pacific water mass formation areas: comparison of Argo floats and World Ocean Atlas 2001. Ocean Sci 2:61–70

Bingham FM, Suga T, Hanawa K (1992) Comparison of upper ocean thermal conditions in the western North Pacific between two pentads: 1938–42 and 1978–82. J Oceanogr 48:405–425

Cannon GA (1966) Tropical waters in the western Pacific Oceans, August–September 1957. Deep Sea Res 13:1139–1148

Cronin MF, Meinig C, Sabine CL, Ichikawa H, Tomita H (2008) Surface mooring network in the Kuroshio Extension. IEEE Syst J 2:424–430

Davis XJ, Rothstein LM, Dewar WK, Menemenlis D (2011) Numerical investigations of seasonal and interannual variability of North Pacific Subtropical Mode water and its implications for Pacific climate variability. J Clim (in press)

de Boyer Montégut C, Madec G, Fischer AS, Lazar A, Iudicone D (2004) Mixed layer depth over the global ocean: an examination of profile data and a profile-based climatology. J Geophys Res 109:C12003. doi:10.1029/2004JC002378

Deser C, Alexander MA, Timlin MS (1996) Upper-ocean thermal variations in the North Pacific during 1970–1991. J Clim 9:1840–1855

Ebuchi N, Hanawa K (2001) Trajectory of mesoscale eddies in the Kuroshio recirculation region. J Oceanogr 57:471–480

Emerson S, Watanabe YW, Ono T, Mecking S (2004) Temporal trends in apparent oxygen utilization in the upper pycnocline of the North Pacific: 1980–2000. J Oceanogr 60:139–147

Endoh T, Jia Y, Richards KJ (2006) Sensitivity of the ventilation process in the North Pacific to eddy-induced tracer transport. J Phys Oceanogr 36:1895–1911

Favorite F, Dodimead AJ, Nasu K (1976) Oceanography of the Subarctic Pacific region, 1960–71. Bull Int North Pac Comm 33:1–187

Freeland H, Roemmich D, Garzoli S, LeTraon P, Ravichandran M, Riser S, Thierry V, Wijffels S, Belbeoch M, Gould J, Grant F, Ignazewski M, King B, Klein B, Mork K, Owens B, Pouliquen S, Sterl A, Suga T, Suk M, Sutton P, Troisi A, Velez-Belchi P, Xu J (2010) Argo—a decade of progress. In: Hall J, Harrison DE, Stammer D (eds) Proceedings of OceanObs'09: sustained ocean observations and information for society, 21–25 September 2009, vol 2. ESA Publication WPP-306, Venice, Italy. doi: 10.5270/OceanObs09.cwp.32

Graham NE (1994) Decadal-scale climate variability in the 1970s and 1980s: observations and model results. Clim Dyn 10:135–159

Gregg MC, Sanford TB (1980) Signatures of mixing from the Bermuda Slope, the Sargasso Sea and the Gulf Stream. J Phys Oceanogr 10:105–127

Gu D, Philander SGH (1997) Interdecadal climate fluctuations that depend on exchanges between the tropics and extratropics. Science 275:805–807

Hanawa K (1987) Interannual variations of the winter-time outcrop area of subtropical mode water in the western North Pacific Ocean. Atmos Ocean 25:358–374

Hanawa K (1996) Interdecadal variability in the North Pacific Ocean: subduction oscillation. Paper presented at WOCE pacific workshop, U.S. WOCE Office, Newport Beach, CA, August 19–23

Hanawa K, Hoshino I (1988) Temperature structure and mixed layer in the Kuroshio region over the Izu Ridge. J Mar Res 46:683–700

Hanawa K, Kamada J (2001) Variability of core layer temperature (CLT) of the North Pacific subtropical mode water. Geophys Res Lett 28:2229–2232

Hanawa K, Sugimoto S (2004) 'Reemergence' areas of winter sea surface temperature anomalies in the world's oceans. Geophys Res Lett 31:L10303. doi:10.1029/2004GL019904

Hanawa K, Talley LD (2001) Mode waters. In: Church J et al (eds) Ocean circulation and climate. Academic Press, London, pp 373–386

Hanawa K, Yoritaka H (2001) North Pacific subtropical mode waters observed in long XBT cross sections along 32.5°N line. J Oceanogr 57:679–692

Hare SR (1996) Low frequency climate variability and salmon production. Ph.D. dissertation, University of Washington, Seattle, WA

Hasumi H, Tatebe H, Kawasaki T, Kurogi M, Sakamoto TT (2010) Progress of North Pacific modeling over the past decade. Deep Sea Res II 57:1188–1200

Hautala SL, Roemmich DH (1998) Subtropical mode water in the Northeast Pacific basin. J Geophys Res 103:13055–13066

Hazeleger W, Drijfhout SS (2000) Eddy subduction in a model of the subtropical gyre. J Phys Oceanogr 30:677–695

Hosoda S, Xie SP, Takeuchi K, Nonaka M (2001) Eastern North Pacific subtropical mode water in a GCM: formation mechanism and salinity effects. J Geophys Res 106:19671–19681

Hosoda S, Xie SP, Takeuchi K, Nonaka M (2004) Interdecadal temperature variations in the North Pacific central mode water simulated by an OGCM. J Oceanogr 60:865–877

Huang RX, Qiu B (1994) Three-dimensional structure of the wind-driven circulation in the subtropical North Pacific. J Phys Oceanogr 24:1608–1622

Inui T, Takeuchi K, Hanawa K (1999) A numerical investigation of the subduction process in response to an abrupt intensification of westerlies. J Phys Oceanogr 29:1993–2015

Isoguchi O, Kawamura H, Oka E (2006) Quasi-stationary jets transporting surface warm waters across the transition zone between the subtropical and the subarctic gyres in the North Pacific. J Geophys Res 111:C10003. doi:10.1029/2005JC003402

Itoh S, Yasuda I (2010a) Characteristics of mesoscale eddies in the Kuroshio-Oyashio Extension region detected from the distribution of the sea surface height anomaly. J Phys Oceanogr 40:1018–1034

Itoh S, Yasuda I (2010b) Water mass structure of warm and cold anticyclonic eddies in the western boundary region of the subarctic North Pacific. J Phys Oceanogr 40:2624–2642

Itoh S, Shimizu Y, Ito S, Yasuda I (2011) Evolution and decay of a warm-core ring within the western subarctic gyre of the North Pacific, as observed by profiling floats. J Oceanogr (in press)

Iwamaru H, Kobashi F, Iwasaka N (2010) Temporal variations of the winter mixed layer south of the Kuroshio extension. J Oceanogr 66:147–153
Jensen TG, Campbell TJ, Allard RA, Small RJ, Smith TA (2011) Turbulent heat fluxes during an intense cold-air outbreak over the Kuroshio Extension Region: results from a high-resolution coupled atmosphere–ocean model. Ocean Dyn 61:657–674
Joyce TM (2011) New perspectives on Eighteen Degree Water formation in the North Atlantic. J Oceanogr (this issue)
Kako S, Kubota M (2007) Variability of mixed layer depth in Kuroshio/Oyashio Extension region: 2005–2006. Geophys Res Lett 34:L11612. doi:10.1029/2007GL030362
Kawabe M (1995) Variations of current path, velocity, and volume transport of the Kuroshio in relation with the large meander. J Phys Oceanogr 25:3103–3117
Kawai H (1972) Hydrography of the Kuroshio Extension. In: Stommel H, Yoshida K (eds) Kuroshio—its physical aspects. University of Tokyo Press, Tokyo, pp 235–352
Kelly KA, Thompson L, Cheng W, Metzger EJ (2007) Evaluation of HYCOM in the Kuroshio Extension region using new metrics. J Geophys Res 112:C01004. doi:10.1029/2006JC003614
Kelly KA, Small RJ, Samelson RM, Qiu B, Joyce T, Cronin M, Kwon YO (2010) Western boundary currents and frontal air-sea interaction: Gulf Stream and Kuroshio Extension. J Clim 23:5644–5667
Kobashi F, Kubokawa A (2011) Review on North Pacific Subtropical Countercurrent and Subtropical Front: role of mode water in ocean circulation and climate. J Oceanogr (this issue)
Kobashi F, Xie SP (2011) Interannual variability of the North Pacific Subtropical Countercurrent: role of local ocean–atmosphere interaction. J Oceanogr (this issue)
Kobashi F, Mitsudera H, Xie SP (2006) Three subtropical fronts in the North Pacific: observational evidence for mode water-induced subsurface frontogenesis. J Geophys Res 111:C09033. doi: 10.1029/2006JC003479
Kobashi F, Xie SP, Iwasaka N, Sakamoto TT (2008) Deep atmospheric response to the North Pacific oceanic subtropical front in spring. J Clim 21:5960–5975
Kouketsu S, Tomita H, Oka E, Hosoda S, Kobayashi T, Sato K (2011) The role of meso-scale eddies in mixed layer deepening and mode water formation in the western North Pacific. J Oceanogr (this issue)
Krémeur AS, Lévy M, Aumont O, Reverdin G (2009) Impact of the subtropical mode water biogeochemical properties on primary production in the North Atlantic: new insights from an idealized model study. J Geophys Res 114:C07019. doi:10.1029/2008 JC005161
Kubokawa A (1997) A two-level model of subtropical gyre and subtropical countercurrent. J Oceanogr 53:231–244
Kubokawa A (1999) Ventilated thermocline strongly affected by a deep mixed layer: a theory for subtropical countercurrent. J Phys Oceanogr 29:1314–1333
Kubokawa A, Inui T (1999) Subtropical countercurrent in an idealized ocean GCM. J Phys Oceanogr 29:1303–1313
Kubokawa A, Xie SP (2002) On steady response of a ventilated thermocline to enhanced Ekman pumping. J Oceanogr 58:565–575
Ladd C, Thompson L (2000) Formation mechanisms for North Pacific central and eastern subtropical mode waters. J Phys Oceanogr 30:868–887
Ladd C, Thompson L (2001) Water mass formation in an isopycnal model of the North Pacific. J Phys Oceanogr 31:1517–1537
Ladd C, Thompson L (2002) Decadal variability of North Pacific central mode water. J Phys Oceanogr 32:2870–2881
Ledwell JR, Watson AJ, Law CD (1993) Evidence for slow mixing across the pycnocline from an open-ocean tracer release experiment. Nature 364:701–703
Lee HC (2009) Impact of atmospheric CO_2 doubling on the North Pacific subtropical mode water. Geophys Res Lett 36:L06602. doi:10.1029/2008GL037075
Liu Q, Hu H (2007) A subsurface pathway for low potential vorticity transport from the central North Pacific toward Taiwan Island. Geophys Res Lett 34:L12710. doi:10.1029/2007GL029510
Luo Y, Liu Q, Rothstein LM (2009) Simulated response of North Pacific mode waters to global warming. Geophys Res Lett 36:L23609. doi:10.1029/2009GL040906
Luyten JR, Pedlosky J, Stommel H (1983) The ventilated thermocline. J Phys Oceanogr 13:292–309
Mantua NJ, Hare SR (2002) The Pacific decadal oscillation. J Oceanogr 58:35–44
Mantua NJ, Hare SR, Zhang Y, Wallace JM, Francis RC (1997) A Pacific interdecadal climate oscillation with impacts on salmon production. Bull Am Meteorol Soc 78:1069–1079
Marshall D (1997) Subduction of water masses in an eddying ocean. J Mar Res 55:201–222
Marshall JC, Nurser AJG, Williams RG (1993) Inferring the subduction rate and period over the North Atlantic. J Phys Oceanogr 23:1315–1329
Masumoto Y (2010) Sharing the results of a high-resolution ocean general circulation model under a multi-discipline framework—a review of OFES activities. Ocean Dyn 60:633–652
Masuzawa J (1969) Subtropical mode water. Deep Sea Res 16: 463–472
Mecking S, Warner MJ (2001) On the subsurface CFC maxima in the subtropical North Pacific thermocline and their relation to mode waters and oxygen maxima. J Geophys Res 106:22179–22198
Mecking S, Langdon C, Feely RA, Sabine CL, Deutsch CA, Min DH (2008) Climate variability in the North Pacific thermocline diagnosed from oxygen measurements: an update based on the U.S. CLIVAR/CO_2 repeat hydrography cruises. Glob Biogeochem Cycles 22:GB3015. doi:10.1029/2007GB003101
Miyazawa Y, Zhang R, Guo X, Tamura H, Ambe D, Lee JS, Okuno A, Yoshinari H, Setou T, Komatsu K (2009) Water mass variability in the western North Pacific detected in a 15-year eddy resolving ocean reanalysis. J Oceanogr 65:737–756
Mizuno K, White WB (1983) Annual and interannual variability in the Kuroshio Current System. J Phys Oceanogr 13:1847–1867
Mori K, Uehara K, Kameda T, Kakehi S (2008) Direct measurements of dissipation rate of turbulent kinetic energy of North Pacific subtropical mode water. Geophys Res Lett 35:L05601. doi: 10.1029/2007GL032867
Nakamura H (1996) A pycnostad on the bottom of the ventilated portion in the central subtropical North Pacific: Its distribution and formation. J Oceanogr 52:171–188
Nishikawa S, Kubokawa A (2007) Mixed layer depth front and subduction of low potential vorticity water in an idealized ocean GCM. J Oceanogr 63:125–134
Nishikawa S, Tsujino H, Sakamoto K, Nakano H (2010) Effects of mesoscale eddies on subduction and distribution of subtropical mode water in an eddy-resolving OGCM of the western North Pacific. J Phys Oceanogr 40:1748–1765
Nitta T, Yamada S (1989) Recent warming of tropical sea surface temperature and its relationship to the Northern Hemisphere circulation. J Meteorol Soc Jpn 67:375–382
Nonaka M, Xie SP (2000) Propagation of North Pacific interdecadal subsurface temperature anomalies in an ocean GCM. Geophys Res Lett 27:3747–3750
Nonaka M, Nakamura H, Tanimoto Y, Kagimoto T, Sasaki H (2006) Decadal variability in the Kuroshio-Oyashio Extension simulated in an eddy-resolving OGCM. J Clim 19:1970–1989
Nonaka M, Xie SP, Sasaki H (2011) Interannual variations in low potential vorticity water and the subtropical countercurrent in an eddy-resolving OGCM. J Oceanogr (this issue)

Ohno Y, Kobayashi T, Iwasaka N, Suga T (2004) The mixed layer depth in the North Pacific as detected by the Argo floats. Geophys Res Lett 31:L11306. doi:10.1029/2004GL019576

Ohno Y, Iwasaka N, Kobashi F, Sato Y (2009) Mixed layer depth climatology of the North Pacific based on Argo observations. J Oceanogr 65:1–16

Oka E (2009) Seasonal and interannual variation of North Pacific subtropical mode water in 2003–2006. J Oceanogr 65:151–164

Oka E, Suga T (2003) Formation region of North Pacific subtropical mode water in the late winter of 2003. Geophys Res Lett 30:2205. doi:10.1029/2003GL018581

Oka E, Suga T (2005) Differential formation and circulation of North Pacific central mode water. J Phys Oceanogr 35:1997–2011

Oka E, Talley LD, Suga T (2007) Temporal variability of winter mixed layer in the mid- to high-latitude North Pacific. J Oceanogr 63:293–307

Oka E, Toyama K, Suga T (2009) Subduction of North Pacific central mode water associated with subsurface mesoscale eddy. Geophys Res Lett 36:L08607. doi:10.1029/2009GL037540

Oka E, Kouketsu S, Toyama K, Uehara K, Kobayashi T, Hosoda S, Suga T (2011a) Formation and subduction of central mode water based on profiling float data, 2003–08. J Phys Oceanogr 41:113–129

Oka E, Suga T, Sukigara C, Toyama K, Shimada K, Yoshida J (2011b) "Eddy-resolving" observation of the North Pacific subtropical mode water. J Phys Oceanogr 41:666–681

Ono T, Midorikawa T, Watanabe YW, Tadokoro K, Saino T (2001) Temporal increases of phosphate and apparent oxygen utilization in the subsurface waters of western subarctic Pacific from 1968 to 1998. Geophys Res Lett 28:3285–3288

Palter JB, Lozier MS, Barber RT (2005) The effect of advection on the nutrient reservoir in the North Atlantic subtropical gyre. Nature 437:687–692

Pan A, Liu Q (2005) Mesoscale eddy effects on the wintertime vertical mixing in the formation region of the North Pacific subtropical mode water. Chin Sci Bull 50:1949–1956

Qiu B (1999) Seasonal eddy field modulation of the North Pacific Subtropical Countercurrent: TOPEX/POSEIDON observations and theory. J Phys Oceanogr 29:2471–2468

Qiu B (2002) The Kuroshio Extension system: its large-scale variability and role in the midlatitude ocean–atmosphere interaction. J Oceanogr 58:57–75

Qiu B (2003) Kuroshio Extension variability and forcing of the Pacific decadal oscillations: responses and potential feedback. J Phys Oceanogr 33:2465–2482

Qiu B, Chen S (2005) Variability of the Kuroshio Extension jet, recirculation gyre and mesoscale eddies on decadal timescales. J Phys Oceanogr 35:2090–2103

Qiu B, Chen S (2006) Decadal variability in the formation of the North Pacific subtropical mode water: oceanic versus atmospheric control. J Phys Oceanogr 36:1365–1380

Qiu B, Chen S (2010a) Interannual-to-decadal variability in the bifurcation of the North Equatorial Current off the Philippines. J Phys Oceanogr 40:2525–2538

Qiu B, Chen S (2010b) Eddy-mean flow interaction in the decadally modulating Kuroshio Extension system. Deep Sea Res II 57:1098–1110

Qiu B, Chen S (2011) Effect of decadal Kuroshio Extension jet and eddy variability on the modification of North Pacific intermediate water. J Phys Oceanogr 41:503–515

Qiu B, Huang RX (1995) Ventilation of the North Atlantic and North Pacific: subduction versus obduction. J Phys Oceanogr 25:2374–2390

Qiu B, Chen S, Hacker P (2004) Synoptic-scale air-sea flux forcing in the western North Pacific: observations and their impact on SST and the mixed layer. J Phys Oceanogr 34:2148–2159

Qiu B, Hacker P, Chen S, Donohue KA, Watts DR, Mitsudera H, Hogg NG, Jayne SR (2006) Observations of the subtropical mode water evolution from the Kuroshio Extension System Study. J Phys Oceanogr 36:457–473

Qiu B, Chen S, Hacker P (2007) Effect of mesoscale eddies on subtropical mode water variability from the Kuroshio Extension System Study (KESS). J Phys Oceanogr 37:982–1000

Qu T, Chen J (2009) A North Pacific decadal variability in subduction rate. Geophys Res Lett 36:L22602. doi:10.1029/2009GL040914

Qu T, Xie SP, Mitsudera H, Ishida A (2002) Subduction of the North Pacific mode waters in a global high-resolution GCM. J Phys Oceanogr 32:746–763

Rainville L, Jayne SR, McClean JL, Maltrud ME (2007) Formation of subtropical mode water in a high-resolution ocean simulation of the Kuroshio Extension region. Ocean Modell 17:338–356

Roden GI (1970) Aspects of the mid-Pacific transition zone. J Geophys Res 75:1097–1109

Roden GI (1972) Temperature and salinity fronts at the boundaries of the subarctic-subtropical transition zone in the western Pacific. J Geophys Res 77:7175–7187

Roemmich D, Boebel O, Desaubies Y, Freeland H, King B, LeTraon PY, Molinari R, Owens WB, Riser S, Send U, Takeuchi K, Wijffels S (2001) Argo: the global array of profiling floats. In: Koblinsky CJ, Smith NR (eds) Observing the oceans in the 21st century. GODAE Project Office, Bureau of Meteorology, Melbourne, pp 248–258

Roemmich D, Johnson GC, Riser S, Davis R, Gilson J, Owens WB, Garzoli SL, Schmid C, Ignaszewski M (2009) Argo: the challenge of continuing 10 years of progress. Oceanography 22:46–55

Saito H, Suga T, Hanawa K, Watanabe T (2007) New type of pycnostad in the western subtropical-subarctic transition region of the North Pacific: transition region mode water. J Oceanogr 63:589–600

Saito H, Suga T, Hanawa K, Shikama N (2011) The transition region mode water of the North Pacific and its rapid modification. J Phys Oceanogr (in press)

Sasaki YN, Schneider N, Maximenko N, Lebedev K (2010) Observational evidence for propagation of decadal spiciness anomalies in the North Pacific. Geophys Res Lett 37:L07708. doi:10.1029/2010GL042716

Sasaki H, Xie SP, Taguchi B, Nonaka M, Hosoda S, Masumoto Y (2011) Interannual variations of the Hawaiian Lee Countercurrent induced by low potential vorticity water ventilation in the subsurface. J Oceanogr (this issue)

Schneider N, Miller AJ, Alexander MA, Deser C (1999) Subduction of decadal North Pacific temperature anomalies: observations and dynamics. J Phys Oceanogr 29:1056–1070

Shimada K, Nomoto M, Yoshida J (2007) Distribution of the density ratio in the North Pacific. La Mer 45:149–158

Sprintall J, Roemmich D (1999) Characterizing the structure of the surface layer in the Pacific Ocean. J Geophys Res 104:23297–23311

Stommel HM (1979) Determination of water mass properties of water pumped down from Ekman layer to the geostrophic flow below. Proc Natl Acad Sci USA 76:3051–3055

Stommel H, Schott F (1977) The beta spiral and the determination of the absolute velocity field from hydrographic station data. Deep Sea Res 24:325–329

Suga T, Hanawa K (1990) The mixed layer climatology in the northwestern part of the North Pacific subtropical gyre and the formation area of subtropical mode water. J Mar Res 48:543–566

Suga T, Hanawa K (1995a) The subtropical mode water circulation in the North Pacific. J Phys Oceanogr 25:958–970

Suga T, Hanawa K (1995b) Interannual variations of North Pacific subtropical mode water in the 137°E section. J Phys Oceanogr 25:1012–1017

Suga T, Hanawa K (1995c) Subtropical mode water south of Honshu, Japan in the spring of 1988 and 1989. J Oceanogr 51:1–19
Suga T, Hanawa K, Toba Y (1989) Subtropical mode water in the 137°E section. J Phys Oceanogr 19:1605–1618
Suga T, Takei Y, Hanawa K (1997) Thermostad distribution in the North Pacific subtropical gyre: the central mode water and the subtropical mode water. J Phys Oceanogr 27:140–152
Suga T, Kato A, Hanawa K (2000) North Pacific tropical water: its climatology and temporal changes associated with the climate regime shift in the 1970s. Prog Oceanogr 47:223–256
Suga T, Motoki K, Hanawa K (2003) Subsurface water masses in the central North Pacific transition region: the repeat section along the 180° meridian. J Oceanogr 59:435–444
Suga T, Motoki K, Aoki Y, Macdonald AM (2004) The North Pacific climatology of winter mixed layer and mode waters. J Phys Oceanogr 34:3–22
Suga T, Aoki Y, Saito H, Hanawa K (2008) Ventilation of the North Pacific subtropical pycnocline and mode water formation. Prog Oceanogr 77:285–297
Suga T, Sato K, Nonaka M, Hosoda S, Ueno H, Shikama N, Kobayashi T, Iwasaka N, Oka E (2011) A revisit to causes of the North Pacific central mode water property changes associated with regime shifts. In: Hall J, Harrison DE, Stammer D (eds) Proceedings of the OceanObs'09: sustained ocean observations and information for society (Annex), 21–25 September 2009. ESA Publication WPP-306, Venice, Italy (in press)
Sugimoto S, Hanawa K (2005a) Remote reemergence areas of winter sea surface temperature anomalies in the North Pacific. Geophys Res Lett 32:L01606. doi:10.1029/2004GL021410
Sugimoto S, Hanawa K (2005b) Why does reemergence of winter sea surface temperature anomalies not occur in eastern mode water areas? Geophys Res Lett 32:L15608. doi:10.1029/2005GL022968
Sugimoto S, Hanawa K (2007) Further evidence for nonreemergence of winter SST anomalies in the North Pacific eastern subtropical mode water area. J Oceanogr 63:625–635
Sugimoto S, Hanawa K (2010) Impact of Aleutian Low activity on the STMW formation in the Kuroshio recirculation gyre region. Geophys Res Lett 37:L03606. doi:10.1029/2009GL041795
Sukigara C, Suga T, Saino T, Toyama K, Yanagimoto D, Hanawa K, Shikama N (2011) Biogeochemical evidence of large diapycnal diffusivity associated with the subtropical mode water of the North Pacific. J Oceanogr 67:77–85
Taguchi B, Xie SP, Schneider N, Nonaka M, Sasaki H, Sasai Y (2007) Decadal variability of the Kuroshio Extension: observations and an eddy-resolving model hindcast. J Clim 20:2357–2377
Takikawa T, Ichikawa H, Ichikawa K, Kawae S (2005) Extraordinary subsurface mesoscale eddy detected in the southeast of Okinawa in February 2002. Geophys Res Lett 32:L17602. doi:10.1029/2005GL023842
Talley LD (1988) Potential vorticity distribution in the North Pacific. J Phys Oceanogr 18:89–106
Taneda T, Suga T, Hanawa K (2000) Subtropical mode water variation in the southwestern part of the North Pacific subtropical gyre. J Geophys Res 105:19591–19598
Tanimoto Y, Iwasaka N, Hanawa K, Toba Y (1993) Characteristic variations of sea surface temperature with multiple time scales in the North Pacific. J Clim 6:1153–1160
Toba Y, Kawamura H, Hanawa K, Otobe H, Taira K (1991) Outbreak of warm water from the Kuroshio south of Japan—a combined analysis of satellite and OMLET oceanographic data. J Oceanogr Soc Jpn 47:297–303
Tomita H, Kako S, Cronin MF, Kubota M (2010) Preconditioning of the wintertime mixed layer at the Kuroshio Extension Observatory. J Geophys Res 115:C12053. doi:10.1029/2010JC006373
Tomosada A (1986) Generation and decay of Kuroshio warm-core rings. Deep Sea Res 33:1475–1486
Toyama K, Suga T (2010) Vertical structure of North Pacific mode waters. Deep Sea Res II 57:1152–1160
Toyama K, Suga T (2011) Roles of mode waters in formation and maintenance of central water in the North Pacific. J Oceanogr (this issue)
Toyoda T, Awaji T, Ishikawa Y, Nakamura T (2004) Preconditioning of winter mixed layer in the formation of North Pacific eastern subtropical mode water. Geophys Res Lett 31:L17206. doi:10.1029/2004GL020677
Toyoda T, Awaji T, Masuda S, Sugiura N, Igarashi H, Mochizuki T, Ishikawa Y (2011) Interannual variability of North Pacific eastern subtropical mode water formation in the 1990s derived from a 4-dimensional variational ocean data assimilation experiment. Dyn Atmos Oceans 51:1–25
Trenberth KE (1990) Recent observed interdecadal climate changes in the Northern Hemisphere. Bull Am Meteorol Soc 71:988–993
Trenberth KE, Hurrell JW (1994) Decadal atmosphereocean variations in the Pacific. Clim Dyn 9:303–319
Tsujino H, Yasuda T (2004) Formation and circulation of mode waters of the North Pacific in a high-resolution GCM. J Phys Oceanogr 34:399–415
Uda M (1963) Oceanography of the subarctic Pacific Ocean. J Fish Res Board Can 20:119–179
Uda M, Hasunuma K (1969) The eastward subtropical countercurrent in the western North Pacific Ocean. J Oceanogr Soc Jpn 25:201–210
Uehara H, Suga T, Hanawa K, Shikama N (2003) A role of eddies in formation and transport of North Pacific subtropical mode water. Geophys Res Lett 30:1705. doi:10.1029/2003GL017542
Watanabe YW, Ono T, Shimamoto A, Sugimoto T, Wakita M, Watanabe S (2001) Probability of a reduction in the formation rate of the subsurface water in the North Pacific during the 1980s and 1990s. Geophys Res Lett 28:3289–3292
Williams RG (1989) The influence of air–sea interaction on the ventilated thermocline. J Phys Oceanogr 19:1255–1267
Williams RG (1991) The role of the mixed layer in setting the potential vorticity of the main thermocline. J Phys Oceanogr 21:1803–1814
Xie SP, Kunitani T, Kubokawa A, Nonaka M, Hosoda S (2000) Interdecadal thermocline variability in the North Pacific for 1958–97: a GCM simulation. J Phys Oceanogr 30:2798–2813
Xie SP, Deser C, Vecchi GA, Ma J, Teng H, Wittenberg AT (2010) Global warming pattern formation: sea surface temperature and rainfall. J Clim 23:966–986
Xie SP, Xu LX, Liu Q, Kobashi F (2011) Dynamical role of mode-water ventilation in decadal variability in the central subtropical gyre of the North Pacific. J Clim 24:1212–1225
Xu LX, Xie SP, Liu Q, Kobashi F (2011) Response of the North Pacific Subtropical Countercurrent and its variability to global warming. J Oceanogr (this issue)
Yamanaka G, Ishizaki H, Hirabara M, Ishikawa I (2008) Decadal variability of the Subtropical Front of the western North Pacific in an eddy-resolving ocean general circulation model. J Geophys Res. 113:C12027. doi:10.1029/2008JC005002
Yasuda I (2003) Hydrographic structure and variability of the Kuroshio-Oyashio transition area. J Oceanogr 59:389–402
Yasuda T, Hanawa K (1997) Decadal changes in the mode waters in the midlatitude North Pacific. J Phys Oceanogr 27:858–870
Yasuda T, Hanawa K (1999) Composite analysis of North Pacific subtropical mode water properties with respect to the strength of the wintertime East Asian monsoon. J Oceanogr 55:531–541
Yasuda T, Kitamura Y (2003) Long-term variability of North Pacific subtropical mode water in response to spin-up of the subtropical gyre. J Oceanogr 59:279–290

Yasuda I, Okuda K, Hirai M (1992) Evolution of a Kuroshio warm-core ring—variability of the hydrographic structure. Deep Sea Res 39:S131–S161

Yoshida S (1964) A note on the variations of the Kuroshio during recent years. Bull Jpn Soc Fish Oceanogr 5:66–69 (in Japanese)

Yoshida T, Hoshimoto M (2006) Heat content change in the surface isothermal layer of a warm core ring in the sea east of Japan. J Oceanogr 62:283–287

Yuan X, Talley LD (1996) The subarctic frontal zone in the North Pacific: characteristics of frontal structure from climatological data and synoptic surveys. J Geophys Res 101:16491–16508

Zhang Y (1996) An observational study of atmosphere–ocean interaction in the northern oceans on interannual and interdecadal time-scales. Ph.D. dissertation, University of Washington, Seattle, WA

Zhang RX, Hanawa K (1993) Features of the water-mass front in the northwestern North Pacific. J Geophys Res 98:967–975

J Oceanogr (2012) 68:21–43
DOI 10.1007/s10872-011-0083-7

SPECIAL SECTION: REVIEW

New developments in mode-water research: Dynamic and climatic effects

Review on North Pacific Subtropical Countercurrents and Subtropical Fronts: role of mode waters in ocean circulation and climate

Fumiaki Kobashi · Atsushi Kubokawa

Received: 13 March 2011 / Revised: 8 September 2011 / Accepted: 9 September 2011 / Published online: 25 November 2011

Abstract A Subtropical Countercurrent (STCC) is a narrow eastward jet on the equator side of a subtropical gyre, flowing against the broad westward Sverdrup flow. Together with theories, recent enhanced observations and model simulations have revealed the importance of mode waters in the formation and variability of North Pacific STCCs. There are three distinct STCCs in the North Pacific, maintained by low potential vorticity (PV) that mode waters carry from the north. Model simulations show that changes in mode water ventilation result in interannual to interdecadal variations and long-term changes of STCCs. STCCs affect the atmosphere through their surface thermal effects, inducing anomalous cyclonic wind curl and precipitation along them. Thus, mode waters are not merely passive water masses but have dynamical and climatic effects. For temporal variability, atmospheric forcings are also suggested to be important in addition to the variability of mode waters. STCCs exist in other oceans and they are also flanked by mode waters on their poleward sides, suggesting that they are maintained by similar dynamics.

F. Kobashi (✉)
Faculty of Marine Technology, Tokyo University of Marine Science and Technology, Etchujima 2-1-6, Koto-ku, Tokyo 135-8533, Japan
e-mail: kobashi@kaiyodai.ac.jp

F. Kobashi
Research Institute for Global Change, Japan Agency for Marine-Earth Science and Technology, Yokosuka, Japan

A. Kubokawa
Faculty of Environmental Earth Science, Hokkaido University, Sapporo, Japan

Keywords Subtropical Countercurrent · Subtropical Front · Mode water · Ocean circulation · Climate

1 Introduction

A Subtropical Countercurrent (STCC) is a narrow eastward jet on the equator side of a subtropical gyre, where the barotropic flow is westward as predicted by the Sverdrup theory. Figure 1a shows long-term mean sea surface dynamic height referenced to 1000 dbar and surface zonal geostrophic velocity relative to 400 dbar, based on a high-resolution temperature climatology (adapted from Kobashi et al. 2006). The surface flow has an eastward component broadly even in the southern half of the Sverdrup subtropical gyre (Fig. 1b). Such a broad surface eastward flow is predicted by the modern wind-driven ocean circulation theory based on the ventilated thermocline theory (Luyten et al. 1983) and potential vorticity (PV) homogenization theory (Rhines and Young 1982; Young and Rhines 1982). Embedded in but distinct from the broad weak eastward flow are narrow eastward jets. They are the STCCs that the present study focuses on.

There are three distinct STCCs in the North Pacific (Kobashi et al. 2006). They each accompany a temperature and density front at subsurface depths of about 100–200 m (Uda and Hasunuma 1969). Figure 1a, c show subsurface fronts with surface eastward vertical shears in the subtropical gyre. One is in the latitudinal range of 21–25°N from 130°E to 175°E, tilting slightly to the north as it extends to the east. To the south, another front is identified in the latitudinal range of 19–21°N from 130°E to 180°, slightly veering to the north east of 160°E. The other front is found in the central basin, along about 26°N between 175°E and 160°W. These three fronts with large eastward

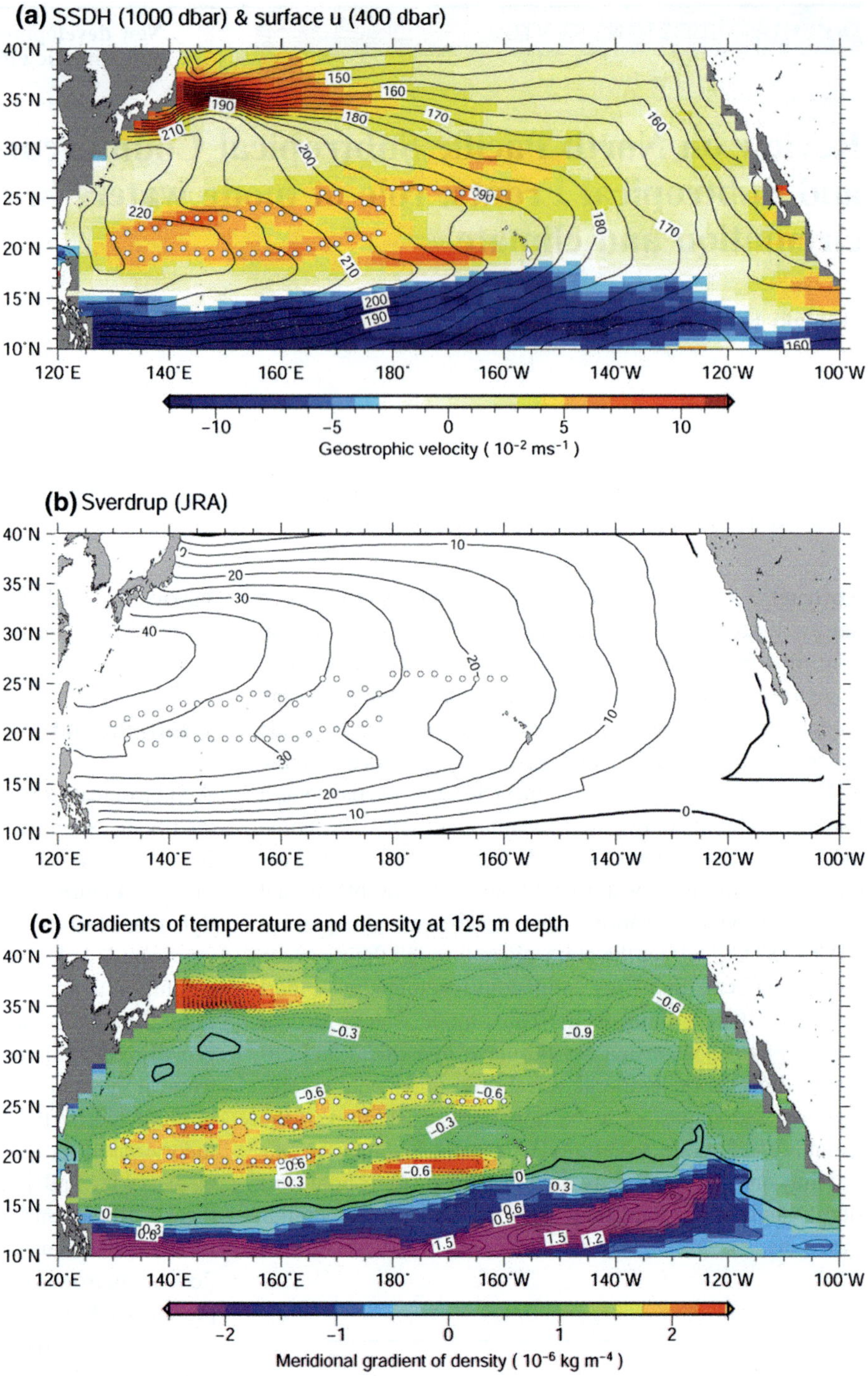

Fig. 1 **a** Long-term mean surface dynamic height referenced to 1000 dbar (contour at 5 cm intervals) and surface zonal geostrophic velocity relative to 400 dbar (*color*). **b** Sverdrup transport streamfunction calculated from long-term mean surface wind stress data of the Japanese 25-year reanalysis (Onogi et al. 2007). **c** Meridional gradients of temperature (contours at 1.0×10^{-5} °C m^{-1} intervals) and density (*color*) at 125 m depth. **a** and **c** are adapted from Kobashi et al. (2006), which are constructed mainly using temperature profiles in the World Ocean Database 2001 (Conkright et al. 2002). *Open circles* in all panels denote the position of STFs defined by Kobashi et al.

current shears exist against the broad southwestward Sverdrup flow (Fig. 1b). In this study, these narrow fronts at subsurface depths and associated currents are called STFs and STCCs, respectively. Following Kobashi et al. (2006), the northern and southern fronts (currents) in the western basin are referred to as the northern and southern STFs (STCCs), respectively, and that in the central basin the eastern STF (STCC).

Another eastward current is found at almost the same latitudes as the southern STCC west of the Hawaiian Islands (Fig. 1a, c), which is called the Hawaiian Lee countercurrent (HLCC). Different from the STCCs, the HLCC is consistent with the Sverdrup flow (Fig. 1b) and driven by Hawaiian Islands-induced wind curls (Xie et al. 2001). In climatology, the HLCC is rather confined in its westward extension up to the dateline (Qiu et al. 1997;

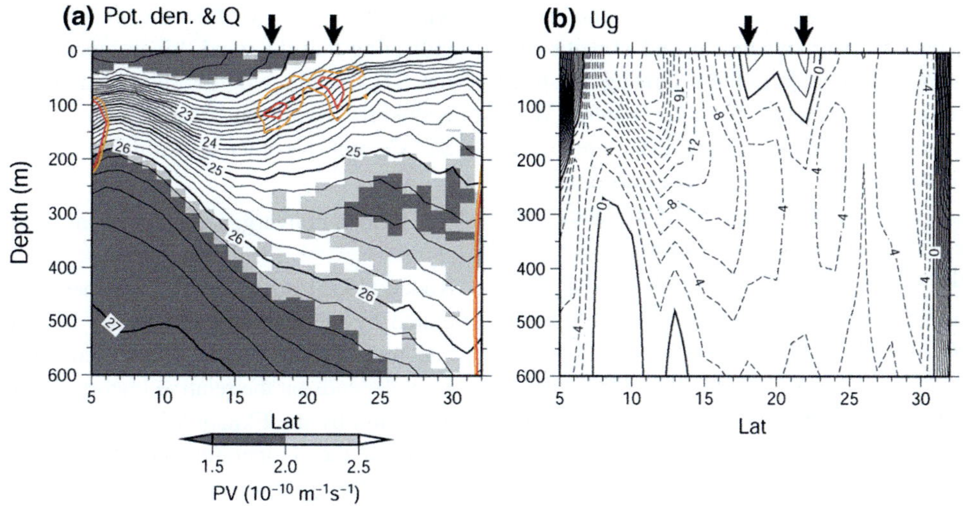

Fig. 2 Mean sections of **a** potential density and **b** zonal geostrophic velocity (referenced to 1000 dbar) along 137°E, calculated from CTD observations in every winter and summer season during 2000–2009, conducted by R/Vs Ryofu Maru and Keifu Maru of the JMA. *Orange* (*red*) contours in **a** denote the meridional gradient of potential density at 2.0 (2.5) × 10^{-6} kg m^{-4}, and *shade* indicates PV computed from $(f/\rho)(\partial\sigma_\theta/\partial z)$, where f is the Coriolis parameter, ρ is in situ water density, σ_θ is potential density, and z is the vertical coordinate. Stations along the sections are located approximately 1° apart in latitude. The locations of the northern and southern STCCs (STFs) are denoted by *arrows*

Yu et al. 2003; Kobashi et al. 2006), and it is separated from the southern STCC. Although the HLCC is sometimes referred to as "STCC" (e.g., Liu et al. 2003), this study focuses on only STCCs against the Sverdrup circulation.

Figure 2 shows mean meridional sections of potential density and zonal geostrophic velocity (referenced to 1000 dbar) along 137°E. They are calculated from 20 sections of conductivity temperature depth (CTD) observations in every winter and summer during 2000 to 2009, conducted by the R/Vs Ryofu Maru and Keifu Maru of the Japan Meteorological Agency (JMA). The data were obtained from the JMA website. The northern and southern STCCs appear as shallow eastward currents around 22°N and 18°N (Fig. 2b), each with an STF of large meridional temperature gradients at subsurface depths (see color contours in Fig. 2a). The main pycnocline is separated in the subtropical gyre into the upper one sloping up to the north and the lower one sloping down to the north. The STFs are associated with a steep northward shoaling embedded in the gradual shoaling of the upper pycnocline. The slope of the upper pycnocline gives rise to an eastward current shear of the STCCs near the surface by the thermal wind relation.

Since the discovery of STCCs by Uda and Hasunuma (1969), the mechanism of STCC formation has been discussed in a number of studies. Recent studies have revealed the importance of mode waters in the formation of STFs and STCCs, in particular, the subtropical mode water (STMW; Masuzawa 1969) and the central mode water (CMW; Nakamura 1996; Suga et al. 1997). These mode waters originate in the Kuroshio Extension region, where a deep surface mixed layer forms in winter due to intense surface cooling caused by warm water of the Kuroshio Extension being in contact with cold and dry winds of the East Asian monsoon. The deep mixed layer water is subducted into the thermocline and is advected by the wind-driven gyre circulation, producing a thick layer of vertically uniform properties in a wide region of the subtropical gyre. Thus, the mode waters are pycnostads characterized by a vertical minimum in PV. Observational studies have shown that STFs appear along the southern flank of mode waters (Uda and Hasunuma 1969; Suga et al. 1989; Chu et al. 2002; Aoki et al. 2002; Kobashi et al. 2006). For example, in Fig. 2a, the northern and southern STFs are located on the southern flank of the STMW that is identified with a pycnostad around 25.0–25.6 σ_θ between the upper and lower pycnoclines in the subtropical gyre. Similarly, the eastern STF is located along the CMW (Aoki et al. 2002; Kobashi et al. 2006).

In this paper, we review the progress in STCC research. Our understanding has been greatly advanced for the past decades with new developments of satellite and Argo profiling float arrays, accumulation of in situ hydrographic observations, improved numerical modeling, and their use to validate and improve theories. Now we recognize that STCCs are anchored by mode waters, and affect the overlying atmosphere and climate through their surface thermal effects. These findings shed light on the important role of mode waters in ocean current formation and climate. Mode waters are traditionally treated as a topic of water mass analysis, in which the main objective is to understand characteristics of mode waters themselves such as formation, circulation, and variability, with not much attention being paid to their dynamical and climatic significance. In this paper, we address the following questions that highlight dynamical effects of mode waters: What causes the formation of STCCs? How are STCCs related to mode waters? How do STCCs affect the atmosphere and climate? What causes variability of STCCs? Does variability in mode waters cause that of STCCs?

The remainder of the paper is organized as follows. In Sect. 2, we explain the mechanisms of STCC formation proposed so far. In Sect. 3, following a review of observations of the three STCCs, we estimate their mean surface current velocities. Then, we show how the three STCCs are related to mode waters, followed by a discussion on the mechanisms of their formation. We also describe STCCs simulated by models, and show that they are rather strong compared with those in observations. Section 4 presents atmospheric and climatic effects of STCCs from observations and models. Seasonal to interdecadal variability and long-term change of STCCs and their mechanisms are discussed in Sect. 5, followed by a summary and implications for future studies in Sect. 6.

2 Generation mechanism

In this section, we review four theories: (1) wind-driven mechanism, (2) Ekman convergence mechanism, (3) shock formation of stationary Rossby wave, and (4) vertical accumulation of low PV fluids. Each theory predicts the formation of one STF with STCC. As will be shown in Sect. 3, the observed three STFs are consistent with the mechanisms (3) and (4), both related to mode waters. The mechanisms (1) and (2) turn out not to be essential for the mean STCCs, but seem important for temporal variability (Sect. 4).

2.1 Wind-driven mechanism

Yoshida and Kidokoro (1967a, b) suggested that STCCs are wind-driven currents. They showed that the Sverdrup streamfunction based on the spring wind-stress curl contains an eastward flow between 20°N and 25°N, roughly along the northern and eastern STCCs. This eastward Sverdrup flow is caused by a small-scale trough of anticyclonic wind stress curl in the subtropical gyre. The trough forms only seasonally, and there is no trough in the annual mean wind-stress curl, whereas the northern and eastern STCCs exist in all seasons in observations (Uda and Hasunuma 1969; Kobashi and Kawamura 2002). On the basis of the ocean general circulation model (GCM) driven by wind stress and surface differential heating, Takeuchi (1984) succeeded in reproducing STCC. The wind stress curl used in the experiment has no trough, indicating that the wind-driven mechanism is not essential for STCCs. As will be shown in Sect. 4, Yoshida and Kidokoro's wind curl trough turns out to be due to the STCCs, not the other way round.

2.2 Ekman convergence mechanism

Roden (1975) proposed the hypothesis that STFs are generated by the Ekman convergence between the westerlies and trades. Whether the Ekman convergence produces a front depends on the relative importance of the temperature advection by the Ekman drift to the air–sea heat exchange terms in the temperature equation as discussed by Welander (1981). If we assume that surface heat exchange is proportional to the temperature difference between atmosphere and ocean with constant coefficient k and the oceanic temperature advection is caused by the zonally uniform meridional Ekman transport $V_e = \tau^x/\rho_0 f$, where τ^x is the zonal component of wind stress, ρ_0 is density, and f is the Coriolis parameter, the equation for the mixed layer temperature T in a steady state is

$$V_e(y)\frac{\partial T}{\partial y} = -k[T - T_a(y)], \qquad (2.1)$$

where y is the northward coordinate and T_a is the atmospheric temperature which is assumed to depend only on y. The derivation of this equation can be found in Welander (1981). If the Ekman transport $V_e(y)$ and atmospheric temperature $T_a(y)$ are linear functions of y, i.e., $V_e = -\alpha y$ and $T_a = T_0 + \frac{\Delta T}{L} y$, and the domain is $-L/2 < y < L/2$ with the boundary conditions $T = T_0 \pm \Delta T/2$ at $y = \pm L/2$, Eq. 2.1 yields

$$T = T_0 + \frac{k}{k-\alpha}\frac{\Delta T}{L}y - \mathrm{sign}(y)\frac{\alpha \Delta T}{2(k-\alpha)}\left(\frac{2|y|}{L}\right)^{k/\alpha}. \qquad (2.2)$$

The meridional gradient of the temperature becomes

$$\frac{dT}{dy} = \frac{k}{k-\alpha}\frac{\Delta T}{L} - \frac{k\Delta T}{L(k-\alpha)}\left(\frac{2|y|}{L}\right)^{k/\alpha - 1}. \qquad (2.3)$$

For $0 < k/\alpha \le 1$, dT/dy is infinite at $y = 0$, and the front becomes weaker as k/α becomes larger. The solution is shown in Fig. 3 for $k/\alpha = 0.5,\ 2,\ 10$. Assuming $d\tau^x/dy = 6 \times 10^{-9}$ (dyn cm^{-3}) based on mean wind-stress distribution, we obtain $\alpha \approx 10^{-4}$ (cm s^{-1}) at the 25°N. On the other hand, the coefficient Q_2 of Haney (1971) is 70–100 (ly day^{-1} K^{-1}), which yields $k \approx 10^{-3}$ (cm s^{-1}). From these values, we can estimate $k/\alpha \approx 10$. In this case, there is no front in Fig. 3. Therefore, the Ekman convergence is likely too weak to generate a significant front.

Takeuchi (1984), introduced in Sect. 2.1, also carried out an ocean GCM experiment with meridional wind stress,

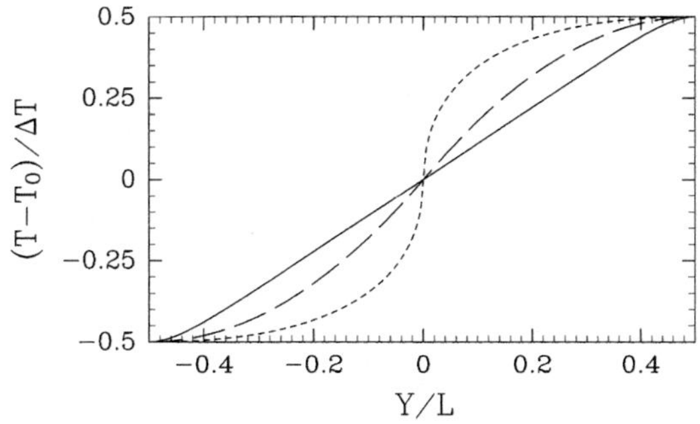

Fig. 3 Meridional distribution of mixed layer temperature given by Eq. 2.2 for $k/\alpha = 0.5$ (*dotted line*), 2.0 (*dashed line*), and 10 (*solid line*)

and obtained STCC very similar to that in the model driven by zonal wind stress. This indicates that the Ekman convergence produced by the westerlies and trades is not essential. In addition, the Ekman convergence occurs between 25°N and 35°N (Roden 1975; Kazmin and Rienecker 1996), slightly to the north of the STCCs, suggesting the same conclusion.

2.3 Shock formation of stationary Rossby wave

The numerical experiments of Takeuchi (1984) strongly suggested that the STCC is wind-driven circulation modified by surface differential heating. Cushman-Roisin (1984) obtained an STF-like structure in a similarity solution to an equation for continuously stratified thermocline (Welander 1959), forced by wind stress and surface differential heating. Although its direct applicability to reality is not clear because it does not satisfy all boundary conditions, the nonlinearity coming from mass conservation produces the STF-like structure. Since the vertical structure is assumed to have a similarity form, the theory may be regarded as that for a single baroclinic mode. Therefore, similar solutions can be obtained in a two-layer system.

Here, we give an explanation of the mechanism of frontogenesis in a two-layer system with some details. The model equation used here is the so-called planetary geostrophic thermocline equation (Dewer 1992) which is based on the hydrostatic balance, geostrophic balance, and mass conservation. The model is driven by the Ekman pumping, in which the total depth H is assumed to be constant, the density difference between two layers is $\Delta\rho$, and mean density is ρ_0. Substituting the geostrophic velocity into the lower layer mass equation, we obtain the equation for the upper layer depth h as

$$\frac{\partial h}{\partial t} - \frac{1}{H} J(\phi, (H-h)/f) - \frac{\beta g'(H-h)h}{f^2 H}\frac{\partial h}{\partial x} = 0, \qquad (2.4)$$

where t is time, x and y are the eastward and northward coordinate, respectively, $J(A,B) = A_x B_y - A_y B_x$ is the Jacobian operator, f is the Coriolis parameter, β is df/dy, g' is the reduced gravity ($=\Delta\rho g/\rho_0$), and

$$\phi = \frac{f^2}{\beta}\int_{x_e}^{x} w_e dx \qquad (2.5)$$

is the Sverdrup function. Here, w_e is the Ekman pumping velocity and x_e is the coordinate of the eastern boundary. The coefficient in front of $\partial h/\partial x$ in Eq. 2.4 is the Rossby wave speed. We assume that the model represents the upper ocean whose bottom corresponds to the main thermocline.

The Rossby wave speed in a two-layer system is the highest when $h/H = 1/2$. If $h/H < 1/2$ and $\partial h/\partial x > 0$ or $h/H > 1/2$ and $\partial h/\partial x < 0$ in the initial condition, the Rossby wave front steepens and eventually forms a shock as it propagates westward (Dewar 1987). In a wind-driven circulation, a similar β-induced steepening mechanism can also cause a steady shock (a "spontaneous shock"; Dewer 1992) as explained below.

Assuming a steady state and using the relation

$$J(A,B) = \left[\left(\frac{\partial A}{\partial \phi}\right)_y \left(\frac{\partial B}{\partial y}\right)_\phi - \left(\frac{\partial A}{\partial y}\right)_\phi \left(\frac{\partial B}{\partial \phi}\right)_y\right] \left(\frac{\partial \phi}{\partial x}\right)_y, \qquad (2.6)$$

where subscripts denote the variables which are kept constant for differentiation, Eq. 2.4 can be transformed from (x, y) to (ϕ, y) coordinates which is called the Sverdrup coordinate system (Kubokawa 1995):

$$\frac{\partial q}{\partial y} - \beta g'(H - fq) q \frac{\partial q}{\partial \phi} = 0. \qquad (2.7)$$

Here $q = (H - h)/f$ is the potential thickness (reciprocal of potential vorticity) of the lower layer. The relation between $x - y$ coordinates and Sverdrup coordinates is shown in Fig. 4. Equation 2.7 is very similar to that for the long Rossby waves; if we set $\phi = 0$ and rewrite Eq. 2.4 using q, the equation for the Rossby wave becomes

$$\frac{\partial q}{\partial t} - \frac{\beta g'(H - fq) q}{fH}\frac{\partial q}{\partial x} = 0. \qquad (2.8)$$

Similarity between Eqs. 2.7 and 2.8 allows for an interpretation that the propagation of q on the $\phi - y$ plane is like that on an $x - t$ plane. Since disturbances are advected southward by the Sverdrup flow, $-y$ corresponds to t, and disturbances propagate in the direction of increasing ϕ (westward) on the $\phi - y$ plane as y decreases. Because of the similarity between Eqs. 2.7 and 2.8, we refer to disturbances in 2.7 as stationary Rossby waves. The propagation speed on the $\phi - y$ plane (characteristic speed) is the highest when $fq/H = 1/2$ which is equivalent to $h/H = 1/2$. When $fq/H > 1/2$ ($h/H < 1/2$), the smaller q is,

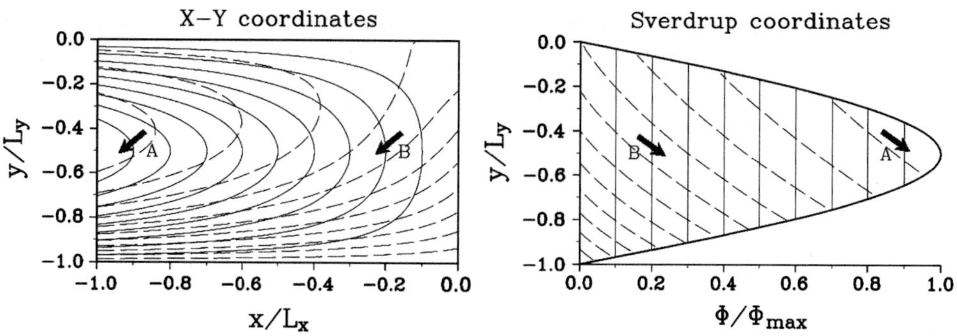

Fig. 4 Relation between the x–y coordinates (*left panel*) and the Sverdrup coordinates (*right panel*). *Solid contours* denote the Sverdrup function, and *dashed contours* the lower layer potential thickness, along which the stationary Rossby wave signals propagate southwestward (*arrows*). The parameters are $\beta L_y/f_0 = 0.5$, $h_e/H = 0.2$ and $\phi_{max}/g'H^2 = 0.2$, where f_0 is the Coriolis parameter at the subpolar–subtropical boundary ($y = 0$) and L_y is the meridional extension of the subtropical gyre, and h_e is the upper layer depth at the eastern boundary. In the present case, the lower layer is at rest. Locations denoted by letters A and B in x–y coordinates correspond to those denoted by the *same letters* in the Sverdrup coordinates. The positive ϕ direction is westward

the higher the speed, and vice versa. Therefore, if q is given along a latitude circle somewhere in the subtropical gyre as satisfying $fq/H > 1/2$ and $\partial q/\partial\phi > 0$ or $fq/H < 1/2$ and $\partial q/\partial\phi < 0$, stationary Rossby waves governed by Eq. 2.7 steepen as y decreases; this behavior is almost the same as that of Rossby waves governed by Eq. 2.8 as time goes.

Curves of constant q in Fig. 4 correspond to trajectories (characteristics) of linear stationary Rossby waves. This example is for a case in which the lower layer is at rest. Since the constant q curve slants southwestward on the $x - y$ plane in the southern half of the subtropical gyre, the eastward vertical shear given by $\partial h/\partial y < 0$ corresponds to $\partial h/\partial x > 0$ ($\partial q/\partial\phi > 0$). Since the sign of $\partial q/\partial\phi$ is unchanged along the characteristics, q in the source region must satisfy the condition that $fq/H > 1/2$ ($h/H < 1/2$) and $\partial q/\partial\phi > 0$ ($\partial q/\partial x < 0$, $\partial h/\partial x > 0$) for the stationary Rossby wave to form a shock with eastward upper-layer current. When the shock forms, it is assumed that a very weak diffusion term which is not explicitly included in the above equations prevents the solution from being multi-valued.

There are four possible sources of stationary Rossby waves and shocks. The first one is the outcrop of the lower layer (Dewer 1992). When the outcrop slants northeastward from the west, q of the subducted water becomes smaller toward east because f is large where the subduction occurs. This structure ($\partial q/\partial x < 0$) steepens as it is advected southward, and a front associated with the eastward current appears in the southern part of the subtropical gyre. The second possible source is the subpolar–subtropical gyre boundary (Dewar 1991, 1992). Although there is a stationary Rossby wave solution on the gyre boundary whose westward propagation velocity locally balances with eastward current weakening toward the eastern boundary, it cannot satisfies the condition, $h/H < 1/2$ and $\partial q/\partial x < 0$, for forming an STCC-like current. Only a shock on the gyre boundary can satisfy this condition. This shock extends into the subtropical gyre, forming an eastward jet in the southern half of the gyre, which is called an "arrested front" (Dewar 1991). This arrested front solution looks like the countercurrent in Cushman-Roisin (1984) which emanates from the gyre boundary, and we can classify these two solutions into the same category, although the model configurations are different from each other and diabatic forcing is imposed in Cushman-Roisin (1984).

The third possible source is the northern part of the western boundary[1] where ϕ changes from 0 to ϕ_{max} (see Fig. 4b). Since $\partial q/\partial\phi > 0$ is equivalent to $\partial q/\partial y < 0$ along the western boundary, when q decreases northward, the stationary Rossby wave emanating from the western boundary can form a shock. The condition $\partial q/\partial y < 0$ implies that the vertical shear is weak or westward there. Furthermore, according to Kubokawa (1995), for the shock to form in the rectangular basin, the vertical shear there must be westward, which seems unrealistic. The last one is the eastern boundary, which has not been discussed as a source of the stationary Rossby wave. In a usual two-layer system, ventilation does not occur at the eastern boundary because the eastern boundary is treated as impermeable and no interlayer flux is allowed. However, in a continuously stratified ocean model with a meridional gradient in the mixed layer, eastern boundary ventilation is inevitable, because density-driven near-surface eastward flows must turn westward below the surface (Sumata and Kubokawa 2001). The eastern boundary ventilation can be incorporated by permitting nonzero baroclinic component of zonal flow at the eastern boundary (Pedlosky 1984). Figure 5 shows a solution with a shock caused by an eastern

[1] The term "western boundary" in this section means the western boundary of the Sverdrup interior region, but not the real western boundary.

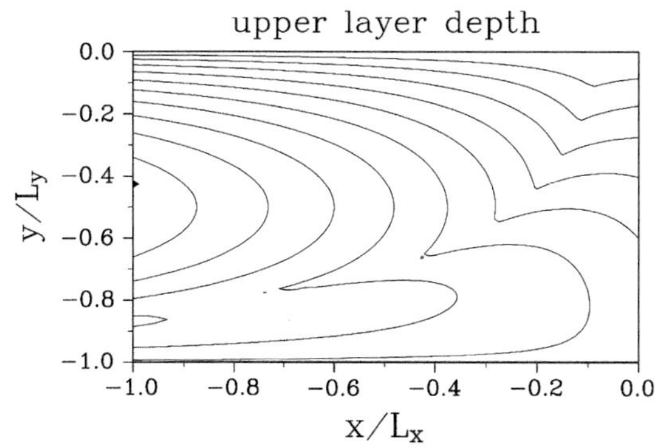

Fig. 5 An example of spontaneous shock solution caused by an eastern boundary ventilation. The *panel* shows the upper layer depth distribution. The parameters are same as those in Fig. 4, but $h_e(y)/H = 0.1 + 0.3\tanh(-2y/L_y)$.

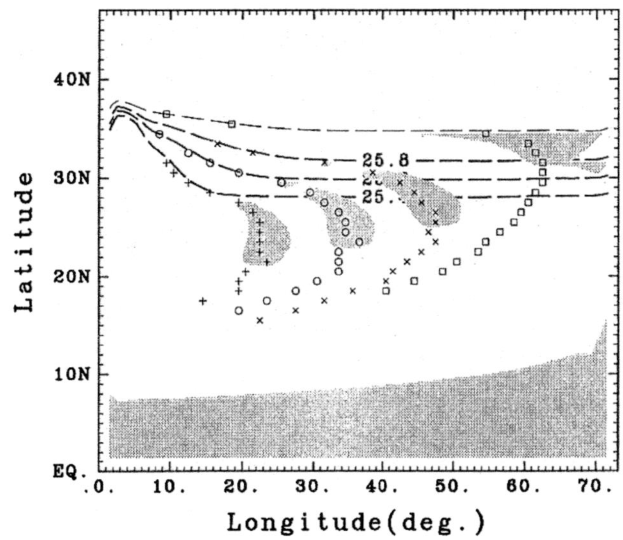

Fig. 6 Trajectories of low PV fluids on four isopycnal surfaces, simulated in an idealized ocean GCM. *Plus symbols*, *open circles*, *multiplication symbols*, and *squares* denote the positions of local minima for given latitudes on the isopycnal surface of 25.4 σ_θ, 25.6 σ_θ, 25.8 σ_θ, and 26.1 σ_θ, respectively. The region with PV lower than 10^{-10} m^{-1} s^{-1} is *shaded*. PV is defined as the same as in Fig. 2. The low PV fluids are vertically stacked up downstream, forming a thick low PV pool, along which STCC appears in the model. From Kubokawa and Inui (1999)

boundary ventilation, in which the thermocline depth at the eastern boundary, $h_e(y)$, is decreasing northward. Since positive $\partial q/\partial y$ along the eastern boundary yields negative $\partial q/\partial x$, the condition for spontaneous shock is satisfied.

In the two-layer model discussed so far, density is constant in each layer. To explore the effects of horizontal density gradient, Kubokawa (1997) presented a theory of a two-level planetary geostrophic model, consisting of equations for vertically integrated density and stratification under the assumptions of geostrophic balance. He discussed the formation of an STCC caused by stationary Rossby waves emanating from the western boundary, and showed that if the density is vertically homogeneous and vertical shear is small at the northwestern subtropical gyre, the stationary Rossby wave can produce a strong countercurrent. Although the mechanism presented by Kubokawa (1997) is essentially the same as that in the two-layer model with western boundary source described above, in the two-level model a stronger countercurrent occurs along the boundary between the regions of weak stratification in the north and strong stratification in the south. Kubokawa (1997) suggested that the northern STF in the real ocean has a similar density structure as in the model, and the homogeneous fluid in the northwestern subtropical gyre may correspond to STMW. The vertical resolution of this model is, however, too coarse; it is therefore desirable to revisit this problem using a multi-layer or a continuously stratified model for a further discussion.

2.4 Vertical accumulation of low PV fluids

Kubokawa and Inui (1999) carried out a numerical experiment similar to that reported by Takeuchi (1984), and pointed out the importance of distribution of low PV fluid subducted from the intersections of outcrops and a mixed-layer-depth front, which is a narrow transition zone separating shallow and deep mixed layers. The mixed-layer-depth front slants slightly northeastward from the western region of the subtropical gyre, whereas outcrops are almost zonal and slant slightly southeastward. As a result, the density of outcrops intersecting the mixed-layer front increases eastward, meaning that the minimum PV fluids on denser isopycnals are formed in the east, whereas those on lighter isopycnals are formed in the west (Fig. 6). Trajectories of these minimum PV fluids on each isopycnal converge on the horizontal plane as they are advected southward, and the low PV fluids are stacked up vertically and form a thick layer of low PV fluid (Fig. 6). This thick low PV layer causes the upper pycnocline to rise, forming a density front and eastward countercurrent on its southern edge (Fig. 2). Since two-layer models cannot reproduce such a situation, Kubokawa (1999) presented a theory using a multi-layer ideal fluid thermocline model. Kubokawa's theory demonstrates that low PV fluids such as mode waters have an important dynamical effect on surface circulation.

Here, we outline this theory. We write the z-coordinate of the bottom of the j-th layer as $z_j(<0)$, numbering the layers upward from the deepest, with layer 1 being the deepest moving layer and z_0 being constant in space and time, to layer N being the uppermost layer. The sea surface is at $z = 0$ and the density of the j-th layer is ρ_j. Under the planetary geostrophic assumption, conservation of the PV in the n-th layer ($n < N$) can be written as

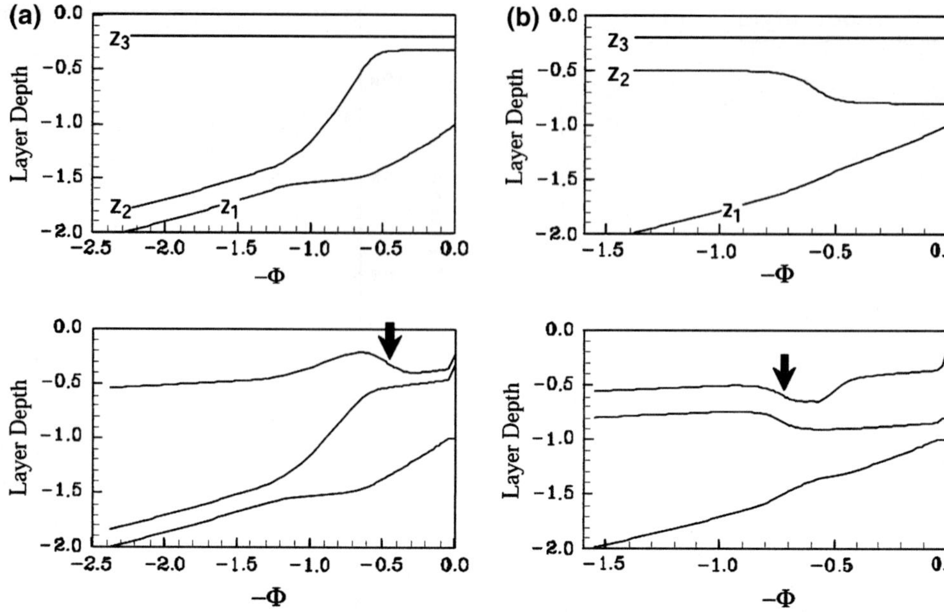

Fig. 7 Two examples of solutions in a 3.5-layer model. Column **a** is for a case in which PV in the deepest moving layer decreases eastward whereas that in the second layer decreases eastward, and column **b** is for a case in which PV in the deepest moving layer decreases westward. The *upper panels* show given zonal distributions of interface depths, z_1, z_2, and z_3 at $f = f_0$, and the *lower panels* show z_1, z_2, and z_3 at $f = 0.8f_0$, calculated by Eqs. 2.12–2.14. The *horizontal axis* $-\phi$ is proportional to $x - x_e$ if w_e is independent of x, i.e., $-\phi = (f^2/\beta)|w_e|(x - x_e)$. z_j values are normalized by z_{01}, and ϕ is normalized by $\gamma_1 z_{01}^2$. γ_j values are assumed to be $\gamma_2/\gamma_1 = \gamma_3/\gamma_1 = 0.5$. *Arrows* are guides for the position of STCC

$$J\left(\sum_{j=1}^{n} \gamma_j z_j, (z_{n+1} - z_n)/f\right) = 0, \tag{2.9}$$

where $\gamma_j = g(\rho_{j-1} - \rho_j)/\rho_0$. For the uppermost layer ($j = N$),

$$-J\left(\sum_{j=1}^{N} \gamma_j z_j, z_N/f\right) = w_e = \frac{\beta}{f^2}\frac{\partial \phi}{\partial x}, \tag{2.10}$$

where ϕ is the Sverdrup function defined by Eq. 2.5. The sum of these equations from $n = 1$ to N yields

$$\frac{\partial}{\partial x}\left(\sum_{j=1}^{N} \frac{\gamma_j}{2} z_j^2\right) = \frac{\partial \phi}{\partial x}, \tag{2.11}$$

which is the Sverdrup relation. Integrating with respect to x from the eastern boundary x_e to x, we obtain

$$\sum_{j=1}^{N} \gamma_j\left(z_j^2 - z_{0j}^2\right) = 2\phi, \tag{2.12}$$

where z_{0j} is the z_j at the eastern boundary.

If the flow in layer 1 is southward, $z_1$is an increasing function of x, and it can be used as a zonal coordinate in place of x. This transformation can be done by the same as that from (x, y) to (ϕ, y) coordinates (see Eq. 2.6), and the equation for the first layer, Eq. 2.9 with $n = 1$, becomes

$$\left(\frac{\partial}{\partial y}\left\{\frac{z_2 - z_1}{f}\right\}\right)_{z_1} = 0, \tag{2.13}$$

where subscript denotes the variable which is kept constant for differentiation. This equation indicates that potential thickness, $(z_2 - z_1)/f$, is constant along lines of constant z_1.

The equation for the second layer becomes

$$\left[\gamma_1 + \gamma_2\left(\frac{\partial z_2}{\partial z_1}\right)_y\right]\left(\frac{\partial}{\partial y}\left\{\frac{z_3 - z_2}{f}\right\}\right)_{z_1} - \gamma_2\left(\frac{\partial z_2}{\partial y}\right)_{z_1}\left(\frac{\partial}{\partial z_1}\left\{\frac{z_3 - z_2}{f}\right\}\right)_y = 0, \tag{2.14}$$

where $\gamma_2\left(\frac{\partial z_2}{\partial y}\right)_{z_1}$ in the second term represents the zonal velocity relative to the first layer. If the velocity in the first layer has a southward component, $z_2 - z_1$ decreases as the fluid column flows southward because of the PV conservation. This implies that interface z_2 tends to deepen southward along the z_1 contour (streamline of the first layer), and this northward inclination of the interface z_2 strengthens the eastward shear. In this way, the current vector tends to rotate counterclockwise with decreasing depth in the subtropical gyre. This tendency is called β-spiral (Stommel and Schott 1977), and implies that the relative location of low PV waters in different layers changes as the water masses flow southward. If low PV fluids in different layers stack vertically, a thick (low PV) layer forms.

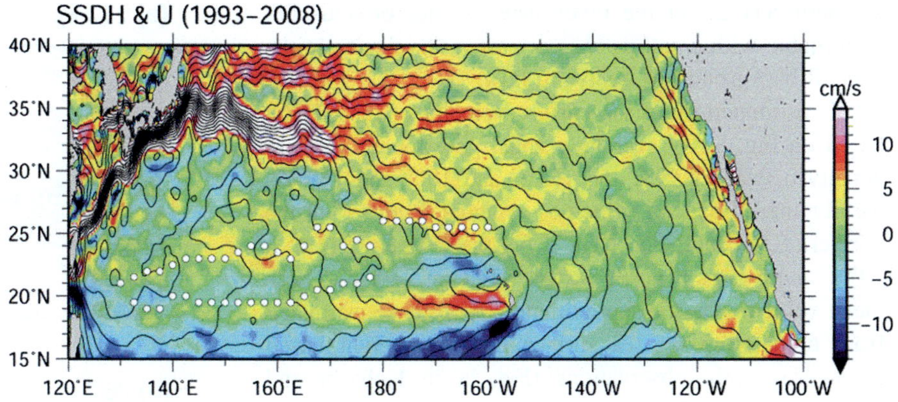

Fig. 8 Absolute mean dynamic height during 1993–2008, computed from an AVISO altimetry product. *Color* indicates eastward geostrophic velocity and *open circles* denote the position of STFs defined by Kobashi et al. (2006)

Figure 7a shows the change of layer thickness distributions in two zonal sections in a 3.5-layer model, in a case that PV in layer 1 decreases eastward whereas that in layer 2 decreases westward in the northern section (upper panel). This PV distribution in the zonal section mimics that in the numerical experiment shown in Fig. 6 (Kubokawa and Inui 1999), in which the PV distribution is caused by the northeastward slanting mixed-layer-depth front. Since the thick low PV fluid in the deep layer moves westward faster than that in the shallower layer, low PV fluid in layer 2 goes on top of that in layer 1 as it is advected southward. In this way, $z_3 - z_1$ increases locally and pushes up the upper interface z_3 (lower panel). This ridge in z_3 extends from northeast to southwest. On the southeastern slope of the ridge in z_3, the eastward current in thermal wind balance occurs (see arrow in Fig. 7a). This is the STCC in Kubokawa (1999). When PV in layer 1 decreases westward whereas PV in layer 2 decreases eastward in the northern zonal section (Fig. 7b), the horizontal positions of high PV regions in layers 1 and 2 get closer as they are advected southward, making a trough in the interface z_3. In this case, an eastward flow occurs along the northwestern slope of the trough (see arrow in Fig. 7b). This case can also occur if the mixed-layer-depth front slants southeastward from the west as shown in Kubokawa (1999).

3 Three STFs

In this section, we first review observations of the three STCCs, and estimate their mean surface current velocities. Then, we show how the three STFs are related to mode waters, followed by a discussion on the mechanisms of their formation. We also show STCCs simulated in models, and discuss their correspondence to the observed STCCs.

3.1 Distribution and mean surface current velocity

Until recently STFs were not fully surveyed. This is partly because of high eddy variability in the STCC regions (e.g., Kawamura et al. 1995; Aoki and Imawaki 1996; Qiu 1999; Kobashi and Kawamura 2001; Hwang et al. 2004). STCCs form a vertically sheared current system with an underlying wind-driven westward flow (Fig. 2b), causing enhanced eddy production through baroclinic instability (Qiu 1999; Kobashi and Kawamura 2002; Noh et al. 2007). The high eddy variability often makes it difficult to describe the details of STCCs from one-time synoptic observations. In addition, because of sparse hydrographic observations, early climatology maps were not able to resolve narrow STCCs, but, instead, they depicted a single broad eastward flow (e.g., Wyrtki 1975; White et al. 1978; Reid and Mantyla 1978), due to strong spatial smoothing of sparse observations.

The three STFs shown by Kobashi et al. (2006) agree in location with previous findings from high resolution observations and climatologies. The northern and southern STFs have been depicted by hydrographic observations (Nitani 1972; Hasunuma and Yoshida 1978; White and Hasunuma 1982; White and Walker 1985; Kaneko et al. 1998; Aoki et al. 2002; Kobashi and Kawamura 2002), and by other observations from surface drifters (Michida 1997), acoustic Doppler current profilers (Ichikawa et al. 2004), and satellite altimetry (Kobashi and Xie 2012). The eastern STF has also been shown from hydrographic observations (Reed 1970; Roden 1980b; White et al. 1978; White and Walker 1985; Aoki et al. 2002; Kobashi and Kawamura 2002).

The three STCCs are identified in absolute mean sea level maps based on satellite altimeter observations (e.g., Niller et al. 2003; Rio and Hernandez 2004; Vianna and Menezes 2010). Figure 8 shows mean ocean dynamic height during 1993–2008, calculated from an altimeter product from Archiving, Validation and Interpretation of Satellite Oceanographic data (AVISO), France. Mean dynamic height used by AVISO is produced by combing satellite altimetry, in situ measurements, and a geoid model (Rio et al. 2011). The northern and eastern STCCs appear clearly along their positions shown by Kobashi et al. (2006), and they are separated around 170–175°E. The southern STCC, on the other hand, is not obvious but identifiable with small patches of eastward velocity in the general background of westward flow.

We estimated the mean surface velocity of the STCCs based on Fig. 8. By averaging zonal geostrophic velocity zonally between 135°E and 165°E and 180° and 160°W, we obtained the mean velocities as about 3 cm s^{-1} at the northern STCC at 23°N, 1 cm s^{-1} at the southern STCC at 19°N, and 5 cm s^{-1} at the eastern STCC at 26.5°N. These mean velocities agree approximately with those estimated from climatology of sea surface dynamic height based on hydrographic observations (Fig. 2a in Kobashi and Kawamura 2002). In contrast, they are remarkably smaller than those from synoptic hydrographic observations, in which surface velocity of greater than 20 cm s^{-1} is reported at the northern and southern STCCs (Uda and Hasunuma 1969; Kaneko et al. 1998; Aoki et al. 2002). The discrepancy is probably due to eddy contamination in synoptic observations, because mesoscale eddies in the STCC regions have large advective speed of about 20 cm s^{-1} at the surface (Kobashi and Kawamura 2001), several times greater than the mean velocity of the STCCs. This indicates that the mean STCCs tend to be masked by eddies at an instantaneous field.

Judging from the sea surface dynamic height contours in Figs. 1a and 8, we note an interesting feature that the surface geostrophic velocity vector points eastward or southeastward, whereas the band of the STCCs extends to the east or slightly to the northeast, which suggests that the current velocity vector does not follow the direction of the STCCs. STCCs in the theories mentioned in Sect. 2 also show a similar feature. On this point, STCCs are not "currents" in the usual sense of the word, but they are narrow bands of eastward-velocity anomaly.

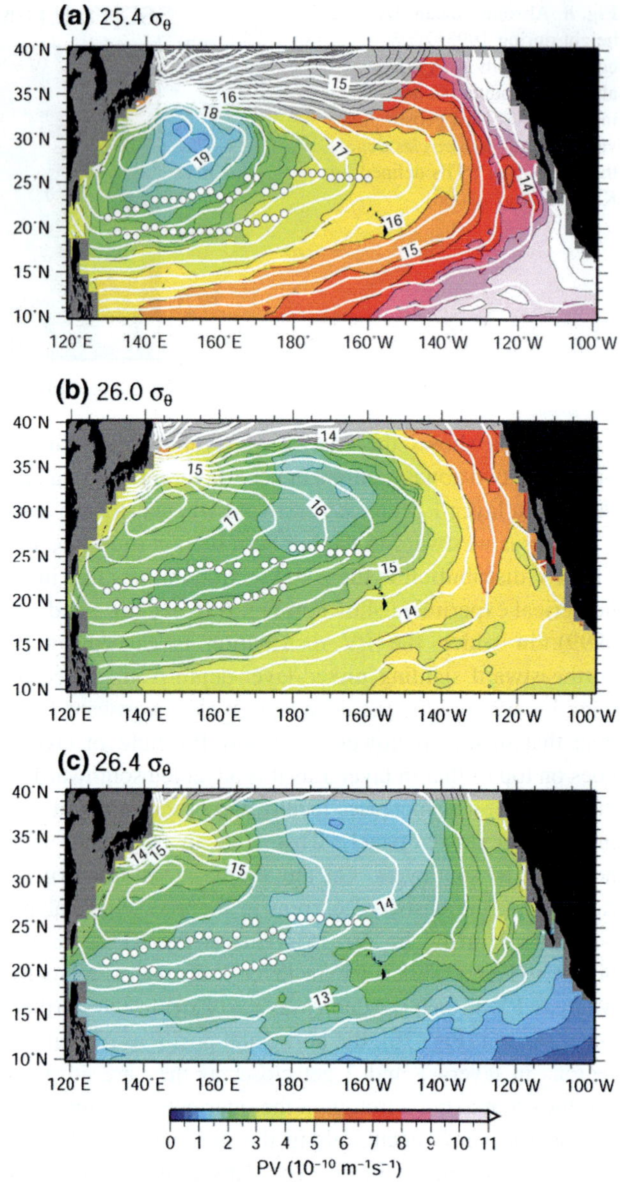

Fig. 9 Map of long-term mean PV on the isopycnal surfaces of **a** 25.4 σ_θ, **b** 26.0 σ_θ, and **c** 26.4 σ_θ. PV is defined as the same as in Fig. 2. Acceleration potential relative to 1000 dbar is shown by *white contours* with an interval of 0.5 m^2 s^{-2}. *Light shade* indicates the area with the winter sea surface density heavier than each isopycnal, and *open circles* denote the STFs. Modified from Kobashi et al. (2006)

3.2 Relation to mode waters

The mechanisms in Sects. 2.3 and 2.4 relate STFs to mode waters. Together with Aoki et al. (2002), Kobashi et al. (2006) revealed a close relationship between the STFs and two mode waters, STMW and CMW, from hydrographic observations. Figure 9 shows climatological mean PV on isopycnal surfaces of 25.4 σ_θ, 26.0 σ_θ, and 26.4 σ_θ with acceleration potential relative to 1000 dbar superimposed. These isopycnal surfaces correspond to those of STMW and the upper and lower portions of CMW. A schematic summary is shown in Fig. 10. Both figures are adapted from Kobashi et al.

The three STFs are all located along the southern flank of low PV waters in the thermocline (Fig. 9). Dynamically, meridional slopes of isopycnal surfaces of the STFs are related to large negative meridional PV gradients along isopycnals in the thermocline below the fronts (Fig. 2a; Kobashi et al. 2006). The northern STF is located on the southern edge of the Kuroshio recirculation, where the negative meridional PV gradient occurs within a narrow density range and is maintained by the low PV core of STMW (Figs. 9a, 10a). On the other hand, the negative PV gradient at the eastern and southern STFs spans over a wide density range. For the eastern STF, the upper and lower portions of CMW, which are advected along the inner and outer parts of the subtropical gyre, respectively (Oka and Suga 2005), converge to the north of the front, forming a thick low PV layer and maintaining the PV gradient at the

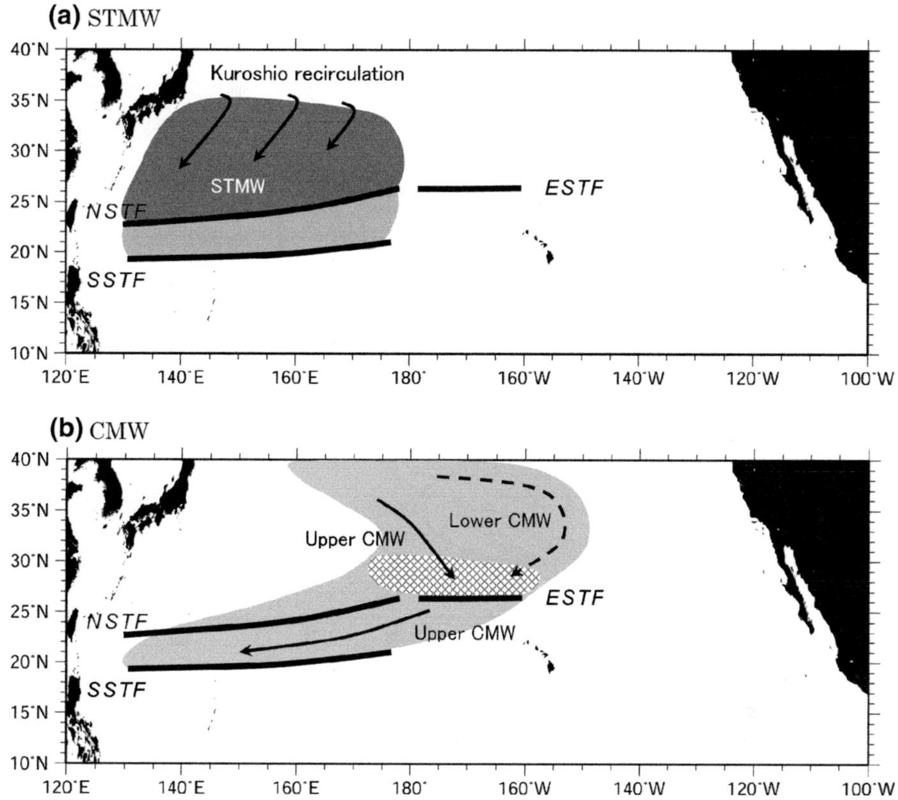

Fig. 10 Schematic of the relationship between STFs and **a** STMW and **b** CMW. The mode waters are represented by *shading*, whereas the northern, southern, and eastern STFs by *thick lines with labels* "NSTF", "SSTF" and "ESTF", respectively. *Arrows* in the *upper panel* indicate the advection of the STMW by the Kuroshio recirculation, whereas *solid and dashed arrows* in the *lower panel* represent the advection of the upper and lower CMWs by the geostrophic gyre flow. The area where the upper and lower CMWs converge is hatched in the *lower panel*. From Kobashi et al. (2006)

front (Figs. 9b, c, 10b). Similarly, at the southern STF, the negative PV gradient is due to a low PV layer to the north, with the southernmost portion of STMW stacked above the upper CMW advected from the central subtropical gyre. The thick low PV waters forces the upper thermocline to shoal, causing surface-intensified STCCs by the thermal wind relation.

The upper CMW extends across the eastern STF to the southwest, almost following streamlines (Fig. 9b), indicating advective spreading by geostrophic flow. On the other hand, the southward spread of low PV across the northern STF in the STMW layer is due to more than just mean flow advection. In Fig. 9a, there are indications that low PV of the STMW south of the northern STF is diffused from a low PV core from the north across the streamlines, possibly by intensive mesoscale eddies in the STCC region. The eddy activity is strongest along the northern STF, and remarkably decreases to the south of the southern STF (Qiu 1999; Kobashi and Kawamura 2002). Eddy transport of STMW across the mean flow is indicated from an analysis of eddy-resolving ocean GCM simulations by Nishikawa et al. (2010). The similar diffusive feature is identified also on maps of PV and streamfunction relative to 2000 dbar produced by Suga et al. (2004) using an isopycnally averaged climatology (their Fig. 5). Thus, the vertical stacking of the mode waters along the southern STF is caused by not just geostrophic advection of the upper CMW but also eddy diffusion of low PV from the STMW core north of the northern STF.

The vertical stacking of low PV waters over a wide density range at the eastern and southern STFs supports the mechanism due to vertical accumulation of low PV fluids in Sect. 2.4 (Kubokawa 1999; Kubokawa and Inui 1999). For the northern STF, the PV gradient is provided mainly by the STMW core to the north advected by the Kuroshio recirculation gyre. As suggested by Aoki et al. (2002), because the STMW resembles a homogeneous water mass in the northwestern subtropical gyre in the theoretical model of Kubokawa (1997), the northern STF could be generated by the stationary Rossby wave mechanism in Sect. 2.3. In addition, the STMW forms progressively eastward on denser isopycnals (Suga and Hanawa 1990), and in the Kuroshio recirculation, lighter STMW stacks above denser STMW (e.g., Bingham 1992; Suga et al. 2004; Oka et al. 2011). The PV gradient at the northern STF occurs in the density range of the STWM (Kobashi et al. 2006). This vertical stacking of the STMW is consistent with Kubokawa's (1999) mechanism. Further examination will be needed to clarify which mechanism is important in the formation of the northern STF.

STCC anchored by mode waters is also simulated by idealized ocean GCMs (Kubokawa and Inui 1999; Endoh et al. 2006) and realistic ocean and coupled GCMs (Yamanaka et al. 2008; Tokinaga et al. 2009a; Xie et al. 2011;

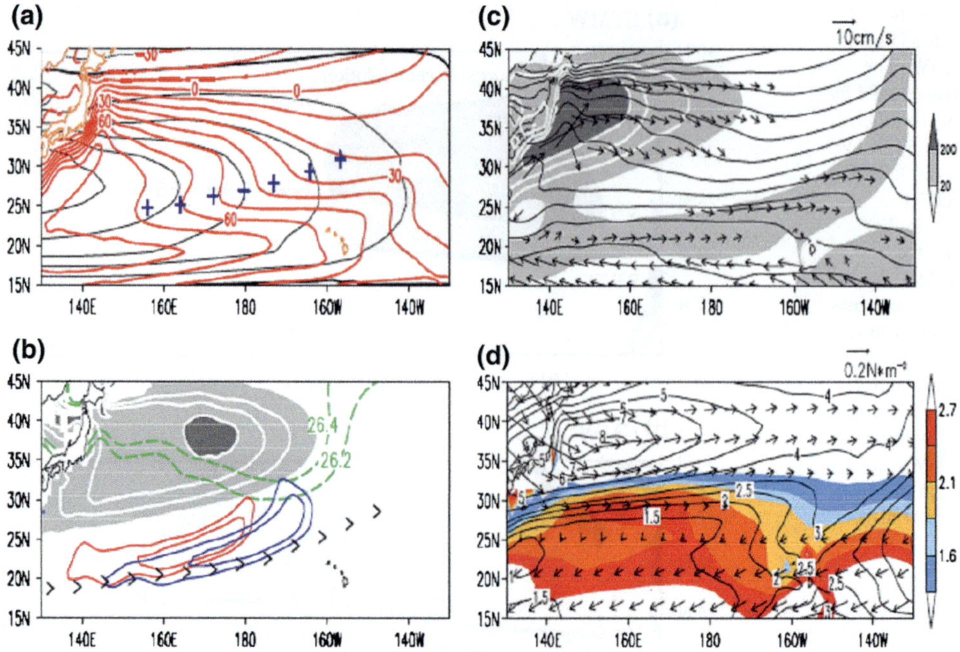

Fig. 11 March climatology from the GFDL coupled model (CM2.1). **a** Annual-mean Sverdrup streamfunction (*black contours* at 20 Sv intervals), SSH (*red contours* at 10 cm intervals), and *plus symbols* denote where a PV minimum is reached in the layer of 25.6–26.6 σ_θ. **b** PV on 26.2 σ_θ (*blue contours*) and 26.4 σ_θ (*blue contours*) isopycnals, along with the mixed layer depth (*light and dark shades* >100 and 400 m; *white contours* at 100 m intervals), surface density (*green*), and ">" symbols denote the axis of STCC. **c** 50 m current velocity, SST (*black contours* at 2°C intervals), and net surface heat flux (*gray shade*; *white contours* at 60 W m^{-2} intervals). **d** Surface wind stress (N m^{-2}), Ekman pumping velocity (*color shade* at 1.0 × 10^{-6} m s^{-1} intervals), and precipitation (*black contours* at 0.5 kg m^{-2} s^{-1}). From Xie et al. (2011)

Xu et al. 2012; Nonaka et al. 2012). Figure 11 displays the result from a 300-year control simulation from the Geophysical Fluid Dynamics Laboratory (GFDL) coupled climate model (CM2.1), shown by Xie et al. (2011). The model reproduces a narrow eastward jet extending from the western Pacific to the northeast of Hawaii, against the Sverdrup flow (Fig. 11a, c). This STCC agrees in location roughly with the observed southern and eastern STCCs (Fig. 1a), with a notable difference. The observed eastern STCC extends nearly zonally, but in the model the band of the eastern STCC tilts northeastward and apparently is continued from the southern STCC. Low PV waters subducted from the intersections of outcrops and a mixed-layer-depth front are advected by the subtropical gyre, forming the tongues of PV minimums on isopycnal surfaces to the southwest (Fig. 11b). They are vertically stacked up, resulting in a thick low PV layer, along which the STCC forms (Fig. 11b), consistent with observations.

Compared with observations, in the coupled climate models of the CM2.1, analyzed by Xie et al. (2011) and Xu et al. (2012), and the Community Climate System Model (CCSM3.0), analyzed by Tokinaga et al. (2009a), the STCC is too strong (compare Figs. 8 with 11c) as a result of too strong mode waters (Xie et al. 2011). Non-eddy-resolving models tend to produce enhanced mode waters (e.g., Ladd and Thompson 2001; Thompson and Cheng 2008).

In contrast, eddy-resolving ocean GCMs, the Meteorological Research Institute Community Ocean Model (MRI.COM) analyzed by Yamanaka et al. (2008), and the Ocean GCM for the Earth Simulator (OFES) by Nonaka et al. (2012) simulate the eastern STCC with a strength comparable to or slightly weaker than that in observations. In these models, the STCC is more confined to the central subtropical gyre as the eastern STCC, and the southern STCC does not appear in those models, probably due to mixing effect of eddies. Low PV tongues subducted from the central subtropical gyre (Fig. 11b) are much more diffused by eddies as they begin their excursion downstream, resulting in the weaker STF and STCC to the west. The eastern STCC in the MRI.COM and OFES extends zonally with no notable northeastward tilt, consistent with observations. The difference from the climate models could be attributable to reproducibility of subduction and circulation of CMW that is responsible for the formation of the eastern STCC.

4 Atmospheric and climatic effects

This section addresses the question: How do STCCs affect atmosphere and climate? We first look at observations in which an SST front plays an important role. We discuss

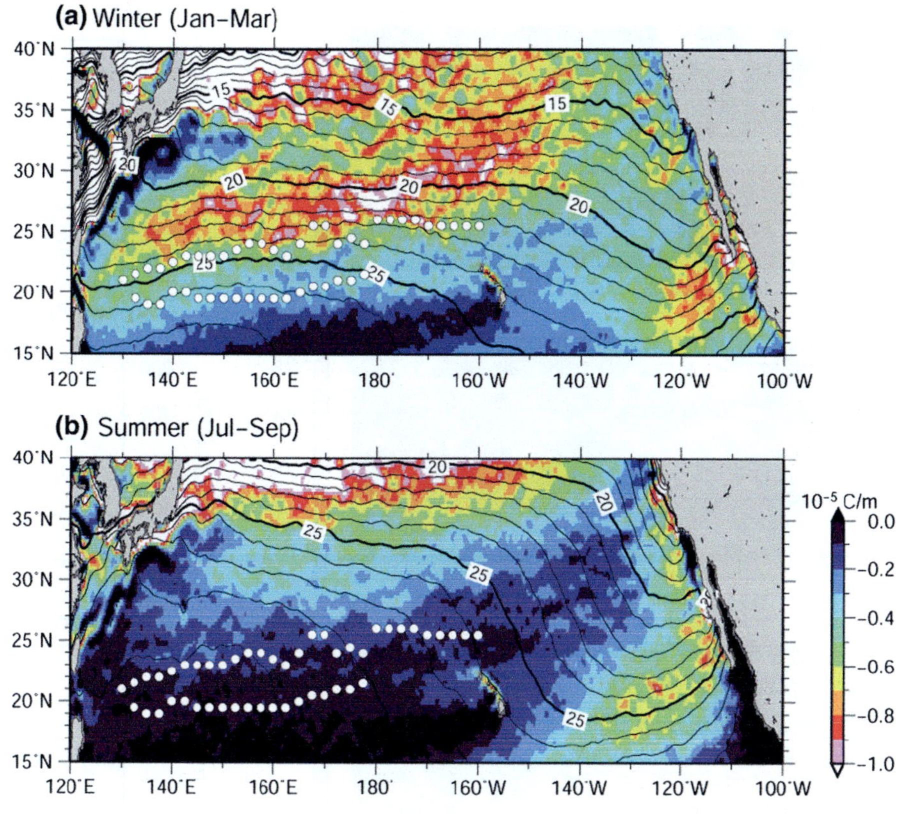

Fig. 12 Seasonal mean SST and its meridional gradient in **a** winter and **b** summer, calculated from an optimally interpolated SST dataset during 1993–2008 (Reynolds et al. 2007). *Open circles* denote the position of STFs defined by Kobashi et al. (2006)

how an SST front is related to subsurface STFs forced by mode waters, and how the SST front affects the atmosphere. Then, we move to results from coupled climate models.

Figure 12 shows seasonal mean SST and its meridional gradient in winter and summer, calculated from an optimally interpolated SST dataset produced from the blend of infrared and microwave satellite observations and in situ measurements (Reynolds et al. 2007). In winter, a pronounced SST front appears in the latitudinal range of 23–35°N, slightly tilting to the northeast, from the western to central subtropical gyre, but it disappears in summer. This SST front is obvious from late fall to early spring (e.g., Roden 1980a). Different from subsurface STFs, the SST front is governed by mixed layer dynamics: its frontogenesis and frontolysis are primarily in response to surface Ekman convergence and surface heat flux (Kazmin and Rienecker 1996; Dinniman and Rienecker 1999). As mentioned in Sect. 2.2, Ekman convergence is too weak to generate a mean STF, but on the seasonal timescale, it could generate an SST front, because k/α changes with season.

To examine the relationship between the SST front and the STFs, we plot the position of the STFs (Kobashi et al. 2006) on Fig. 12. The winter SST front is located along the north of the northern and eastern STFs, and its magnitude rapidly decreases to the south of the STFs. The feature is similar in early spring. This relationship strongly suggests that the SST front is affected not only by Ekman convergence but also by the subsurface STFs. Probably the STFs anchor the SST front above, through an entrainment process in which subsurface temperature anomalies are released to the surface when deep mixed layer develops. Further investigations are needed to reveal the effects of the STFs on the SST front.

A recent finding is that the SST front exerts significant influence on the atmosphere and climate in the western subtropical gyre. From an analysis of QuikSCAT and Tropical Rainfall Measuring Mission (TRMM) satellite observations and reanalysis data from JRA-25, Kobashi et al. (2008) found that, in April to May along the SST front, surface wind stress curls turn weakly cyclonic on the general background of anticyclonic curls that drive the subtropical gyre (Fig. 13). The cyclonic wind curl corresponds to a trough of wind curl shown by Yoshida and Kidokoro (1967a, b), though rather confined to the western subtropical gyre compared with the Yoshida and Kidokoro's trough (see Sect. 2.1). The cyclonic curl anomaly is accompanied by a band of high column water vapor slightly to the south (Fig. 13c), indicating a deep structure of the moist layer. On the weather timescale, the cyclonic curl anomaly is associated with low pressure systems of a subsynoptic scale in space. In the lows, convective rain takes place, with deep upward motion moistening the entire

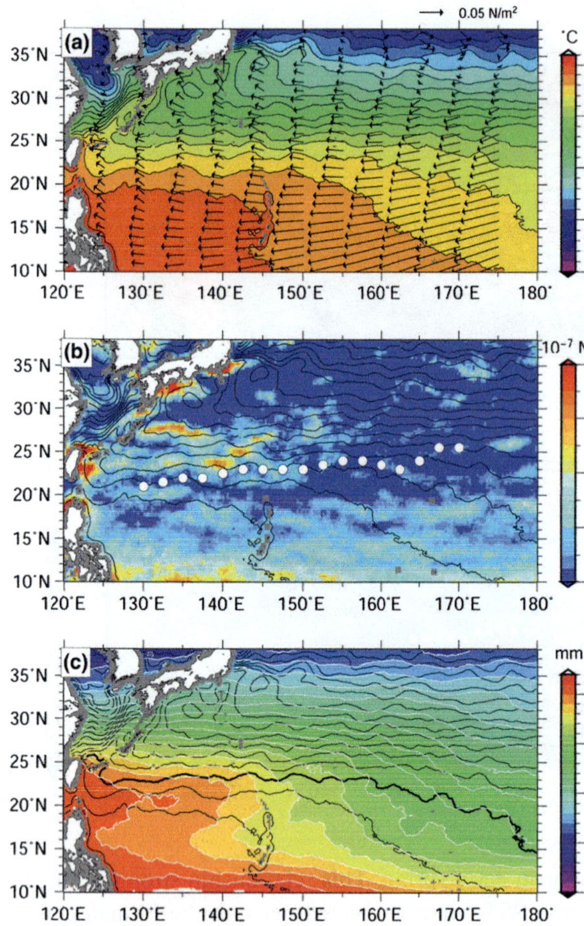

Fig. 13 May climatology from QuikSCAT and TRMM satellite observations: **a** vector wind stress, **b** wind stress curl, and **c** columnar water vapor, along with SST in contours. *Open circles* in **b** denote the position of the northern STF defined by Kobashi et al. (2006). The *thick contour* in **c** marks a 27°C isotherm. Along the SST front near 25°N, a local trough of easterly trade winds (**a**) forms cyclonic wind curl anomaly (**b**). Modified from Kobashi et al. (2008)

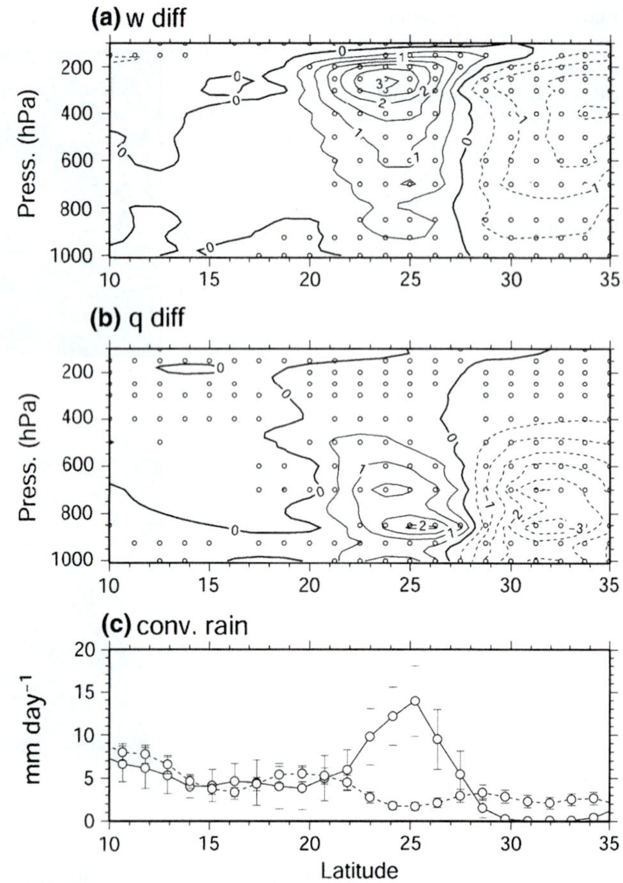

Fig. 14 Composite differences of **a** upward vertical wind (cm/s) and **b** specific humidity (g/kg) at 142.875°E, between the case when cyclonic wind curl appears at the surface and the others. *Open circles* in **a** and **b** show grid points of significant difference with confidence greater than 95%. The vertical wind speed is computed from pressure vertical velocity by the approximate relationship for synoptic scales under hydrostatic approximation. **c** Meridional profiles of convective rain rate (mm/day) in the cyclonic curl (*solid line*) and the other (*dashed line*) composites, along with 95% confidence intervals. These are calculated from JRA-25 data. Along the SST front around 25°N, convective rain takes place with strong upward motion moistening the entire troposphere. Modified from Kobashi et al. (2008)

troposphere (Fig. 14). The lows are enhanced by condensational heating and grow on the baroclinicity anchored by the SST front, resulting in the formation of cyclonic wind curls on intraseasonal timescales.

Deep atmospheric responses have been recognized in other regions of strong mid-latitude western boundary currents; the Kuroshio Extension (Tokinaga et al. 2009b), the Gulf Stream (Minobe et al. 2008), and the Agulhas Return Current (Liu et al. 2007). Distinct from these western boundary currents, the SST front in the STCC regions is not so strong, but features SST greater than 27°C on its southern flank in April to May (Fig. 13c). This high SST is conducive to deep convection, favoring the formation of cyclonic wind curl anomaly (Kobashi et al. 2008).

The cyclonic wind curl anomaly in May is accompanied by enhanced precipitation. Kobashi et al. (2008) suggested that this increase in precipitation appears to correspond to the so-called pre-Baiu/Meiyu front, which has been independently identified from previous meteorological studies (Kato and Kodama 1992; Tanaka 1992; Wang and LinHo 2002). The pre-Baiu/Meiyu front is manifested as a cloud and rain band prior to the onset of the Baiu/Meiyu that is one of the most remarkable events in the East Asia summer monsoon. The SST front seems to anchor this pre-Baiu/Meiyu band, a hypothesis that needs further investigation.

Atmospheric effects of STCCs are demonstrated in coupled model simulations based on the CCSM3.0 and CM2.1, by Tokinaga et al. (2009a) and Xie et al. (2011), respectively. In these models, the eastern STCC creates an SST ridge via thermal advection (Fig. 11c), forming cyclonic surface wind curls and local maximum of precipitation along the STCC (Fig. 11d). Xie et al. (2011)

suggest that the wind curl response provides a positive feedback on SST via the Sverdrup flow advection mechanism (Liu 1993). The cyclonic wind curl forces a northward Sverdrup flow that acts to strength the SST warming, and the SST warming intensifies the cyclonic wind curl anomalies. This positive feedback is suggested to enhance the coupling between the eastern STCC and the atmosphere.

The atmospheric effect shown by the models differs from that by the observations. In the observations, it results from enhanced atmospheric baroclinicity by the SST front, whereas in the models it is caused by SST anomaly due to thermal advection by the STCC. Recent careful analyses of in situ surface meteorology data and satellite observations find a local peak of net surface heat flux from ocean to atmosphere along the eastern STCC (F. Kobashi and S.-P. Xie, personal communication), suggesting a similar atmospheric response to that of the models. The effect of SST front observed in the western subtropical gyre has not been reported so far from models. Further investigations are needed to clarify the whole picture of atmospheric effects of STCCs and their feedback into the ocean.

The results from the models reveal that mode waters affect atmosphere and climate through surface current formation. This mechanism is different from a previously recognized re-emergence mechanism (e.g., Alexander et al. 1999; Hanawa and Sugimoto 2004; Sugimoto and Hanawa 2005), in which mode waters serve as an isolated heat reservoir memorizing wintertime ocean–atmosphere interaction, modulating local or remote SST by releasing their heat anomalies when deep surface mixed layer develops and entrains mode waters. Mode waters that re-emerge are usually located at subsurface depths in the seasonal thermocline. In contrast, the new mechanism indicates that mode waters that are subducted into the permanent thermocline change surface thermal structure via the surface current formation, affecting atmosphere and climate.

5 Variability and its mechanism

This section addresses the questions: What causes variability of STCCs? Does variability in mode waters cause that of STCCs? Several mechanisms have been suggested, which emphasize the importance of mode waters, atmospheric forcings, and ocean internal dynamics. We review variability of STCCs and its mechanism on seasonal to decadal and longer timescales.

5.1 Seasonal variations

The three STCCs exhibit different seasonal variations. Climatologies of SSH based on hydrographic observations and altimeter observations show that the northern STCC is strong in late winter to spring and weak in fall, whereas the eastern STCC is strong in summer to fall and weak in spring (White et al. 1978; Kobashi and Kawamura 2002; Kobashi and Xie 2012). The southern STCC appears with eastward velocity only in spring to summer (Kobashi and Kawamura 2002), and in winter and fall it is identified as a meridional minimum of westward flow (Kobashi and Xie 2012). Similar seasonal variations of the eastern STCC are also simulated by the OFES (Nonaka et al. 2012).

Mechanisms have not been clarified yet, though there are some implications from observations and numerical simulations. Observations indicate that the strong northern STCC in late winter is caused by a strong SST front (Fig. 12a) with a deep vertical extent within mixed layer depth, whereas that in spring is due to a strengthened subsurface front (Kobashi and Kawamura 2002; Kobashi and Xie 2012). Kobashi and Xie suggest the importance of cyclonic wind stress curl in May along the northern flank of the northern STCC (Fig. 13b). They conducted composite analyses using altimeter-derived surface velocity data, showing that in years with strong cyclonic curl anomaly, the seasonal cycle of the northern STCC is clear with a peak in June, whereas in other years, it is not. In contrast, performing numerical experiments using an idealized ocean GCM, Takeuchi (1986) reproduced seasonal variations similar to those of the northern STCC, suggesting that seasonal change in basin-scale wind stress is important for the seasonal variations of the model STCC.

In climatology, the three STCCs are anchored by mode waters (Sect. 3.2). The STMW, which maintains the northern and southern STCCs, has clear seasonal change in distribution and PV, because of the production of thick STMW in winter and subsequent advection by the Kuroshio recirculation and considerable dissipation within a year or so (Suga and Hanawa 1995). It has not been examined yet whether such seasonal changes of the STMW affect STCC variations. Another possible mechanism is annual Rossby waves (e.g., Mysak 1983; Cummins et al. 1986) that could modulate mode waters and thus induce STCC variations. Seasonal variations could be a superposition of currents forced by atmospheric forcings and mode waters.

5.2 Interannual variations

Interannual variations of the northern STCC are investigated by Qiu and Chen (2010) and Kobashi and Xie (2012). From analysis of satellite observations and JMA repeat hydrographic surveys along 137°E, Qiu and Chen (2010) found interannual changes in mesoscale eddy variability in the northern STCC region, and showed that the vertical shear of the northern STCC tends to be larger in eddy-rich years than in eddy-weak years. They suggested that the intensity of baroclinic instability controls the interannual changes in the eddy activity. Furthermore, they showed a correlation between the eddy activity and the meridional Ekman convergence associated with basin-scale surface winds, suggesting that the Ekman convergence affects SST frontogenesis along the STCC, causing change in the vertical shear and thus the eddy activity. Their results indicate that Ekman convergence is a cause of interannual variations in the STCC.

On the other hand, Kobashi and Xie (2012) emphasized the importance of a zonal band of cyclonic wind curl in May along the northern flank of the northern STCC (Fig. 13b). From analyses of satellite altimeter data and Argo profiling floats, Kobashi and Xie (2012) found a positive correlation of interannual variations between the cyclonic wind stress curl and the SST front in May, suggesting a positive feedback between them. Furthermore, the cyclonic wind curl is negatively correlated with the SSH and SST below. The strong (weak) cyclonic wind anomaly elevates (depresses) the thermocline and causes the fall (rise) in the SSH and SST, accelerating (decelerating) the northern STCC to the south. The cyclonic wind curl takes a peak in May, followed by the peak of the STCC in June with about 1 month lag. The response of the thermocline to the wind forcing occurs probably locally along the band of the cyclonic wind rather than via Rossby wave propagation. Their results suggest that the wind forcing in May causes the subsequent variations in the northern STCC.

Qiu and Chen (2010) and Kobashi and Xie (2012) both indicate a central role of atmospheric forcings that are not important in the formation of mean STCCs (Sects. 2.1 and 2.2). Their atmospheric forcings are different, but their results are not mutually exclusive. Because eddy production in the northern STCC is seasonally dependent, caused by baroclinic instability in late winter (Qiu 1999; Kobashi and Kawamura 2002), the interannual change in the eddy activity is considered to represent the STCC strength in late winter. Therefore, the result of Qiu and Chen suggests that the Ekman convergence is important for variations in the late winter STCC. Indeed, Kobashi and Xie (2012) suggested that interannual variations of the SST front in late winter are correlated with meridional Ekman convergence forcing. In addition, they found a significant positive correlation between the SST front in late winter and the cyclonic wind curl in May, suggesting that the strong SST front in late winter provides favorable conditions for the development of a strong SST front in spring, affecting the formation of the cyclonic wind curl in May.

Other mechanisms of interannual variations are proposed from an analysis of a long hindcast simulation of the OFES. Nonaka et al. (2012) found that interannual variations of the eastern STCC in the model occur almost simultaneously along a zonal band of the STCC, with a vertically coherent change in current velocity down to at least 1000 m depth. These properties resemble those for zonal jets or striations examined by Richards et al. (2006), suggesting one possibility that they are oceanic internally induced variability, likely caused by geostrophic turbulence. In addition, they found that interannual variations in the near-surface vertical shear of the STCC, which are deviations from the vertically uniform change, are negatively correlated with those in PV of CMW carried to the north of the STCC, indicating that lower PV mode water induces intensified vertical shear of the STCC. They showed that variations in PV of the mode water explain about 20% of interannual variance in zonal velocity of the STCC.

STCC variability due to mode waters has not yet been confirmed from observations. Kobashi and Xie (2012) did a preliminary comparison between changes in STMW thickness and those in strength of the northern STCC using 8-year Argo profiling data in the western subtropical gyre. Apparently, there seems to be no correspondence between them. Although not at STCCs, mode water-induced variations are demonstrated at HLCC by Sasaki et al. (2012). As mentioned in Sect. 1, HLCC is a wind-driven current west of the Hawaiian Islands. Analyzing satellite and Argo float observations, they found that enhanced eastward HLCC in 2003 and 2005 coincided with anomalous southward intrusion of CMW and eastern subtropical mode water (Hautala and Roemmich 1998) to the north of the HLCC. They showed that the intrusion of the mode waters increases the slope of the upper pycnocline, causing the strong HLCC at the surface. An elaborate analysis as done by Nonaka et al. (2012) will be needed to evaluate how interannual variations of STMW and CMW (Suga and Hanawa 1995; Yasuda and Hanawa 1997, 1999; Taneda et al. 2000; Hanawa and Kamada 2001; Hanawa and Yoritaka 2001; Oka 2009) affect STCCs.

5.3 Decadal to longer timescale variations

Mode water-induced variations are dominant on timescales longer than interannual (Nonaka et al. 2012). Model studies showed that changes in ventilation path and core PV of mode water in response to slow variations in wind bring about large subsurface temperature anomalies in the central subtropical gyre (Xie et al. 2000; Kubokawa and Xie 2002; Hosoda et al. 2004). Such changes in mode waters induce variations of STCCs on longer timescales.

Long hindcast simulations by eddy-resolving ocean GCMs indicate that the eastern STCC is intensified in the late 1970s with a strong southward intrusion of CMW to the north of the STCC, and is weakened in the early 1990s

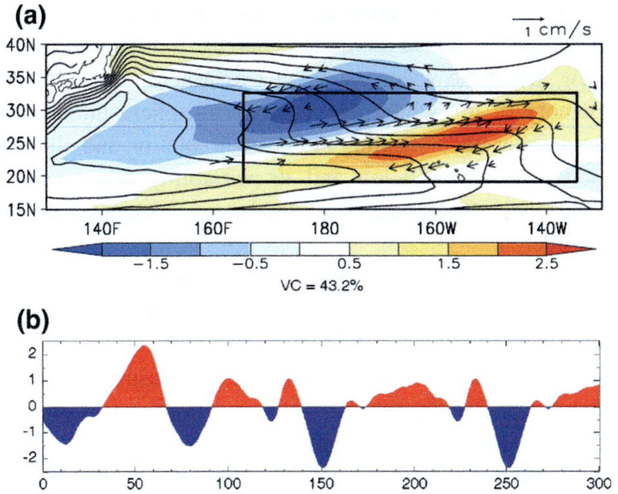

Fig. 15 **a** First empirical orthogonal function (EOF) mode of SSH (*color shade* in cm) in the central-gyre region (*box*), and **b** its principal component, calculated from a 300-year control simulation by the GFDL coupled model (CM2.1). Mean SSH (*black contours* at 10 cm intervals) and regression of 50 m current velocity upon the principal component are superimposed in **a**. From Xie et al. (2011)

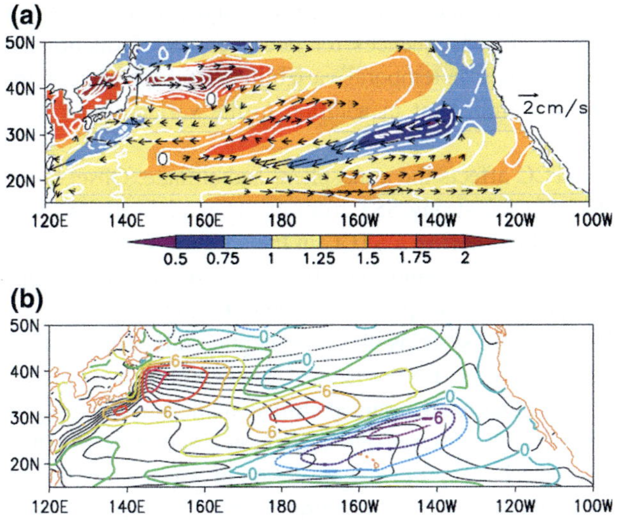

Fig. 16 50-year (2050–2000) changes in March in the A1B global warming simulation by the GFDL climate model (CM2.1): **a** SST (°C), 50 m current velocity (cm s^{-1}), and net surface heat flux (*white contours* at 5 W m^{-2}); SSH change (contours in cm) superimposed on **b** the climatology (*black contours*). From Xie et al. (2011)

when the ventilation of CMW is weak (Yamanaka et al. 2008; Nonaka et al. 2012). Yamanaka et al. related this interdecadal variability to changes in mode water formation and basin-scale wind fields, suggesting that the strong intrusion of the mode water is caused by much more production of CMW due to intensification of the westerly in the mid-1970s (Yasuda and Hanawa 1997), whereas the weak ventilation is associated with the increased warm water advection by the Kuroshio Extension (Hanawa and Kamada 2001) and the eastward shift of mode water formation area, both due to the spin up of the subtropical gyre in a delayed response to the wind intensification.

Changes in mode water ventilation leave characteristic signatures in spatial patterns of surface circulation and SST (Xie et al. 2011; Xu et al. 2012). The interdecadal variability of the eastern STCC is characterized by a northeast-slanted dipole of SSH changes in the central basin (Fig. 15), whose pattern indicates northeastward and southwestward current anomalies by geostrophic balance. The change in the advection of mean SST field produces northeast-slanted bands in SST anomaly fields. Such a pattern with slanted bands is characteristic of SST warming in the North Pacific and North Atlantic subtropical gyres in future projections (Fig. 16; Xie et al. 2010). The global warming reduces the subduction of mode waters in the Kuroshio Oyashio Extension region (Luo et al. 2009; Xu et al. 2012), decelerating the eastern STCC (Xu et al. 2012) and thus causing the slanted bands of changes in surface circulation and SST (Fig. 16). Mode water ventilation is an important mechanism for interdecadal variations as well as changes under global warming.

There is no observational study so far on the decadal variation of STCC. Argo observations have started in 2000, and a large amount of hydrographic data that resolving STCCs and mode waters has been accumulated for the past several years. In the near future, Argo observations will enable us to examine how STCCs are affected by decadal fluctuations in STMW and CMW (e.g., Yasuda and Hanawa 1997; Ladd and Thompson 2002; Suga et al. 2003; Qiu and Chen 2006; Qu and Chen 2009; Sugimoto and Hanawa 2010; Oka and Qiu 2012).

6 Summary and other implications

We have reviewed the progress in studies of the North Pacific STCCs and STFs. Theoretical, observational, and model simulation studies for the past decade have revealed the importance of mode waters in the formation and variability of STCCs. There are three distinct STCCs accompanying subsurface STFs in the North Pacific. They are each anchored by low PV that STMW or CMW carries from the north. The three STCCs support the formation mechanism proposed by Kubokawa (1999) and Kubokawa and Inui (1999), in which the vertical stacking of low PV fluids causes STCCs. Shock formation of stationary Rossby waves (Kubokawa 1997) is also suggested to be important in the formation of the northern STCC. Eddy-resolving ocean models and coupled climate models show that mode waters not only cause the STCCs but also induce variations on interannual to interdecadal timescales and long-term changes in response to global warming. These results demonstrate that mode water is a dynamical tracer with important consequences for ocean current formation and variability. For temporal variability, atmospheric forcings are also important in addition to the variability of mode waters.

STCC affects the atmosphere and climate. Observations find that the strong SST front in spring along the northern flank of the northern STCC in the western subtropical gyre affects surface wind curl, precipitation, and water vapor content with a deep vertical extent, through generations of low pressure systems of a subsynoptic scale. This observed effect of the SST front has not been confirmed by models. On the other hand, coupled climate models tend to produce another type of ocean–atmosphere interaction, in which the eastern STCC causes the advection of warm water to form an SST anomaly in the central subtropical gyre and affect the atmosphere.

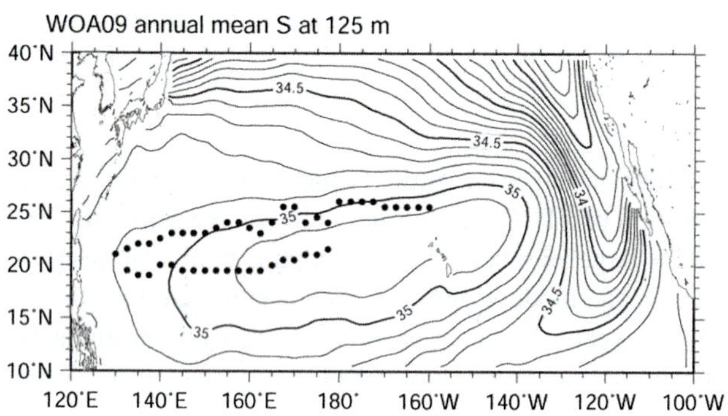

Fig. 17 Annual-mean salinity distribution at depth of 125 m, derived from WOA09. *Black dots* denote the position of STFs defined by Kobashi et al. (2006), the same as those in Fig. 1

These results represent dynamical and climatic effects of mode waters, which are quite different from the traditional view that these water masses act as a heat reservoir with memory of the wintertime ocean–atmosphere interaction. For a fuller understanding of ocean circulation and climate, mode water ventilation and its variability and coupling with the atmosphere are important elements to be considered.

Finally, we discuss some implications for future study. Early observational studies noticed that the northern STF is located not only along STMW to the north but also along the tropical water to the south (Uda and Hasunuma 1969). The North Pacific tropical water is a water mass characterized as a subsurface salinity maximum in the subtropical gyre, which results from the subduction of surface saline water formed by intensive surface evaporation. Previous studies have not paid much attention to salinity effects on STFs, because they have described STFs only from temperature data, using a constant salinity value (White et al. 1978) and salinity estimated from climatological mean temperature and salinity relation (White and Walker 1985; Kobashi and Kawamura 2002; Kobashi et al. 2006).

We plot the position of the STFs (Kobashi et al. 2006) over an annual-mean subsurface salinity map using World Ocean Atlas 2009 data (WOA09; Antonov et al. 2010). Figure 17 is the result, showing that the northern and eastern STFs are located along the northern flank of the tropical water. Indeed, analyses of climatology data (Yuan and Talley 1996) and synoptic observations (Roden 1980b) indicate that STFs are almost collocated with the salinity front. Because temperature and salinity have opposing effects on density, the salinity front weakens the density

Fig. 18 Magnitude of horizontal gradients of density at depth of 125 m, calculated from annual mean climatology of WOA09

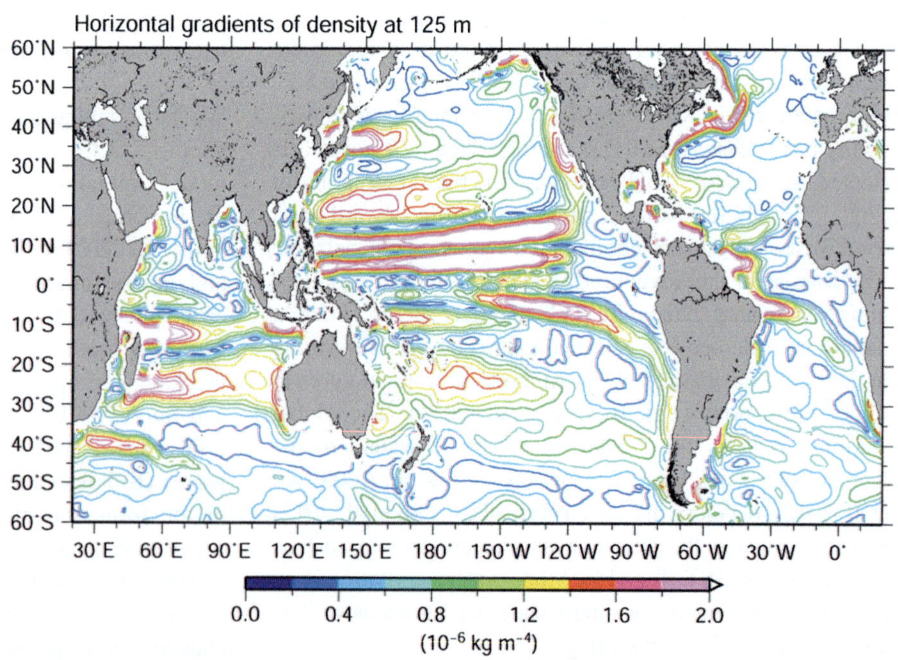

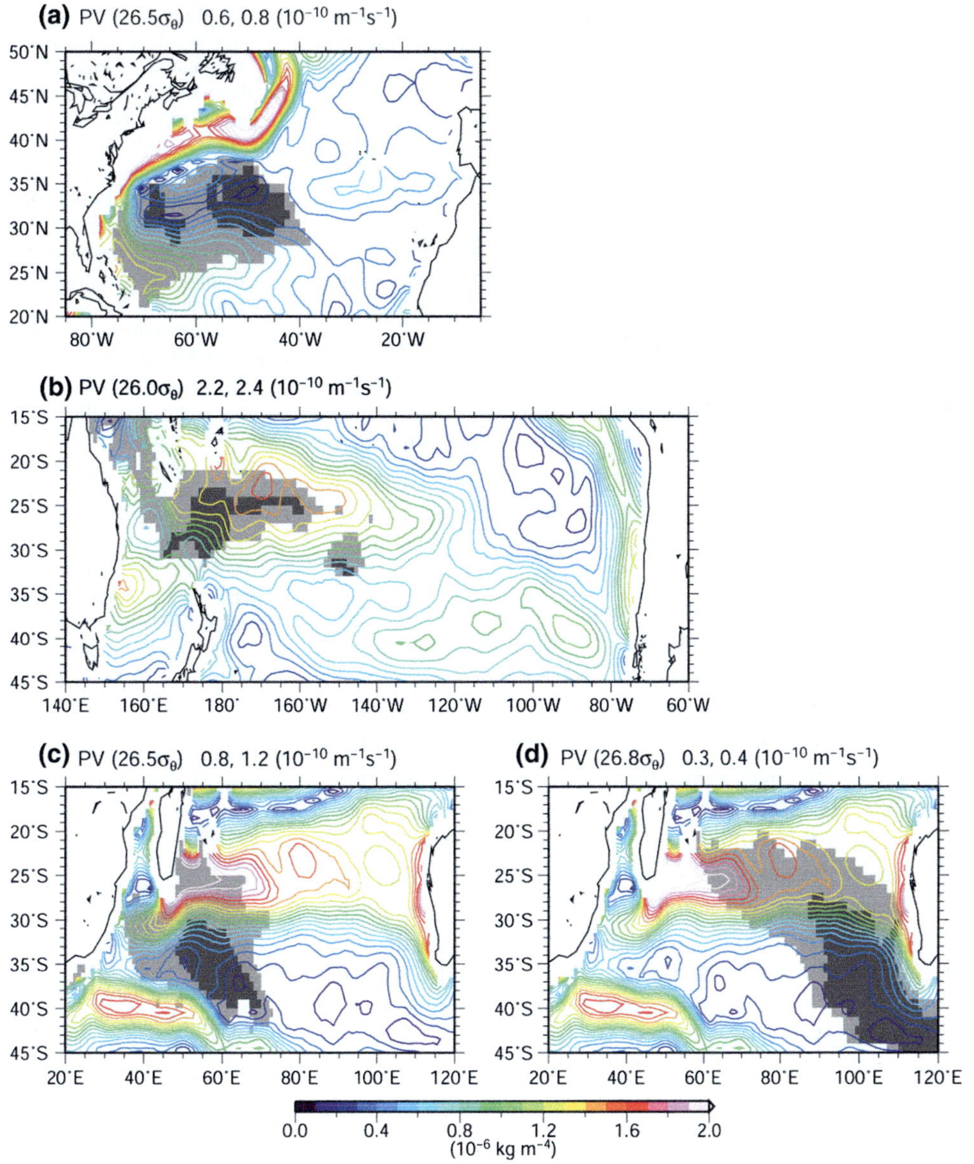

Fig. 19 Isopycnal PV lower than **a** 0.6, 0.8×10^{-10} m^{-1} s^{-1} on 26.5 σ_θ, **b** 2.2, 2.4×10^{-10} m^{-1} s^{-1} on 26.0 σ_θ, **c** 0.8, 1.2×10^{-10} m^{-1} s^{-1} on 26.5 σ_θ, and **d** 0.3, 0.4×10^{-10} m^{-1} s^{-1} on 26.8 σ_θ is highlighted with *light* and *dark shades*, calculated from annual mean climatology of WOA09. PV is defined as the same as in Fig. 2. *Color contours* are the same as in Fig. 18. Low PV water corresponds to **a** North Atlantic STMW, **b** South Pacific STMW, **c** Indian Ocean STMW, and **d** Southeast Indian subantarctic mode water

STFs. It is still unclear how large interannual variability of the tropical water (Suga et al. 2000) affects STCC variations. Accumulation of Argo observations will make it possible to quantitatively estimate salinity effects on STCC variations.

STCC is commonly observed in other oceans, though our knowledge is very limited with regard to the formation, variability, and atmospheric effect. Figure 18 shows the global map of the magnitude of horizontal density gradient at subsurface depth. The magnitude was calculated using $\sqrt{(\partial\rho/\partial x)^2+(\partial\rho/\partial y)^2}$, where ρ is annual mean density computed from WOA09 (Antonov et al. 2010; Locarnini et al. 2010). Strong spatial smoothing of WOA09 data is not ideal for the analysis on narrow bands of STF, but it yields the overall distribution of a broad feature of STF. Indeed, the two bands of STF in the western North Pacific are smoothed out and appear as one broad front in Fig. 18.

In the North Atlantic subtropical gyre, a relatively strong front appears near 25°N east off the Bahamas, extending to the east in the western Sargasso Sea. This front is a subsurface STF associated with eastward current at the surface (e.g., Reid 1978; Olson et al. 1984; Halliwell et al. 1994). In the South Pacific, a prominent zonal band with large density gradients is seen in the latitudes of 20–30°S to the west of 120°W. This front corresponds to an STF, and accompanies a shallow eastward current called the south tropical countercurrent (Merle et al. 1969; Qiu and Chen 2004). A similar broad front is evident in the South Indian Ocean, where the front between 20°S and 30°S extends from the east of Madagascar to the west of Australia. An eastward current associated with this front was recently identified as the South Indian Ocean countercurrent (Siedler et al. 2006; Palastanga et al. 2007). In the South Atlantic, although STF and its associated countercurrent were reported around 28°S in the western subtropical gyre (Tsuchiya 1985; Me'mery et al. 2000), its signature is not clear in Fig. 16, probably due to the paucity of data and strong smoothing in WOA09.

These STFs seem to be associated with low PV mode water on the poleward side, similar to the North Pacific. With reference to the horizontal distribution of mode waters illustrated by Hanawa and Talley (2001), we examine the possible connection between the STFs and mode waters. Figure 19 displays the isopycnal PV of mode waters, together with the density gradient in Fig. 18. The PV is calculated using WOA09. The STF in the North Atlantic is located along the southern flank of North Atlantic STMW (Worthington 1959), very similar to the northern STF in the North Pacific (Fig. 19a). In the South Pacific, the STF associated with south tropical countercurrent is almost collocated with the South Pacific STMW (Roemmich and Cornuelle 1992; Tsubouchi et al. 2007), and the meridional peak of the front is located slightly to the north of the STMW (Fig. 19b). In the Indian Ocean, the STF seems to be related to two different mode waters (Fig. 19c, d). To the west of the front, the STF is located along the northern flank of the Indian Ocean STMW (Gordon et al. 1987; Tsubouchi et al. 2010), whereas, to the east, it is found along the northern flank of the Southeast Indian subantarctic mode water (Thompson and Edwards 1981). The close relationship in spatial distribution between STFs and mode waters is suggestive of a role of mode waters in eastward current formation. More comprehensive analysis is necessary to reveal the impact of mode waters on global surface circulation and climate.

Acknowledgments We thank Prof. Shang-Ping Xie for many fruitful discussion and comments. Constructive comments from anonymous reviewers greatly helped improve the manuscript. This study is partially supported by Grants-in-Aid for Scientific Research of the Ministry of Education, Culture, Sports, Science and Technology, Japan [20740267, 22106007, 23340139]. The altimeter product was produced by Ssalto/Duacs and distributed by AVISO, with support from CNES. The SST dataset was obtained from the NOAA website.

References

Alexander MA, Deser C, Timlin MS (1999) The reemergence of SST anomalies in the North Pacific Ocean. J Clim 12:2419–2433

Antonov JI, Seidov D, Boyer TP, Locarnini RA, Mishonov AV, Garcia HE, Baranova OK, Zweng MM, Johnson DR (2010) World Ocean Atlas 2009. Salinity, vol 2. In: Levitus S (ed) NOAA Atlas NESDIS 69. US Government Printing Office, Washington, DC

Aoki S, Imawaki S (1996) Eddy activities of the surface layer in the western North Pacific detected by satellite altimeter and radiometer. J Oceanogr 52:457–474

Aoki Y, Suga T, Hanawa K (2002) Subsurface subtropical fronts of the North Pacific as inherent boundaries in the ventilated thermocline. J Phys Oceanogr 32:2299–2311

Bingham FM (1992) Formation and spreading of subtropical mode water in the North Pacific. J Geophys Res 97:11177–11189

Chu PC, Li R, You X (2002) Northwest Pacific subtropical countercurrent on isopycnal surface in summer. Geophys Res Lett 29:1842. doi:10.1029/2002GL014831

Conkright ME, Antonov JI, Baranova O, Boyer TP, Garacia HE, Gelfeld R, Johnson D, Locarnin RA, Murphy PP, O'Brien TD, Smolyar I, Stephens C (2002) World Ocean Database 2001. Introduction, vol 1. In: Levitus S (ed) NOAA Atlas, NESDIS 42. US Government Printing Office, Washington, DC

Cummins PF, Mysak LA, Hamilton K (1986) Generation of annual Rossby waves in the North Pacific by the wind stress curl. J Phys Oceanogr 16:1179–1189

Cushman-Roisin B (1984) On the maintenance of the subtropical front and its associated countercurrent. J Phys Oceanogr 14:1179–1190

Dewar WK (1987) Planetary shock waves. J Phys Oceanogr 17:470–482

Dewar WK (1991) Arrested fronts. J Mar Res 49:21–52

Dewer WK (1992) Spontaneous shocks. J Phys Oceanogr 22:505–522

Dinniman MS, Rienecker MM (1999) Frontogenesis in the North Pacific oceanic frontal zones: a numerical simulation. J Phys Oceanogr 29:537–559

Endoh T, Jia Y, Richards KJ (2006) Sensitivity of the ventilation process in the North Pacific to eddy-induced tracer transport. J Phys Oceanogr 36:1895–1911

Gordon AL, Lutjeharms JRE, Grundlingh ML (1987) Stratification and circulation at the Agulhas retroflection. Deep Sea Res 34:565–599

Halliwell GR Jr, Peng G, Olson DB (1994) Stability of the Sargasso Sea subtropical frontal zone. J Phys Oceanogr 24:1166–1183

Hanawa K, Kamada J (2001) Variability of core layer temperature (CLT) of North Pacific subtropical mode water. Geophys Res Lett 28:2229–2232

Hanawa K, Sugimoto S (2004) 'Reemergence' areas of winter sea surface temperature anomalies in the world's oceans. Geophys Res Lett 31:L10303. doi:10.1029/2004GL019904

Hanawa K, Talley LD (2001) Mode waters. In: Siedler G, Church J (eds) Ocean circulation and climate. International geophysics series. Academic, San Diego, pp 373–386

Hanawa K, Yoritaka H (2001) North Pacific subtropical mode water observed in long cross section along 32.5°N line. J Oceanogr 57:679–692

Haney RL (1971) Surface thermal boundary condition for ocean circulation models. J Phys Oceanogr 1:241–248

Hasunuma K, Yoshida K (1978) Splitting the subtropical gyre in the western North Pacific. J Oceanogr Soc Japan 34:160–172

Hautala SL, Roemmich DH (1998) Subtropical mode water in the Northeast Pacific basin. J Geophys Res 106:19671–19681

Hosoda S, Xie S-P, Takeuchi K, Nonaka M (2004) Interdecadal temperature variations in the North Pacific central mode water simulated by an OGCM. J Oceanogr 60:865–877

Hwang C, Wu C-R, Kao R (2004) TOPEX/Poseidon observations of mesoscale eddies over the subtropical countercurrent: kinematic characteristics of an anticyclonic eddy and a cyclonic eddy. J Geophys Res 109:C08013. doi:10.1029/2003JC002026

Ichikawa K, Gohda N, Arai M, Kaneko A (2004) Monitoring surface velocity from repeated ADCP observations and satellite altimetry. J Oceanogr 60:365–374

Kaneko I, Takatsuki Y, Kamiya H, Kawae S (1998) Water property and current distributions along the WHP-P9 section (137°–142°E) in the western North Pacific. J Geophys Res 103:12959–12984

Kato K, Kodama Y (1992) Formation of the quasi-stationary Baiu front to the south of the Japan Islands in early May of 1979. J Meteor Soc Jpn 70:631–647
Kawamura H, Sawa Y, Sakaida F (1995) Satellite observations of 3–6 months variation in the Kuroshio and the subtropical front. Umi Sora 71:9–15 (in Japanese with English abstract and legends)
Kazmin AS, Rienecker MM (1996) Variability and frontogenesis in the large-scale oceanic frontal zones. J Geophys Res 101:907–921
Kobashi F, Kawamura H (2001) Variation of sea surface height at periods of 65–220 days in the subtropical gyre of the North Pacific. J Geophys Res 106:26817–26831
Kobashi F, Kawamura H (2002) Seasonal variation and instability nature of the North Pacific subtropical countercurrent and the Hawaiian Lee countercurrent. J Geophys Res 107:3185. doi: 10.1029/2001JC001225
Kobashi F, Xie SP (2012) Interannual variability of the North Pacific subtropical countercurrent: role of local ocean–atmosphere interaction. J Oceanogr. doi:10.1007/s10872-011-0048-x
Kobashi F, Mitsudera H, Xie S-P (2006) Three subtropical fronts in the North Pacific: observational evidence for mode water-induced subsurface frontogenesis. J Geophys Res 111:C09033. doi:10.1029/2006JC003479
Kobashi F, Xie S-P, Iwasaka N, Sakamoto TT (2008) Deep atmospheric response to the North Pacific oceanic subtropical front in spring. J Clim 21:5960–5975
Kubokawa A (1995) Stationary Rossby waves and shocks on the Sverdrup coordinate. J Oceanogr 51:207–224
Kubokawa A (1997) A two-level model of subtropical gyre and subtropical countercurrent. J Oceanogr 53:231–244
Kubokawa A (1999) Ventilated thermocline strongly affected by a deep mixed layer: a theory for subtropical countercurrent. J Phys Oceanogr 29:1314–1333
Kubokawa A, Inui T (1999) Subtropical countercurrent in an idealized ocean GCM. J Phys Oceanogr 29:1303–1313
Kubokawa A, Xie S-P (2002) On steady response of a ventilated thermocline to enhanced Ekman pumping. J Oceanogr 58:565–575
Ladd C, Thompson L (2001) Water mass formation in an isopycnal model of the North Pacific. J Phys Oceanogr 31:1517–1537
Ladd C, Thompson L (2002) Decadal variability of North Pacific central mode water. J Phys Oceanogr 32:2870–2881
Liu Z (1993) Interannual positive feedbacks in a simple extratropical air–sea coupling system. J Atmos Sci 50:3022–3028
Liu Q, Wang S, Wang Q, Wang W (2003) On the formation of subtropical countercurrent to the west of the Hawaiian Islands. J Geophys Res 108:3167. doi:10.1029/2002JC001366
Liu WT, Xie X, Niiler PP (2007) Ocean–atmosphere interaction over Agulhas Extension meanders. J Clim 20:5784–5797
Locarnini RA, Mishonov AV, Antonov JI, Boyer TP, Garcia HE, Baranova OK, Zweng MM, Johnson DR (2010) World Ocean Atlas 2009. Temperature, vol 1. Levitus S (ed) NOAA Atlas NESDIS 68. US Government Printing Office, Washington, DC
Luo Y, Liu Q, Rothstein LM (2009) Simulated response of North Pacific mode waters to global warming. Geophys Res Lett 36:L23609. doi:10.1029/2009GL040906
Luyten JR, Pedlosky J, Stommel H (1983) The ventilated thermocline. J Phys Oceanogr 13:292–309
Masuzawa J (1969) Subtropical mode water. Deep Sea Res 16:436–472
Me'mery L, Arhan M, Alvarez-Salgado XA, Messias M-J, Mercier H, Castro CG, Rios AF (2000) The water masses along the western boundary of the south and equatorial Atlantic. Prog Oceanogr 47:69–98
Merle J, Rotschi H, Voituriez B (1969) Zonal circulation in the tropical western South Pacific at 170°E. Bull Jpn Soc Fish Oceanogr. Special Number (Prof. Uda's Commemorative Papers), 91–98
Michida Y (1997) Surface current field in the area of subtropical countercurrent observed with surface drifters. Rep Hydro Res 33:21–30 (in Japanese with English abstract)
Minobe S, Kuwano-Yoshida A, Komori N, Xie S-P, Small RJ (2008) Influence of the Gulf Stream on the troposphere. Nature 452:206–209
Mysak LA (1983) Generation of annual Rossby waves in the North Pacific. J Phys Oceanogr 13:1908–1923
Nakamura H (1996) A pycnostad on the bottom of the ventilated portion in the central subtropical North Pacific: its distribution and formation. J Oceanogr 52:171–188
Niller PP, Maximenko NA, McWilliams JC (2003) Dynamically balanced absolute sea level of the global ocean derived from near-surface velocity observations. Geophys Res Lett 30(22):2164. doi:10.1029/2003GL018628
Nishikawa S, Tsujino H, Sakamoto K, Nakano H (2010) Effects of mesoscale eddies on subduction and distribution of subtropical mode water in an eddy-resolving OGCM of the western North Pacific. J Phys Oceanogr 40:1748–1765
Nitani H (1972) Beginning of the Kuroshio. Stommel H, Yoshida K (eds) Kuroshio—its physical aspects. University of Tokyo Press, Tokyo, pp 129–163
Noh Y, Yim BY, You SH, Yoon JH, Qiu B (2007) Seasonal variation of eddy kinetic energy of the North Pacific subtropical countercurrent simulated by an eddy-resolving OGCM. Geophys Res Lett 34:L07601. doi:10.1029/2006GL029130
Nonaka M, Xie SP, Sasaki H (2012) Interannual variations in low potential vorticity water and the subtropical countercurrent in an eddy-resolving OGCM. J Oceanogr. doi:10.1007/s10872-011-0042-3
Oka E (2009) Seasonal and interannual variation of North Pacific subtropical mode water in 2003–2006. J Oceanogr 65:151–164
Oka E, Qiu B (2012) Progress of North Pacific mode water research in the past decade. J Oceanogr. doi:10.1007/s10872-011-0032-5
Oka E, Suga T (2005) Differential formation and circulation of North Pacific central mode water. J Phys Oceanogr 35:1997–2011
Oka E, Suga T, Sukigara C, Toyama K, Shimada K, Yoshida J (2011) "Eddy-resolving" observation of the North Pacific subtropical mode water. J Phys Oceanogr 41:666–681
Olson DB, Schott FA, Zantopp RJ, Leaman KD (1984) The mean circulation east of the Bahamas as determined by a recent measurement program and historical XBT data. J Phys Oceanogr 14:1470–1487
Onogi K, Tsutsui J, Koide H, Sakamoto M, Kobayashi S, Hatsushika H, Matsumoto T, Yamazaki N, Kamahori H, Takahashi K, Kadokura S, Wada K, Kato K, Oyama R, Ose T, Mannoji N, Taira R (2007) The JRA-25 reanalysis. J Meteor Soc Jpn 85:369–432
Palastanga V, van Leeuwen PJ, Schouten MW, de Ruijter WPM (2007) Flow structure and variability in the subtropical Indian Ocean: instability of the South Indian Ocean countercurrent. J Geophys Res 112:C01001. doi:10.1029/2005JC003395
Pedlosky J (1984) Cross-gyre ventilation of the subtropical gyre: an internal mode in the ventilated thermocline. J Phys Oceanogr 14:1172–1178
Qiu B (1999) Seasonal eddy field modulation of the North Pacific subtropical countercurrent: TOPEX/Poseidon observations and theory. J Phys Oceanogr 29:2471–2486
Qiu B, Chen S (2004) Seasonal modulations in the eddy field of the South Pacific Ocean. J Phys Oceanogr 34:1515–1527

Qiu B, Chen S (2006) Decadal variability in the formation of the North Pacific subtropical mode water: oceanic versus atmospheric control. J Phys Oceanogr 36:1365–1380

Qiu B, Chen S (2010) Interannual variability of the North Pacific subtropical countercurrent and its associated mesoscale eddy field. J Phys Oceanogr 40:213–225

Qiu B, Koh DA, Lumpkin C, Flament P (1997) Existence and formation mechanism of the North Hawaiian Ridge current. J Phys Oceanogr 27:431–444

Qu T, Chen J (2009) A North Pacific decadal variability in subduction rate. Geophys Res Lett 36:L22602. doi:10.1029/2009GL040914

Reed RK (1970) On subtropical currents and thermal structure in the central North Pacific. J Oceanogr Soc Japan 26:183–184

Reid JL (1978) On the mid-depth circulation and salinity field in the North Atlantic Ocean. J Geophys Res 83:5063–5067

Reid JL, Mantyla AW (1978) On the mid-depth circulation of the North Pacific Ocean. J Phys Oceanogr 8:946–951

Reynolds RW, Smith TM, Liu C, Chelton DB, Casey KS, Schlax MG (2007) Daily high-resolution blended analyses for sea surface temperature. J Clim 20:5473–5496

Rhines PB, Young WR (1982) A theory of the wind-driven circulation. Part I: mid-ocean gyres. J Mar Res 40(Suppl):559–596

Richards KJ, Maximenko NA, Bryan FO, Sasaki H (2006) Zonal jets in the Pacific Ocean. Geophys Res Lett 33:L03605. doi: 10.1029/2005GL024645

Rio M-H, Hernandez F (2004) A mean dynamic topography computed over the world ocean from altimetry, in situ measurements, and a geoid model. J Geophys Res 109:C12032. doi: 10.1029/2003JC002226

Rio M-H, Guinehut S, Larnicol G (2011) New CNES-CLS09 global mean dynamic topography computed from the combination of GRACE data, altimetry, and in situ measurements. J Geophys Res 116:C07018. doi:10.1029/2010JC006505

Roden GI (1975) On the north Pacific temperature, salinity, sound velocity fronts, and the relation to the wind and energy flux fields. J Phys Oceanogr 5:557–571

Roden GI (1980a) On the variability of surface temperature fronts in the western Pacific, as detected by satellite. J Geophys Res 85:2704–2710

Roden GI (1980b) On the subtropical frontal zone north of Hawaii during winter. J Phys Oceanogr 10:342–362

Roemmich D, Cornuelle B (1992) The subtropical mode waters of the South Pacific Ocean. J Phys Oceanogr 22:1178–1187

Sasaki H, Xie SP, Taguchi B, Nonaka M, Hosoda S, Masumoto Y (2012) Interannual variations of the Hawaiian Lee countercurrent induced by low potential vorticity water ventilation in the subsurface. J Oceanogr. doi:10.1007/s10872-011-0074-8

Siedler G, Rouault M, Lutjeharms JRE (2006) Structure and origin of the subtropical South Indian Ocean countercurrent. Geophys Res Lett 33:L24609. doi:10.1029/2006GL027399

Stommel H, Schott F (1977) The beta spiral and the determination of the absolute velocity field from hydrographic station data. Deep Sea Res 24:325–329

Suga T, Hanawa K (1990) The mixed layer climatology in the northwestern part of the North Pacific subtropical gyre and the formation area of subtropical mode water. J Mar Res 48: 543–566

Suga T, Hanawa K (1995) The subtropical mode water circulation in the North Pacific. J Phys Oceanogr 25:958–970

Suga T, Hanawa K, Toba Y (1989) Subtropical mode water in the 137°E section. J Phys Oceanogr 19:1605–1618

Suga T, Takei Y, Hanawa K (1997) Thermostad distribution in the North Pacific subtropical gyre: the central mode water and the subtropical mode water. J Phys Oceanogr 27:140–152

Suga T, Kato A, Hanawa K (2000) North Pacific tropical water: its climatology, temporal changes associated with the climate regime shift in the 1970s. Prog Oceanogr 47:223–256

Suga T, Motoki K, Hanawa K (2003) Subsurface water masses in the central North Pacific transition region: the repeat section along the 180° meridian. J Oceanogr 59:435–444

Suga T, Motoki K, Aoki Y, Macdonald AM (2004) The North Pacific climatology of winter mixed layer and mode waters. J Phys Oceanogr 34:3–22

Sugimoto S, Hanawa K (2005) Remote reemergence areas of winter sea surface temperature anomalies in the North Pacific. Geophys Res Lett 32:L01606. doi:10.1029/2004GL021410

Sugimoto S, Hanawa K (2010) Impact of Aleutian Low activity on the STMW formation in the Kuroshio recirculation gyre region. Geophys Res Lett 37:L03606. doi:10.1029/2009GL041795

Sumata H, Kubokawa A (2001) Numerical study of eastern boundary ventilation and its effects on the thermocline structure. J Phys Oceanogr 31:3002–3019

Takeuchi K (1984) Numerical study of the subtropical front and the subtropical countercurrent. J Oceanogr Soc Jpn 40:371–381

Takeuchi K (1986) Numerical study of the seasonal variations of the subtropical front and the subtropical countercurrent. J Phys Oceanogr 16:919–926

Tanaka M (1992) Intraseasonal oscillation and the onset and retreat dates of the summer monsoon over East, Southeast Asia and the western Pacific region using GMS high cloud amount data. J Meteor Soc Jpn 70:613–629

Taneda T, Suga T, Hanawa K (2000) Subtropical mode water variation in the northwestern part of the North Pacific subtropical gyre. J Geophys Res 105:19591–19598

Thompson L, Cheng W (2008) Water masses in the Pacific in CCSM3. J Clim 21:4514–4528

Thompson RORY, Edwards RJ (1981) Mixing and water-mass formation in the Australian subantarctic. J Phys Oceanogr 11:1399–1406

Tokinaga H, Xie S-P, Kobashi F, Tanimoto Y (2009a) Local and remote influences of the Kuroshio Extension on the atmosphere. US CLIVAR Var 7:1–4

Tokinaga H, Tanimoto Y, Xie S-P, Sampe T, Tomita H, Ichikawa H (2009b) Ocean frontal effects on the vertical development of clouds over the western North Pacific: in situ and satellite observations. J Clim 22:4241–4260

Tsubouchi T, Suga T, Hanawa K (2007) Three types of South Pacific subtropical mode water: the relation to the large-scale circulation of the South Pacific subtropical gyre and their temporal variability. J Phys Oceanogr 37:2478–2490

Tsubouchi T, Suga T, Hanawa K (2010) Indian Ocean subtropical mode water: its water characteristics and spatial distribution. Ocean Sci 6:41–50

Tsuchiya M (1985) Evidence of a double-cell subtropical gyre in the South Atlantic Ocean. J Mar Res 43:57–65

Uda M, Hasunuma K (1969) The eastward subtropical countercurrent in the western North Pacific Ocean. J Oceanogr Soc Jpn 25:201–210

Vianna ML, Menezes VV (2010) Mean mesoscale global ocean currents from geodetic pre-GOCE MDTs with a synthesis of the North Pacific circulation. J Geophys Res 115:C02016. doi: 10.1029/2009JC005494

Wang B, Lin Ho (2002) Rainy season of the Asian–Pacific summer monsoon. J Clim 15:386–398

Welander P (1959) An advective model of ocean thermocline. Tellus 11:309–318

Welander P (1981) Mixed layer and fronts in simple ocean circulation model. J Phys Oceanogr 11:148–152

White WB, Hasunuma K (1982) Quasi-stationary banded structure in the mean zonal geostrophic current regimes of the western North Pacific. J Mar Res 40:1035–1046
White WB, Walker AE (1985) The influence of the Hawaiian Archipelago upon the wind-driven subtropical gyre in the western North Pacific. J Geophys Res 90:7061–7074
White WB, Hasunuma K, Solomon H (1978) Large-scale seasonal, secular variability of the subtropical front in the western North Pacific from 1954 to 1974. J Geophys Res 83:4531–4544
Worthington LV (1959) The 18° water in the Sargasso Sea. Deep Sea Res 5:297–305
Wyrtki K (1975) Fluctuations of the dynamic topography in the Pacific Ocean. J Phys Oceanogr 5:450–459
Xie S-P, Kunitani T, Kubokawa A, Nonaka M, Hosoda S (2000) Interdecadal thermocline variability in the North Pacific for 1958–1997: a GCM simulation. J Phys Oceanogr 30:2798–2813
Xie S-P, Liu WT, Liu Q, Nonaka M (2001) Far-reaching effects of the Hawaiian Islands on the Pacific Ocean–atmosphere system. Science 292:2057–2060
Xie S-P, Deser C, Vecchi GA, Ma J, Teng H, Wittenberg AT (2010) Global warming pattern formation: sea surface temperature and rainfall. J Clim 23:966–986
Xie S-P, Xu L-X, Liu Q, Kobashi F (2011) Dynamical role of mode-water ventilation in decadal variability in the central subtropical gyre of the North Pacific. J Clim 24:1212–1225
Xu L-X, Xie S-P, Liu Q, Kobashi F (2012) Response of the North Pacific subtropical countercurrent and its variability to global warming. J Oceanogr. doi:10.1007/s10872-011-0031-6
Yamanaka G, Ishizaki H, Hirabara M, Ishikawa I (2008) Decadal variability of the subtropical front of the western North Pacific in an eddy-resolving ocean general circulation model. J Geophys Res 113:C12027. doi:10.1029/2008JC005002
Yasuda T, Hanawa K (1997) Decadal changes in the mode waters in the midlatitude North Pacific. J Phys Oceanogr 27:858–870
Yasuda T, Hanawa K (1999) Composite analysis of North Pacific subtropical mode water properties with respect to the strength of the wintertime East Asian monsoon. J Oceanogr 55:531–541
Yoshida K, Kidokoro T (1967a) A subtropical countercurrent in the North Pacific—an eastward flow near the subtropical convergence. J Oceanogr Soc Jpn 23:88–91
Yoshida K, Kidokoro T (1967b) A subtropical countercurrent (II)—a prediction of eastward flows at lower subtropical latitudes. J Oceanogr Soc Jpn 23:231–236
Young RW, Rhines PB (1982) A theory of the wind-driven circulation II. Circulation models and western boundary layers. J Mar Res 40:849–872
Yu Z, Maximenko N, Xie SP, Nonaka M (2003) On the termination of the Hawaiian Lee countercurrent. Geophys Res Lett 30(5):1215. doi:10.1029/2002GL016710
Yuan X, Talley LD (1996) The subarctic frontal zone in the North Pacific: characteristics of frontal structure from climatological data and synoptic surveys. J Geophys Res 101:16491–16508

J Oceanogr (2012) 68:45–52
DOI 10.1007/s10872-011-0029-0

SPECIAL SECTION: REVIEW

New developments in mode-water research: Dynamic and climatic effects

New perspectives on eighteen-degree water formation in the North Atlantic

Terrence M. Joyce

Received: 6 January 2011 / Revised: 9 April 2011 / Accepted: 9 April 2011 / Published online: 7 June 2011

Abstract In this report, eighteen-degree water (EDW) formation will be discussed, with emphasis on advances in understanding emerging within the past decade. In particular, a recently completed field study of EDW (CLIMODE) is suggesting that EDW formation within a given winter can have at least two different dominant physics and distinct locations: one type formed in the northern Sargasso Sea, largely away from the strong flows of the Gulf Stream where 1D physics may apply, and a second type formed along the southern flank of the Gulf Stream, in a region where the background vorticity of the flow and cross-frontal mixing play key roles in the convective formation process.

Keywords Mode water · Air–sea interaction · Gulf Stream

1 Introduction

The subtropical-mode water of the North Atlantic Ocean was first identified as having remarkably uniform temperature (17.9 ± 0.3°C) and salinity (36.50 ± 0.10 psu) properties by Worthington (1959), who named it eighteen-degree water (EDW), a name which has persisted over time. It is very evident in volumetric atlases from hydrographic data (Wright and Worthington 1970) and here using the modern gridded WOCE-era electronic atlas (Gouretski and Koltermann 2004, herein Fig. 1). One distinct "mode" of this 2D histogram of waters warmer than 7°C in the N. Atlantic is clearly the EDW identified by Worthington. Mode waters as a generic entity (Hanawa and Talley 2000) exist in all the major ocean basins and are pervasive water masses that can be detected throughout the year both locally, where they are formed, and in locations far from the formation site(s). Worthington identified the Sargasso Sea as the location where EDW could be readily found and the northern reaches of the Sargasso Sea, just south of the separated Gulf Stream, as the region of formation. Formation has generally been determined by where the mode water is seen to outcrop at the ocean surface in late winter. Worthington refined his ideas about EDW in a later study (Worthington 1976) where he associated the formation with the large mean negative oceanic heat flux to the atmosphere. At the time, information about ocean/atmosphere exchange was limited to rather coarse summaries from available shipboard data collected over time, and was clearly affected by limited sampling during periods of maximum exchange. These early summaries by Budyko (1963) and Bunker (1976) painted the first basin-wide picture of air–sea exchange. Worthington argued that the mean heat-loss region over the separated Gulf Stream (GS) was responsible for transforming subtropical waters into EDW, whereas Warren (1972) argued that the region where EDW was "formed", near 35N, 60W was one of zero net heat loss—leading him to conclude that EDW formation was the process by which the locally formed seasonal thermocline was removed during winter: something that occurs in the absence of strong advection and requiring no net heat loss when averaged over the year. Confounding these early ideas about EDW formation was the problem inherent in estimating heat exchange over the ocean, where short-wave and long-wave radiation measurements were few, and estimates of turbulent exchange

T. M. Joyce (✉)
Woods Hole Oceanographic Institution, 360 Woods Hole Rd., Mail Stop 21, Woods Hole, MA 02543, USA
e-mail: tjoyce@whoi.edu

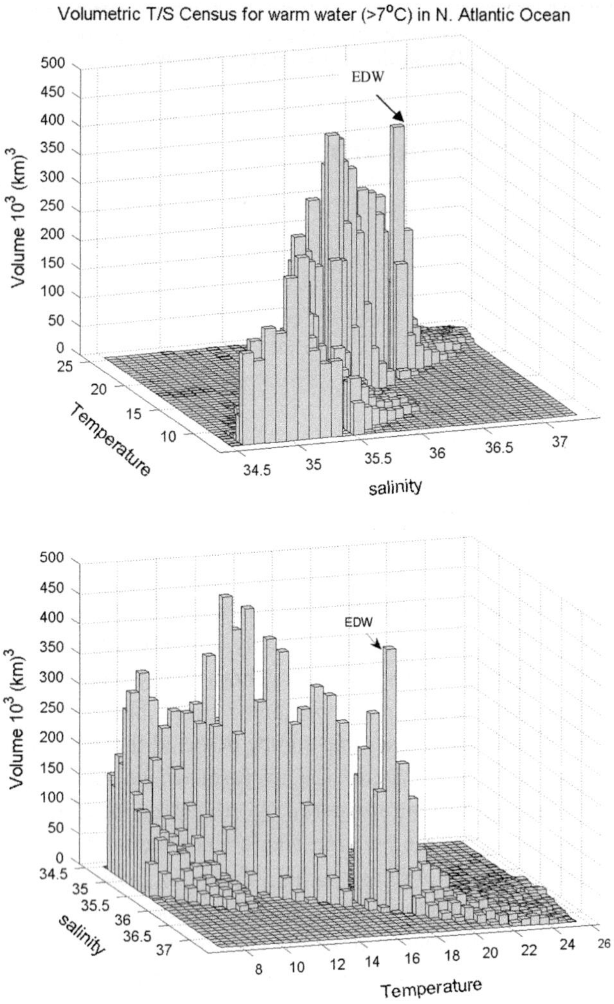

Fig. 1 Two different views of a 2D histogram showing the volume (in 10^3 km^3) of N. Atlantic waters having temperatures >7°C, based on the WOCE Global Hydrographic Climatology (Gouretski and Koltermann 2004). The total volume of EDW bounded by (36.4–36.6 psu, 17.5–18.5°C) is estimated to be 912 × 10^3 km^3

of sensible and evaporative heat were limited by both data and models of turbulent exchange over the ocean.

Walin's (1982) ideas about direct estimation of water mass formation using air–sea exchange data were a major boost to the field, yet application to EDW formation (Speer and Tziperman 1992; Maze et al. 2009) yielded greatly different estimates of EDW formation. This has prompted a concerted effort to collect field observations and conduct modeling studies to better understand the different physical processes in the ocean associated with water mass formation (advection, subduction, mixing) and better parameterize turbulent exchanges with the atmosphere in the northern Sargasso Sea: this experiment was called CLIMODE (CLIvar MOde water Dynamics Experiment; The Climode Group 2009). In this update, an attempt will be made to summarize what we have learnt about EDW since the review of Hanawa and Talley (2000), including the recently published results from CLIMODE. However, much of the CLIMODE analysis has yet to appear in print and it is expected that this review would be quite different if written a few years from now. It is hoped that enough of the recent thinking will show where advances are to be expected and will serve as a current assessment for EDW, to be contrasted with major contemporary efforts (reported elsewhere in this collection) on mode waters found in other ocean basins.

2 EDW definitions

To begin, it is instructive to see how one might define EDW in order to then estimate how much of it there is and where it is formed. Worthington (1976, Table 6) noted that although there were approximately 2512 × 10^3 km^3 of EDW (waters with temperatures between 17 and 19°C) in the N. Atlantic, if one looked in the North American Basin (west of the mid-Atlantic Ridge), the volume was reduced to 1763 × 10^3 km^3. Siedler et al. (1987) have since shown that a substantial amount of mode water with a temperature of ca. 18°C is formed every winter in the North African Basin, near the island of Madeira. This type of subtropical mode water was called Madeira mode water. Unlike its counterpart in the Sargasso Sea, by the end of summer heating, there is little evidence of a homogeneous thermostad, or region of reduced vertical temperature gradient: it has vanished into the seasonal thermocline (Siedler et al. 1987; Fig. 9). If one confines oneself to the region west of the Mid-Atlantic Ridge one can limit the contribution of this eastern basin water mass from the census.

The EDW volume can further be refined by estimating the anomaly resulting from the reduced temperature gradient of EDW imbedded in the permanent thermocline. Worthington (1976; Table 8) did this for the western North Atlantic by subtracting the volume of 16–17 and 19–20°C water from that of 17–19°C water in his census and found that the EDW volume for the western N. Atlantic was further reduced by approximately 50% to 889 × ^{3} km^3: thus for one author alone, published estimates of EDW volume differ by a factor of 3, depending on its definition. The last EDW volume estimate is quite close to one based on T/S criteria presented below. Others have tried looking at anomalies of EDW by limiting it to those waters having a temperature gradient less than some limiting value (Kwon and Riser 2004) or some limiting planetary potential vorticity appropriate to the thermocline. Here we use both temperature and salinity constraints and find, on the basis of Fig. 1, the volume of EDW between 17.5 and 18.5°C with bounding salinities of 36.4 and 36.6 psu is 912 × 10^3 km^3; so addition of a salinity criterion and

Atlantic numbers substantially. As a check on the new climatological data, the total volume was calculated with the WOCE-era atlas without salinity constraints and estimated for 7–19°C water to be 2237 × 10^3 km^3, within the error bars of the Forget et al. (2011) figure of 2360 × 10^3 km^3, and slightly less than Worthington's value given above. Clearly, a restricted temperature definition and use of salinity to define the EDW volume (this work) makes a substantial difference. Limiting the spatial domain (western basin) and requiring the local temperature gradient to be less than some reference value are the two principle issues that need further resolution in the literature in the definition of EDW volume. Resulting EDW volume estimates are clearly dependent on the definitions and thus arises the potential for conflicting claims on volumes, formation rates, and ventilation timescales, as we will see later.

Talley and Raymer (1982) first pointed out that there were spatial variations of EDW within a given winter and that there was substantial year-to-year variability in the thickness and other properties at Bermuda. Joyce et al. (2000) showed that interannual variations of EDW PV, as observed at Bermuda, were well-correlated with the NAO (with no lag/lead), suggesting that changes in air–sea forcing as reflected in major climate indices were important to newly formed EDW observed at Bermuda. Because EDW is not locally formed near Bermuda in most winters, its time of arrival there has been estimated as a few to several months after formation by using oxygen (Jenkins 1982) or PV (Talley and Raymer 1982; Phillips and Joyce 2007). We will return later to the issue of spatial variations of EDW formed in a given winter with some new data drawn from CLIMODE. Because newly formed EDW is highly saturated with dissolved oxygen, Jenkins and Goldman (1985) were able to study biological processes affecting EDW oxygen consumption after subduction by calibrating change against "true" age determined from tritium/helium-3.

As was shown by Schmitz (1996; Fig. 1-21) the southern re-circulation gyre of the GS occupies the region bounded by a box with lat/lon corners at (40N, 50W) and (30N, 55W) to the east and (35N, 75W) and (25N, 75W) to the west. This roughly agrees with the float-based circulation on the 26.5 kg m^{-3} surface by Kwon and Riser (2004) reproduced here (Fig. 2). The float trajectories were used to map the geostrophic pressure on an isopycnal associated with EDW. Data for this circulation map are drawn from a period from 1998 to 2002. The 60 cm contour (Fig. 2) roughly delineates the southern GS re-circulation gyre identified by Schmitz (1996; Fig. 1-21). Clearly the mean circulation is defining or being defined by the presence of EDW, which is largely contained within this region, on the basis of anomaly calculations using a reference vertical temperature gradient. The WOCE mean climatology atlas was used to examine subtle water mass variations of EDW in the Sargasso Sea (Fig. 2, right panel). One can see that saltier (and also warmer—not shown)

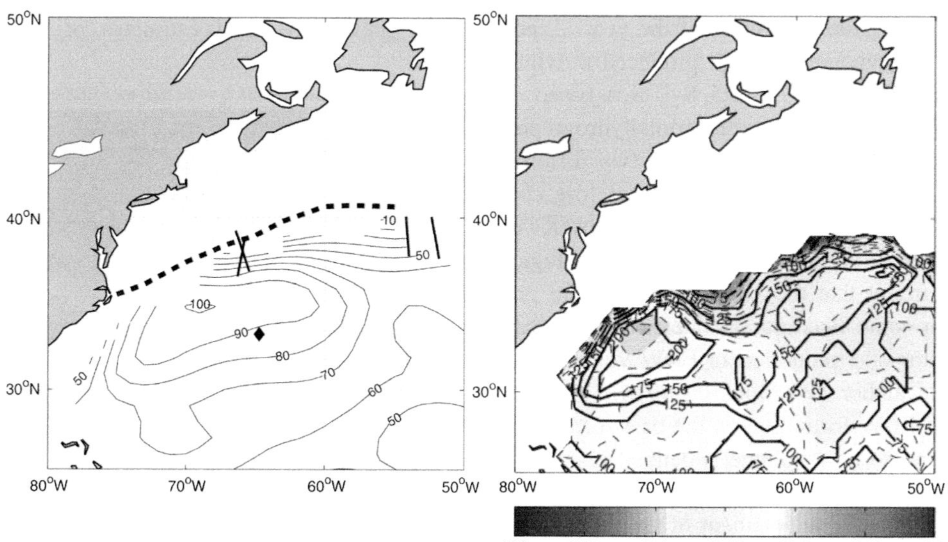

Fig. 2 *Left panel* from Kwon and Riser (2004), the geostrophic pressure on the $\sigma_\theta = 26.5$ kg m^{-3} potential density surface (in cm), based on mean float velocities, is plotted with hydrographic lines from the CLIMODE winter cruise of the Knorr in 2007 (*thick black lines*), the north wall of the Gulf Stream defined by the 200 m temperature = 15°C from 1955–2008 (*heavy dashed line*), and the location of Bermuda (*black diamond*). *Right panel* from the WOCE mean climatology, the thickness of the 17.5–18.5°C temperature surface (*solid black contours*) has been plotted for all values exceeding 50 m. The color represents the average salinity (psu) in this layer, and the contour interval (*dashed lines*) is 0.01. The thickness and salinity reveal a climatological pattern of EDW variation in the Sargasso Sea

varieties of EDW are found within the recirculation center, whereas fresher (and colder) varieties surround them, with a clear tendency for fresher values along the northern boundary in proximity to the Gulf Stream. Because this representation includes all seasons, the wintertime properties are obscured, especially the thickness, which is reduced because of restratification.

3 Annual EDW budgets and formation

EDW formation budgets using the "Walin" method have been nicely exemplified by Forget et al. (2011) for the 3-year period 2004–2006, inclusive. Various data sets (e.g. SST, SSH, Argo floats) have been assimilated in a 1° × 1° lat/lon resolution numerical model that has 16-month, overlapping annual cycles for the above period. Estimates for air–sea exchange have been examined with the inverse machinery and "adjusted" when necessary to reduce model/data misfits. The resulting annual EDW cycle has been compared with more traditional, float-based estimates of EDW volumes during the same period. The region considered is the entire N. Atlantic as reflected in surface areas and volumes in various temperature classes with 1°C resolution; there was no attempt to stratify the calculation using salinity or surface salinity forcing. The amount of EDW formed annually during the winter months is estimated to be 8.6 ± 1.8 Svy (1 Svy = 31.5 × 10^3 km^3), where they have been able to account for the effects of subgridscale mixing in the model, which actually acts to increase EDW formation during winter, although reducing EDW formation during the remainder of the year. A previous estimate of the amount of EDW produced in winter (Kwon and Riser 2004) of 3.5 ± 0.5 Svy was based on increases in EDW volume during winter from hydrographic data alone. Although the error bars of the two different estimates do not overlap, they are remarkably close considering that the EDW definition criteria chosen by Kwon and Riser (2004) focus on the western N. Atlantic region and satisfy a constraint that the vertical temperature gradient be less than some maximum value, which we saw earlier gave a factor of 3 reduction in EDW volume estimated by a single author (Worthington 1976). Yet both cited works define EDW formation on the basis of volume increases in winter—so how volume is defined is inherent to estimates of annual formation. Whereas the Kwon and Riser (2004) calculation is independent of air–sea forcing, relying only on changes in EDW volume based on hydrography, the Forget et al. (2011) calculation is not. So we need to address the sensitivity of "formation" rates to the input air–sea fluxes.

Wintertime air–sea heat loss over the region of EDW formation can be o(300 W m^{-2}) with uncertainties of o(10%). Joyce et al. (2011) have shown that the maximum of winter heat and water loss in the N. Atlantic is located exactly over the location of the meandering, separated GS, with the warm core of the GS defining the region of largest ocean–atmosphere exchange in winter. During the period of CLIMODE (Feb/Mar 2007, Fig. 3) the winter mean SST field indicates that after traversing the region from 75W to 55W, the warm surface core water of the GS eventually disappears. If one were to choose the 17.5 and 18.5°C isotherms for Walin-type formation analysis, it would be clear (Fig. 3) that the domain would be multiply connected and that 10% errors in the surface heat flux together with the sensitivity in flux products to spatially dependent air–sea temperature and humidity differences could lead to substantial differences in water mass formation calculations. Forget et al. (2011) show that, depending on the flux product used, EDW volume increases during winter range from 5 to 13 Svy based on the Walin-type calculation alone. Because their EDW calculation was constrained by ocean data, air–sea exchange forcing is adjusted to give a better model/data fit and is, therefore, given a reality check that the unconstrained range of estimates lacks. The annual EDW formation rate of ca. 13 Svy by Speer and Tziperman (1992; Table 1) is at an extreme limit of this Walin-type calculation, although the authors made an effort to do the formation calculation in density rather than temperature space, taking into account freshwater forcing. Presumably a constrained estimate of this calculation using Argo float salinities and newer flux estimates might be obtained in the near future. For the present, it would therefore seem that the range of EDW formation of 3.5–8.6 Svy between the two previously cited estimates of formation based on

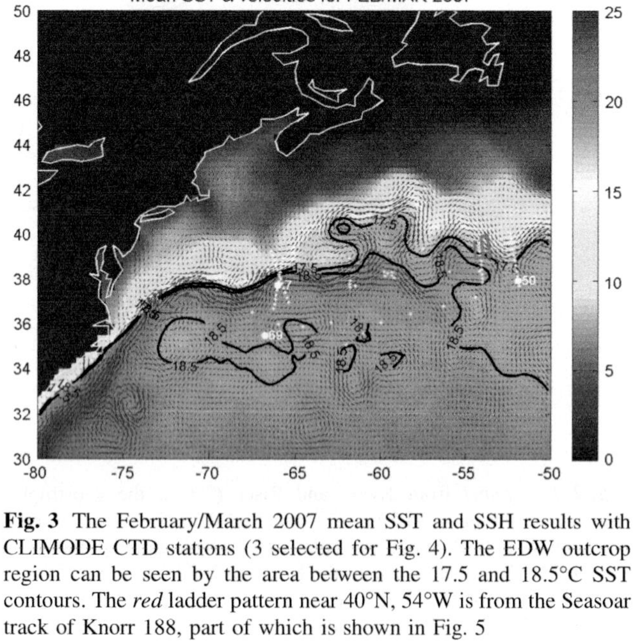

Fig. 3 The February/March 2007 mean SST and SSH results with CLIMODE CTD stations (3 selected for Fig. 4). The EDW outcrop region can be seen by the area between the 17.5 and 18.5°C SST contours. The *red* ladder pattern near 40°N, 54°W is from the Seasoar track of Knorr 188, part of which is shown in Fig. 5

temperature must be because of the definition of EDW. Forget et al. (2011) estimate that the principle agents for dissipation of EDW are air–sea exchange in the warm months, leading to formation of a surface thermocline, and cross-frontal mixing of EDW with higher PV waters, principally found to the north of the Gulf Stream. One final comment related to air–sea exchanges, the CLIMODE program offered some unique extensions for improving the parameterization of fluxes using bulk formulae. With high winds and large air–sea temperature differences, and with clear problems with the algorithms at wind speeds above 15 m s^{-1} (The Climode Group 2009), the expected changes to bulk formulae should improve future estimation of turbulent air–sea fluxes for both heat and water.

4 Multiple types of EDW formed annually

Examination of the outcrop window for EDW (Fig. 3) during CLIMODE indicates two possible scenarios for EDW formation. The first is in the region from 73–63 W at a latitude near 35N. Here in the northern Sargasso Sea and well south of the GS flow, EDW can be seen at the surface. The vertical characteristics of the water are portrayed (Fig. 4) at 3 selected CTD stations from the Knorr cruise in 2007. Station 7, taken in the "upstream" region within the warm surface core of the GS, shows a strong stratification near the surface and evidence of tropical salinity maximum water (Worthington 1976; Wüst 1936) with elevated salinity at a pressure of ca. 200 dbars. Below this layer, a region of reduced stratification, "old EDW" can be seen with oxygen saturation values below 90%. In the northern Sargasso Sea (Station 69), a homogeneous water mass, "new EDW", can be seen between 100 and 400 dbar pressures with elevated oxygen content, a temperature slightly above 18°C, and a salinity of approximately 36.6 psu. A warmer, less dense surface layer has capped the EDW, probably because of restratification in the presence of lighter surrounding waters. This is also suggested by the lack of an outcrop of the EDW layer near the location of the station (Fig. 3). The salinity of this EDW looks approximately like that of the vertically mixed version of the incoming waters, exemplified by station 7. In contrast, the downstream station near the GS (station 50) is located in the other EDW outcrop zone which stays close to the GS until approximately 55 W where it erupts from the front and fills the region to the south. This EDW is colder and fresher. Joyce et al. (2011) argue that formation of EDW within the frontal region entrains cooler, fresher waters across the GS front from the north and results in a lower salinity, and a slightly colder version of EDW. Both types of EDW are highly saturated with oxygen and are different types of EDW formed in the same year, as was first noted by Talley and Raymer (1982), who speculated

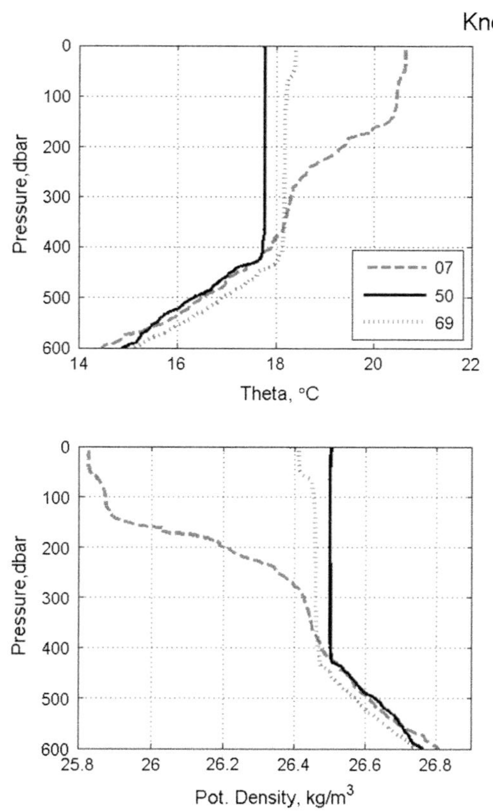

Fig. 4 Selected CTD stations from CLIMODE in 2007. Station 7 (*red curves*) is from the upstream inflow region of the GS just south of the strongest zonal flow. Stations 50 (*black line*) and 69 (*blue dotted line*) are from the regions of newly ventilated EDW along the GS front (50) and in the northern Sargasso Sea (69). The four panels show temperature, salinity, potential density, and percentage oxygen saturation for the stations. Station locations are indicated in *white* on Fig. 3

that EDW variations occurred west to east as the EDW further circulated under the strong surface cooling and thus became progressively cooler. According to Joyce et al. (2011), the high-heat loss region of the GS front is also one in which evaporation exceeds precipitation, making air–sea exchange a candidate for increasing, not decreasing, EDW salinity in the region. By invoking the need for lateral mixing across the GS front ($\kappa \sim 100\ m^2\ s^{-1}$), Joyce et al. (2011) were able to satisfy both the downstream advective budgets for salinity and heat for the EDW formed in the GS, clearly indicating that EDW formed along the GS would necessarily be fresher (and colder) than that found forming further to the south and west. The cooling and freshening of the northern version of EDW is affected by cross-frontal exchange and this is reflected in the mean property changes in the hydrography (Fig. 2, right panel). Forget et al. (2011) did not include salinity in their budget calculations, but found that cross-frontal mixing in winter could actually enhance EDW production. With salinity, it would have been clear that this new component of EDW formation introduced lower salinity water into the region and affected mean EDW properties of temperature and salinity in the northern Sargasso Sea.

It would seem that the two different types of EDW observed in 2007 are the end points of the disparate formation hypotheses of Worthington (1959, 1976) and Warren (1972): frontal formation under the strong air–sea cooling region (Worthington) and formation in the northern Sargasso Sea away from strong advection and cooling (Warren). Thus, the mean properties of EDW are saltier (Fig. 2, right panel) and warmer within the center of the southern recirculation gyre, where Warren argued for one non-advective endpoint for EDW formation. Worthington's (1972) circulation schematic for northward surface flow into the formation region and southward spreading of new EDW away from the region are reasonable if interpreted in density space rather than depth space: lighter waters advect northward and eastward in the GS and are cooled and (slightly) freshened. They become detached from the GS and spread southward to the east of 60W where the southern re-circulation gyre detrains fluid southward from the GS (Fig. 2) and spreads this fresher and cooler version of EDW around the edges of the recirculation gyre. How much of the EDW formation occurs in the two regions remains to be determined. Joyce et al. (2011) estimated that ca. 1.8–2.9 Svy were formed in the GS flow during 2007, which amounts to either 47–82% of the total according to Kwon and Riser (2004), or 21–34% if one uses the Forget et al. (2011) estimate for annual EDW formation. Other CLIMODE data sets, for example Argo floats having salinity sensors, should be able to address this point also.

The formation of EDW within the GS represents a new paradigm for mode water formation, one in which the strong lateral and vertical shear of the flow alters convective processes and allows for symmetric and inertial instabilities (Joyce et al. 2009). Regions of negative potential vorticity are still formed by cooling, but along front wind stress is also an important contributor to the forcing as denser fluid is advected by winds equatorward across the front (Thomas 2005). The forced, slantwise convection can have lateral scales of only a few km and is not easily observed except in sub-mesoscale resolving surveys using devices such as Seasoar (Fig. 5) during CLIMODE. The entire region between 30 and 110 km along the undulating track is filled with new EDW, but closer inspection (see Joyce et al. 2009) reveals some of the distinctive fine structure of the formation: note the subtle changes in oxygen and fluorescence along the track as one passes through waters recently near the surface (elevated values of these tracers) and waters likely to be upwelling from below (lower tracer concentrations). The entire ventilation process is different from 1D convection and is poorly parameterized in models not resolving the sub-mesoscale, and poorly sampled in most ocean observations.

5 Some concluding thoughts and challenges

Were the formation rates and volumes of EDW well established, one could take the ratio of the two to estimate the ventilation rate for EDW: how much of it is ventilated in a typical year. An independent estimate of this was attempted by Jenkins (1982) using tritium/helium-3 dating at Bermuda: the EDW estimate is difficult to extract from his figures because the age change is large with potential density, and the potential density resolution, limited by discrete sampling bottles, is not good for low potential densities (for example, for EDW): an upper limit is 8.5 years (for $\sigma_\theta = 26.5$) from Jenkins (1982; Table 1) and 3 years (Jenkins and Goldman 1985; Fig. 9, depth ca. 230 m), which defines a range nearly identical to that between Kwon and Riser's (2004) estimate of 3.6 years and the Forget et al. (2011) value of 8.7 years. Thus, independent tracer ages do not seem to offer a way out of this conundrum at present. Although we have identified at least two types of EDW formed within a given year, we have not yet ascertained which of the two types is dominant in the annual budget.

The issue of the fate of EDW after formation has not been addressed yet with CLIMODE data. One might expect some revised mean circulation update on that offered in Fig. 2, left panel, and some assessment of the role that eddies play in the spreading and homogenization of EDW from the formation region. One can also expect some future

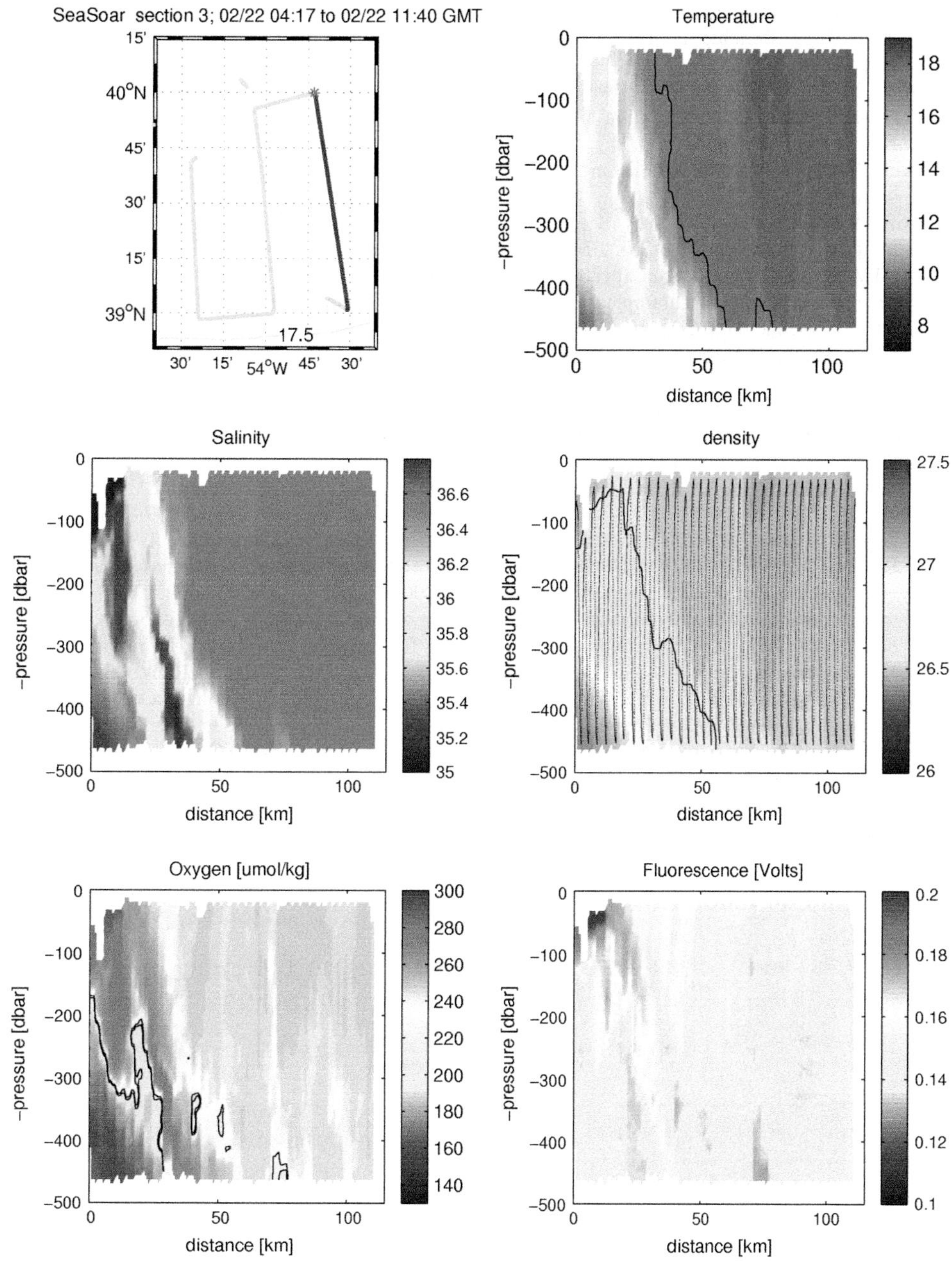

Fig. 5 EDW formation within the GS is seen on this Seasoar track (*blue line*, *upper left*) with the north on the left hand side of the other panels which show temperature (°C, *upper right*), salinity (psu, *middle left*), σ_θ (kg m^{-3}, *middle right*) with the Seasoar track shown as *thin black lines*, dissolved oxygen (μmol kg^{-1}, *lower left*), and chlorophyll fluorescence (uncalibrated volts, *lower right*)

assessment of how representative the CLIMODE period is compared with past winters. Joyce et al. (2011) noted that air–sea exchange during the winter of 2007 was larger over the region than the mean of the previous 19 years. This has not been translated into EDW formation or circulation at this stage and thus represents another aspect of EDW dynamics that may yet emerge from our recent field program. Finally, we should expect that comparative dynamics will emerge for EDW and its close kin, North Pacific subtropical mode water, found immediately to the south of the separated Kuroshio.

Acknowledgments Thanks are due to all my colleagues in the CLIMODE project, to the National Science Foundation, for its support to the author (OCE 0959387), to the editors of this special section of Journal of Oceanography for inviting this contribution, and to Jane Dunworth-Baker, who prepared many of the figures in this manuscript.

References

Budyko MI (1963) Atlas of heat balance of the world. Glav. Geofiz. Obo, Moscow (in Russian). Translation by I.A. Donehoo, Weather Bureau, WB/T106, Washington, DC

Bunker AF (1976) Computations of surface energy flux and annual air–sea interaction cycles of the North Atlantic Ocean. Mon Weather Rev 104:1122–1140

Forget G, Maze G, Buckley M, Marshall J (2011) Estimated seasonal cycle of North Atlantic eighteen degree water volume. J Phys Oceanogr 41:269–286. doi:10.1175/2010JPO4257.1

Gouretski VV, Koltermann KP (2004) WOCE global hydrographic climatology. Berichte des Bundesamtes für Seeschifffahrt und Hydrographie, 35/2004

Hanawa H, Talley L (2000) Mode waters. In: Siedler J, Church J, Gould J (eds) Ocean circulation and climate. Academic Press, New York, pp 373–386

Jenkins WJ (1982) Tritium and 3He in the Sargasso Sea. J Mar Res 40(Supplement):533–569

Jenkins WJ, Goldman JC (1985) Seasonal cycling and primary production in the Sargasso Sea. J Mar Res 43:465–491

Joyce TM, Deser C, Spall MA et al (2000) On the relation between decadal variability of subtropical mode water and the North Atlantic oscillation. J Clim 13:2550–2569

Joyce TM, Thomas LN, Bahr F (2009) Wintertime observations of subtropical mode water formation within the Gulf Stream. Geophys Res Lett 36:L02607. doi:10.1029/2008GL035918

Joyce TM, Thomas LN, Dewar WK, Girton JB (2011) Eighteen degree water formation within the Gulf Stream: a new paradigm arising from CLIMODE. Deep Sea Res II (submitted)

Kwon Y-O, Riser SC (2004) North Atlantic subtropical mode water: a history of ocean-atmosphere interaction 1961–2000. Geophys Res Lett 31:L19307. doi:10.1029/2004GL021116

Maze G, Forget G, Buckley M, Marshall J, Cerovecki I (2009) Using transformation and formation maps to study the role of air–sea heat fluxes in North Atlantic eighteen degree water formation. J Phys Oceanogr 39:1818–1835

Phillips HE, Joyce TM (2007) Bermuda's tale of two time series: hydrostation S and BATS. J Phys Oceanogr 37:554–571

Schmitz WJ Jr (1996) On the world ocean circulation, vol I. Some global features/North Atlantic circulation. WHOI Technical Report WHOI-96-03

Siedler G, Kuhl A, Zenk W (1987) The Madeira mode water. J Phys Oceanogr 17:1561–1570

Speer K, Tzipermann E (1992) Rates of water mass formation in the North Atlantic Ocean. J Phys Oceanogr 22:93–104

Talley LD, Raymer ME (1982) Eighteen degree water variability. J Mar Res 40(Supplement):757–775

The Climode Group (2009) Observing the cycle of convection and restratification over the Gulf Stream system and the subtropical gyre of the North Atlantic ocean: preliminary results from the CLIMODE field campaign. Bull Am Meteorol Soc 90:1337–1350

Thomas LN (2005) Destruction of potential vorticity by winds. J Phys Oceanogr 35:2457–2466

Walin G (1982) On the relation between sea-surface heat flow and thermal circulation in the ocean. Tellus 34:187–195

Warren BA et al (1972) Insensitivity of subtropical mode water characteristics to meteorological fluctuations. Deep Sea Res 19:1–19

Worthington LV (1959) The 18° water in the Sargasso Sea. Deep Sea Res 5:297–305

Worthington LV (1972) Negative oceanic heat flux as a cause of water-mass formation. J Phys Oceanogr 2:205–211

Worthington LV (1976) On the North Atlantic circulation. Johns Hopkins oceanographic studies. The Johns Hopkins University Press, Baltimore

Wright WR, Worthington LV (1970) The water masses of the North Atlantic Ocean: a volumetric census of temperature and salinity. Ser Atlas Mar Environ 19:8 (7 plates)

Wüst G (1936) Oberflächensalzgehalt, Verdunstung, und Niederschlag auf dem Weltmeere. Lünderkundliche Forschung. Festschrift Norbert Krebs, Stuttgart, pp 347–359

J Oceanogr (2012) 68:53–62
DOI 10.1007/s10872-011-0086-4

SPECIAL SECTION: ORIGINAL ARTICLE

New developments in mode-water research: Dynamic and climatic effects

Mixed layer depth front and subduction of low potential vorticity water under seasonal forcings in an idealized OGCM

Shiro Nishikawa · Atsushi Kubokawa

Received: 28 February 2011 / Revised: 1 September 2011 / Accepted: 2 September 2011 / Published online: 25 November 2011

Abstract The mixed layer depth (MLD) front and subduction under seasonal variability are investigated using an idealized ocean general circulation model (OGCM) with simple seasonal forcings. A sharp MLD front develops and subduction occurs at the front from late winter to early spring. The position of the MLD front agrees with the curve where $DT_s/Dt = \partial T_s/\partial t + \boldsymbol{u}_g \cdot \nabla T_s = 0$ is satisfied (t is time, $\boldsymbol{u}_g$ is the upper-ocean geostrophic velocity, T_s is the sea surface temperature (SST), and ∇ is the horizontal gradient operator), indicating that thick mixed-layer water is subducted there parallel to the SST contour. This is a generalization of the past result that the MLD front coincides with the curve $\boldsymbol{u}_g \cdot \nabla T_s = 0$ when the forcing is steady. Irreversible subduction at the MLD front is limited to about 1 month, where the beginning of the irreversible subduction period agrees with the first coincidence of the MLD front and $DT_s/Dt = 0$ in late winter, and the end of the period roughly corresponds to the disappearance of the MLD front in early spring. Subduction volume at the MLD front during this period is similar to that during 1 year in the steady-forcing model. Since the cooling of the deep mixed-layer water occurs only in winter and SST can not fully catch up with the seasonally varying reference temperature of restoring, the cooling rate of SST is reduced and the zonal gradient of the SST in the northwestern subtropical gyre is a little altered in the seasonal-forcing case. These effects result in slightly lower densities of subducted water and the eastward shift of the MLD front.

Keywords Mixed layer depth front · Seasonal variation · Subduction · Idealized OGCM

1 Introduction

The winter mixed layer tends to be shallow toward the equator and deep toward the pole in the subtropical gyre region. It is known that there is a front-like region where the mixed layer becomes abruptly deep toward the pole, which is referred to as the mixed layer depth (MLD) front or the mixed layer front. The sharp MLD front structure is often seen in ocean general circulation model (OGCM) simulations (e.g., Williams et al. 1995; Tsujino and Yasuda 2004). It is located at about 30°N in the North Pacific subtropical gyre and about 30–40°N in the North Atlantic subtropical gyre. Observational data sets such as the World Ocean Atlas also imply a similar MLD front structure in the North Pacific (e.g., Huang and Qiu 1994; Suga et al. 2004), although it is less sharp than that in models. Oka and Suga (2003) showed the existence of a sharp MLD front structure at about 30°N in the North Pacific using Argo profiling float data. Several studies suggested that the MLD front has an important role in the formation of the mode waters such as the subtropical mode water (STMW) and central mode water (CMW) in the North Pacific (e.g., Xie et al. 2000; Tsujino and Yasuda 2004).

The MLD distribution is one of the boundary conditions of the ventilated thermocline theory (Luyten et al. 1983) and therefore it plays an important role in the determination of the upper thermocline structure. Williams (1991)

S. Nishikawa (✉)
Data Research Center for Marine-Earth Sciences,
Japan Agency for Marine-Earth Science and Technology,
3173-25 Showa-machi, Kanazawa-ku,
Yokohama 236-0001, Japan
e-mail: snishika@jamstec.go.jp

A. Kubokawa
Faculty of Environmental Earth Science,
Hokkaido University, Sapporo 060-0810, Japan

formulated the role as the relation between the potential vorticity (PV) of the subducted water and mixed layer quantities (their Eq. 4). Kubokawa and Inui (1999) and Kubokawa (1999) showed that the MLD front is an important source of low PV water on isopycnal surfaces using an idealized OGCM and the ventilated thermocline model. They pointed out that the low PV waters are formed around the intersection of the MLD front and the outcrops in isopycnal surfaces. These low PV waters form a thick low PV pool in the western subtropical gyre, lift the isopycnals in the surface layer, and cause a subtropical countercurrent to the south of the thick low PV pool (Kubokawa and Inui 1999; Kubokawa 1999).

Nishikawa and Kubokawa (2007, hereafter NK07) studied the formation mechanism of the MLD front and the subduction process at the MLD front using an idealized OGCM with steady forcings. Figure 1a shows the distribution of the MLD in their model. They showed that the MLD front (30–40°N) in their model occurs where $\boldsymbol{u}_g \cdot \nabla T_s = 0$, where $\boldsymbol{u}_g$ is the upper ocean geostrophic velocity, T_s is the sea surface temperature (SST), and ∇ is the horizontal gradient operator. The curve $\boldsymbol{u}_g \cdot \nabla T_s = 0$ gives the boundary between the deep mixed layer region ($\boldsymbol{u}_g \cdot \nabla T_s < 0$) and the subduction region ($\boldsymbol{u}_g \cdot \nabla T_s > 0$). On this curve, the deep mixed-layer water is subducted parallel to the SST contours to go beneath a thin stratified surface layer (Fig. 1b). Nishikawa and Kubokawa explained that the MLD front structure is formed as a result of this subduction process along the outcrops.

Subduction in reality occurs only from late winter to midsummer. Stommel (1979) argued that only the water which has the property of the late-winter mixed layer enters the main thermocline. The mixed layer becomes deepest in late winter and therefore waters that subduct afterwards enter not the main thermocline but the seasonal thermocline. On the basis of this idea, NK07's steady-forcing model used a thermal condition of late winter. Stommel's hypothesis ("mixed-layer demon") has been examined by several studies (e.g., Liu and Pedlosky 1994; Williams et al. 1995) and found to be largely correct. However, details of how seasonal variation of forcings affects the position of the MLD front and subduction at the front are not well known.

The three main problems of this study are the following: First, how is the position of the MLD front determined under seasonal variability? Second, when, where, and how is low PV water subducted at the MLD front in a seasonally varying field? Third, what differences arise in the formation of the MLD front and the volume and density of subducted water at the MLD front between the steady and seasonally varying fields? To address these problems, we add a simple seasonality to NK07's model. Seasonal forcing introduces local variation in the upper ocean and, as we will see, the inclusion of $\frac{\partial T_s}{\partial t}$ generalizes NK07's mechanism in a straightforward manner.

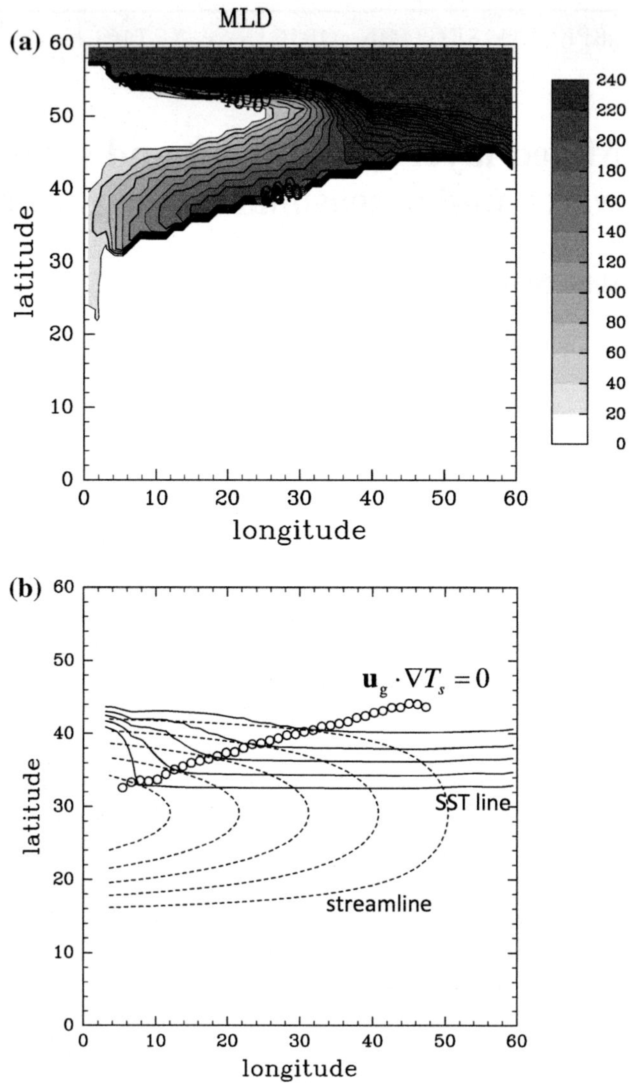

Fig. 1 **a** MLD distribution in the steady-forcing model of NK07. Here, MLD is defined as the thickness of surface layer where vertical density gradient $(-\frac{\partial \sigma_\theta}{\partial z})$ is less than 2.0×10^{-4} kg m^{-4}. Contours larger than 500 m are omitted. **b** Schematic diagram which shows the relation among SST contours (*solid contours*), streamlines (*dashed contours*), and the position of $\mathbf{u}_g \cdot \nabla T_s = 0$ (*circles*) from NK07

The rest of the paper is organized as follows. The model is briefly described in Sect. 2. After the annual cycle of the mixed layer is outlined in Sect. 3, the above problems are explored in Sect. 4. A summary is given in Sect. 5.

2 Model description

The model used in this study is the CCSR Ocean Component Model (COCO) developed at the University of

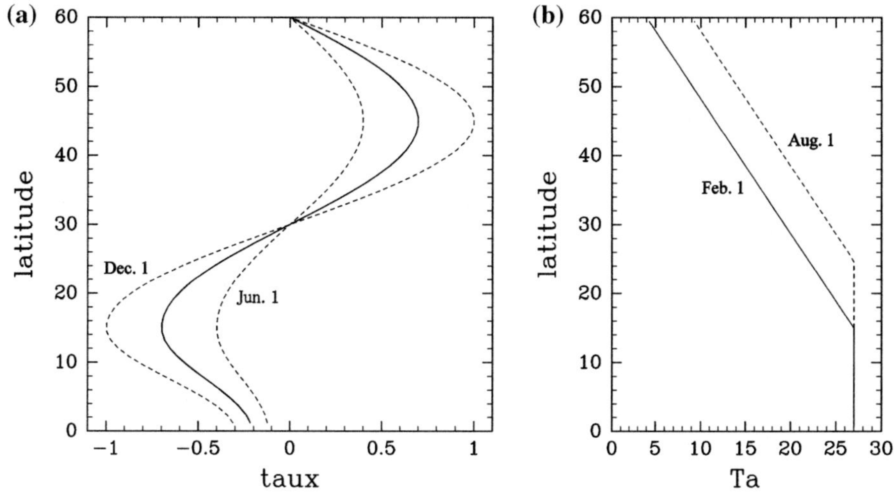

Fig. 2 **a** Meridional profile of τ_x[dyne cm^{-2}]. *Solid line* denotes annual mean, which is the same as that used in NK07. *Dashed lines* denote the maximum (December 1) and minimum (June 1). **b** Meridional profile of T_a[$^\circ$C]. *Solid line* denotes the minimum (February 1), which is the same as that used in NK07. *Dashed lines* denote the maximum (August 1)

Tokyo (Hasumi 2000). This model is the same as NK07's; we describe it briefly here and refer the reader to NK07 for details. The model domain is a rectangular basin with no bottom topography which is 60° wide zonally, extends from the equator to 60°N meridionally, and has a constant depth of 3000 m. The horizontal resolution is 1.2°. There are 30 levels in the vertical with grid spacing increasing from 10 m near the surface to 400 m near the bottom. The vertical resolutions in the upper layers are 10, 20, and 20 m at depths of 0–50 m and 25 m at depths of 50–300 m. A simple convective adjustment is used for the formation of the surface mixed layer. Since the MLD front discussed here is associated with winter-time convection, wind-induced mixing is not included. Salinity is kept constant at 34.9 psu and therefore potential density σ_θ is a function of temperature only.

The model is driven by wind stress and surface heat flux both with seasonal variation (Fig. 2). The surface wind stress, (τ_x, τ_y), is given by

$$\tau_x = \begin{cases} \tau_0(t)[0.35\cos(\frac{\pi}{15}\varphi) - 0.65] & \text{for} \quad 0 < \varphi \le 15 \\ \tau_0(t)\cos(\frac{\pi}{30}\varphi + \frac{\pi}{2}) & \text{for} \quad 15 < \varphi \le 60, \end{cases} \tag{1}$$

$$\tau_y = 0, \tag{2}$$

where $\tau_0(t)$[dyne cm^{-2}] is a time-dependent amplitude given by

$$\tau_0 = 0.7 + 0.3\cos\left(\frac{2\pi}{t_0}(t - t_1)\right), \tag{3}$$

φ is the latitude, t is the time, t_0 is 1 year, and t_1 corresponds to December 1. In this model, 1 model year is 360 days and 1 model month is 30 days for each month. The functional form (1) is based on a zonal average of the Hellerman and Rosenstein (1983) data. The meridional profile of τ_x is shown in Fig. 2a. The annual mean of τ_x is the same as that used in NK07. The Sverdrup circulation driven by (1) consists of three gyres and the gyre boundaries are located at 15°N and 44°N (see Fig. 2 of NK07).

The surface heat flux, F_H[W m^{-2}], is a restoring-type flux given by

$$\frac{F_H}{\rho_0 C_p} = \gamma[T_a(\varphi, t) - T_s)], \tag{4}$$

where T_s is the surface temperature of the model, T_a is a reference temperature, ρ_0 is a constant density, and C_p is the specific heat of seawater [$\rho_0 = 1026.952$ kg m^{-3} and $C_p = 3986.0$ J kg^{-1} K^{-1}]. The restoring coefficient γ is taken to be 100 cm day^{-1}. The reference temperature $T_a(\varphi, t)$[$^\circ$C] is given by

$$T_a = \begin{cases} 27 & \text{for} \quad 0 < \varphi \le \varphi_1(t) \\ 27 - \frac{23}{45}(\varphi - 15) + T_1(t) & \text{for} \quad \varphi_1(t) < \varphi \le 60, \end{cases} \tag{5}$$

where the temperature $T_1(t)$[°C] and the latitude $\varphi_1(t)$ are as follows:

$$T_1 = \Delta T\left\{1 - \cos\left(\frac{2\pi}{t_0}(t - t_2)\right)\right\} \tag{6}$$

$$\varphi_1 = \frac{45}{23}T_1 + 15. \tag{7}$$

The amplitude ΔT is taken to be 2.5°C and t_2 corresponds to February 1. The functional form (5) is based on the zonal mean SST of the World Ocean Atlas 94 (Levitus and Boyer 1994). The profile of T_a at the coldest time of the year (February 1) is the same as that used in NK07. The meridional profile of T_a is shown in Fig. 2b.

The model is integrated for 83 years from the same initial condition as that of NK07 and the upper, wind-driven circulation is fully spun up. Outputs (snapshots) taken every 5 days from the last 3 years are used for analysis. In this study, MLD is defined as the thickness of surface layer where vertical density gradient ($-\frac{\partial\sigma_\theta}{\partial z}$) is less than

2.0×10^{-4} kg m^{-4}. This definition is somewhat different from that used in NK07. With a traditional definition using density difference, a very thin, stratified layer just below the sea surface is ignored and the mixed layer is defined as thicker than this thin stratified layer. In the present study, this thin layer should be resolved because it marks the poleward edge of subduction as will be seen below. The new criterion using density gradient is therefore more suitable.

3 Outline of annual cycle

This section briefly describes the annual cycle of the upper ocean thermal structure.

Figure 3 presents the snapshots of the MLD and surface heat flux on January 1, March 6, and March 26 in the 82nd model year. There is an MLD front, which extends between 30°N and 40°N in the subtropical gyre. The mixed layer is shallow in the western part of the subpolar gyre (30°E, 44–54°N), where the main thermocline is shallow. Figure 4 depicts the time–latitude diagrams at 25.8°E of the MLD and SST for 1 year. From December to February, the mixed layer gradually deepens and the MLD front gradually moves to the south. The surface heat flux is negative (ocean being cooled) in the whole subtropical gyre (Fig. 3d). From February to early March, the MLD attains its maximum in the subtropical gyre. The MLD front comes to the southernmost position in early March (Fig. 3b). The surface heat flux is positive in the central subtropical gyre and in the western subpolar gyre (Fig. 3e). From March to April, the deep mixed layer region in the northwest of the subtropical gyre rapidly shrinks and the MLD front retreats to the north. Most of the MLD front structure disappears in mid to late April. An important feature is that the MLD front is sharper in March to April (Figs. 3c, 4a) than in other seasons (Figs. 3a, 4a), which is discussed in detail in the next section.

4 MLD front and subduction under seasonal forcings

4.1 Position of the MLD front

To investigate the relation among SST, MLD front, and subduction in the subtropical gyre, we use the following quantity:

$$\frac{DT_s}{Dt} \equiv \frac{\partial T_s}{\partial t} + \boldsymbol{u}_g \cdot \nabla T_s, \tag{8}$$

generalizing $\boldsymbol{u}_g \cdot \nabla T_s$ used in NK07, including temporal change. When $\frac{DT_s}{Dt} < 0$, the surface density increases along the geostrophic flow, so that the stratification tends to be unstable and the mixed layer will develop. When $\frac{DT_s}{Dt} > 0$, surface density decreases along the geostrophic flow and hence the surface layer will be stabilized. Therefore, the location of $\frac{DT_s}{Dt} = 0$ is expected to give the boundary between a stably stratified region and a vertically mixed region. The quantity $\frac{DT_s}{Dt}$ is also proportional to the vertical component of the PV flux and is related to subduction rate (see, Marshall and Nurser 1992; Marshall et al. 2001). Positive $\frac{DT_s}{Dt}$ in the subtropical gyre, where Ekman pumping is downward, corresponds to subduction, while negative $\frac{DT_s}{Dt}$ corresponds to obduction. In the following calculation of $\frac{DT_s}{Dt}$, we use the velocity at $z = -40.0$ m as $\boldsymbol{u}_g$ since the Ekman flow is negligible at this depth (the vertical viscosity coefficient adopted in the present model, $A_V = 1.0$ cm^2 s^{-1}, gives an Ekman layer of only a few meters thick).

Figure 4b shows the seasonal cycle of the sign of $\frac{DT_s}{Dt}$ at 25.8°E. From autumn to winter, $\frac{DT_s}{Dt}$ is negative in the whole subtropical gyre implying the deepening of the mixed layer. The front-like structure in MLD appearing from autumn to midwinter (Fig. 3a) would be maintained simply because the seasonal thermocline is being eroded to expose the previous year's deep mixed layer. From spring to summer $\frac{DT_s}{Dt}$ is positive everywhere, and only from late winter to early spring does the boundary between the two regions ($\frac{DT_s}{Dt} = 0$) occur in the subtropical gyre. The MLD front in this season (Fig. 3b, c) is much sharper than that in midwinter (Fig. 3a) as mentioned above, and it is the counterpart of the MLD front in NK07.

Figure 5 shows the distribution of $\frac{DT_s}{Dt}$ on March 26. In the northwest of the subtropical gyre, where $\frac{DT_s}{Dt} < 0$, the deep mixed layer is maintained by local surface cooling and subsurface advection of warm water. In the rest of the subtropical gyre, where $\frac{DT_s}{Dt} > 0$, PV flux is downward and the mixed layer is shallow. The position of the MLD front agrees well with the position of $\frac{DT_s}{Dt} = 0$ (Figs. 4b, 5a, b), where the thick mixed-layer water in the north is subducted under the thin stratified surface layer as shown later, and this subduction maintains the sharp MLD front as in NK07. Note that $\frac{\partial T_s}{\partial t} > 0$ (Fig. 5b) accelerates the subduction as the thin stratified surface layer rapidly covers the water column, resulting in the northward migration of the MLD front (Fig. 4a).

4.2 Subduction at the MLD front

To explore how water is subducted at the MLD front under seasonal variability, we compute the trajectories of water parcels using snapshots of the model velocity field taken every 5 days, with the same method of particle tracking as

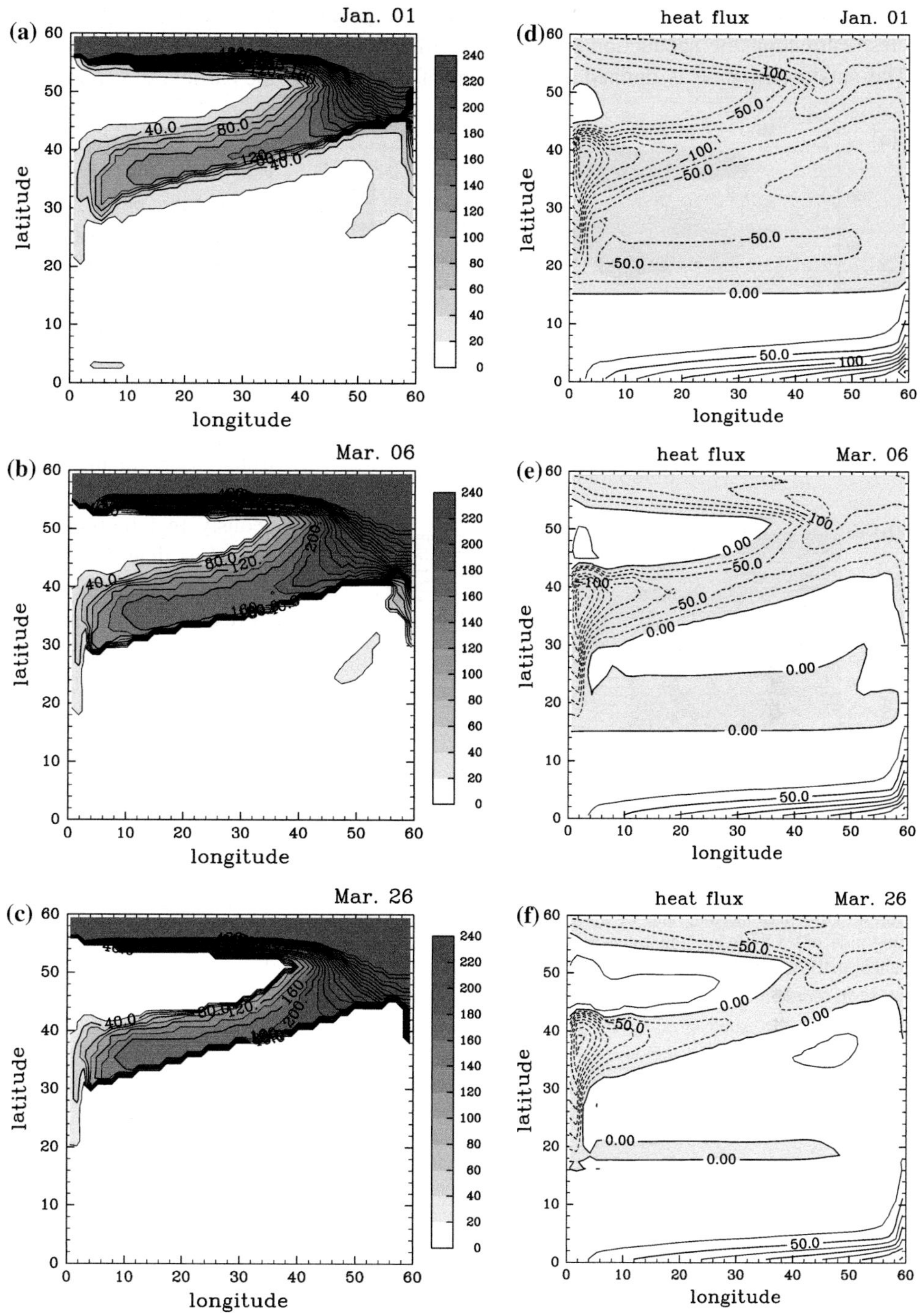

Fig. 3 Horizontal distributions of **a–c** MLD [m] and **d–f** surface heat flux [Wm^{-2}] for January 1 (*top*), March 6 (*middle*), and March 26 (*bottom*). Contour intervals of MLD and heat flux are 20 m and 25 Wm^{-2}, respectively. Contours larger than 500 m are omitted in the *left panels*. *Shading* in the *right panels* indicates negative values (ocean being cooled)

in NK07. Figure 6a shows a trajectory over 2.5 years of particles released on January 6 of the 81st model year. Figure 6b and c show that this particle is subducted around day 90 (April 5) first, reentrained into the mixed layer in the next winter around day 390 (February 5), and is irreversibly subducted on a day between days 421–425 (March 6–10). Figure 6b shows that $\frac{DT_s}{Dt} = 0$ at the subduction, indicating that the trajectory of the water parcel is parallel to the local SST contour then and there; and that $\frac{DT_s}{Dt} > 0$ after crossing the MLD front. As the tall water column in the deep mixed layer moves toward the MLD front, it remains tall because $\frac{DT_s}{Dt} < 0$; after it crosses the MLD front, only its top part gets stratified by the surface heating ($\frac{DT_s}{Dt} > 0$) and the remaining low PV water column goes under this newly formed, thin mixed layer. This is the subduction of this water column. This subduction process makes the sharp MLD front structure there, which is a generalization of NK07's result.

We calculate the timing of subduction using the particles, estimating the subduction period at the MLD front. The subduction time is defined as the moment at which

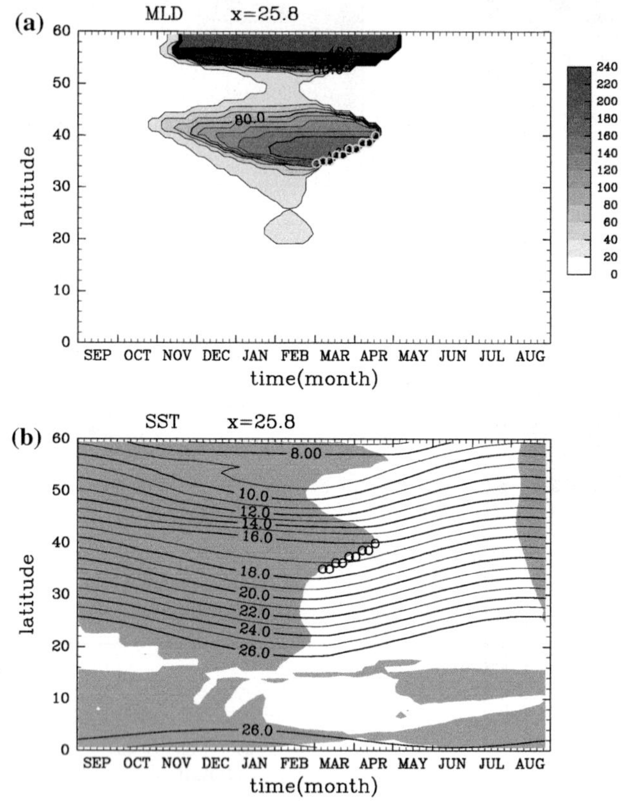

Fig. 4 Time–latitude diagrams of **a** MLD [m] and **b** SST [°C] at the longitude of 25.8° from September in the 81st year. *Circles* in both panels mark the position of the MLD front which this study focuses on. In **b**, *shaded region* denotes $\frac{DT_s}{Dt} < 0$ and *unshaded region* denotes $\frac{DT_s}{Dt} > 0$. In the subtropical gyre 15–44°N the *unshaded area* corresponds to subduction region

water is irreversibly transfered from the mixed layer into the main thermocline. For example, the subduction date of the particle shown in Fig. 6 is March 6–10 in the 82nd year (days 421–425 in Fig. 6c). We calculate the subduction dates of 864 particles which start every 5 days in the 81st year from the 12 release points shown in Fig. 6a (72 samples per point). The release points and the integration period (2.5 years) are chosen so that all particles are subducted at the MLD front. Figure 7 shows the statistics of the subduction times, indicating that the irreversible subduction at the MLD front occurs for 30–40 days from the end of February to early April. The beginning of the subduction period corresponds to the beginning of the coincidence of the MLD front and $\frac{DT_s}{Dt} = 0$, when the MLD front comes to the southernmost position (Fig. 4a). The coincidence occurs on about March 1, almost simultaneously in longitude. Therefore, the number of the subducted particles is sharply increased in March (Fig. 7). The subduction period ends from early to mid April, roughly corresponding to the disappearance of the MLD front. A substantial amount of the water subducted in April is expected to be reentrained into the deep mixed layer in the next winter. The reentrainment in the next winter would affect the transition from irreversible subduction to temporal subduction and lead to the relatively smooth reduction of the number of irreversibly subducted particles in April (Fig. 7).

Subduction of low PV water at the MLD front is quantitatively evaluated using the subduction rate,

$$S = -\frac{\partial h}{\partial t} - \boldsymbol{u} \cdot \nabla h - w, \tag{9}$$

with the subduction period estimated above and the integration region shown in Fig. 8b (the shaded region), where

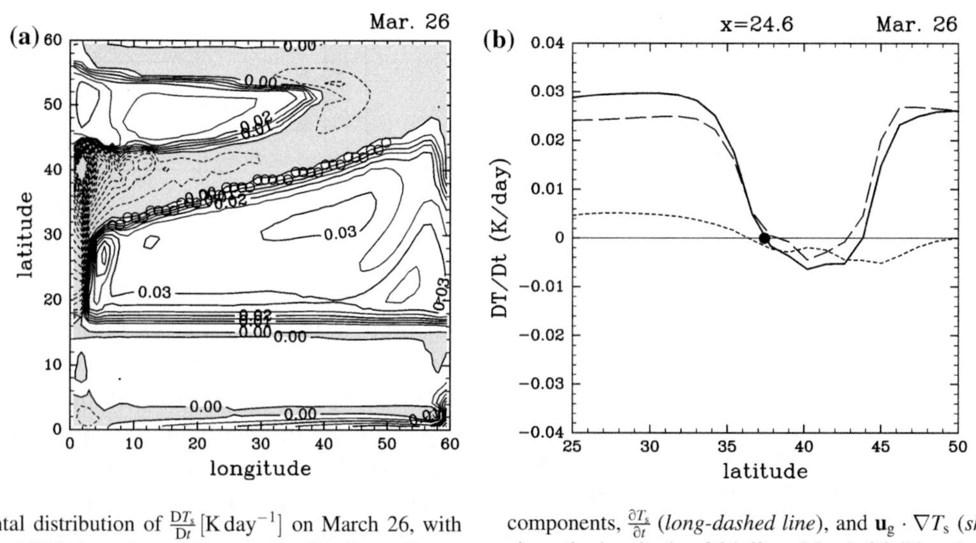

Fig. 5 **a** Horizontal distribution of $\frac{DT_s}{Dt}$ [K day^{-1}] on March 26, with the location of the MLD front denoted by *circles*. *Shading* indicates negative values. **b** Meridional profiles of $\frac{DT_s}{Dt}$ (*solid line*) and its components, $\frac{\partial T_s}{\partial t}$ (*long-dashed line*), and $\mathbf{u}_g \cdot \nabla T_s$ (*short-dashed line*), along the longitude of 24.6° on March 26. The *closed circle* denotes the location of the MLD front

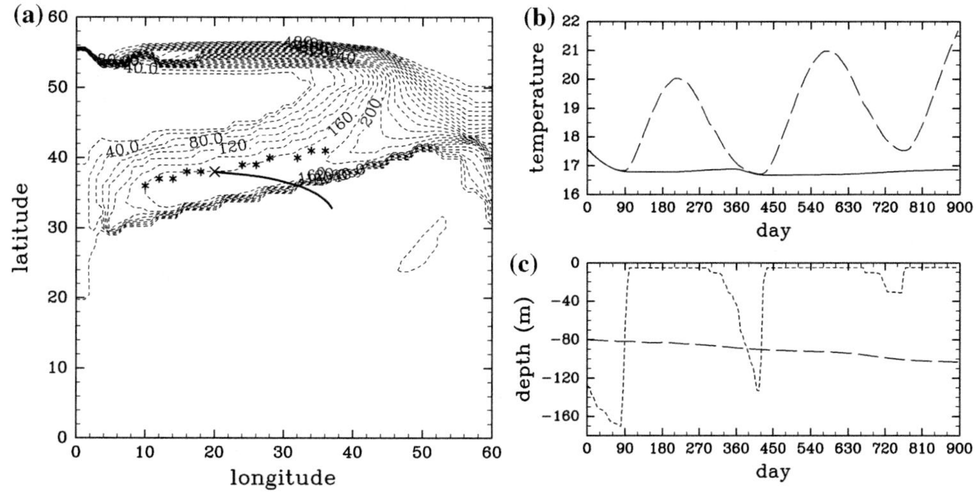

Fig. 6 **a** The trajectory (*solid line*) of a particle seeded at the point denoted by the cross (×) on January 6 of the 81st model year. *Dashed contours* indicate the MLD when the particle is subducted at the MLD front (March 6). The *asterisks* (∗) denote the other release points of particles to calculate subduction time in Fig. 7. **b** SST (*long-dashed line*) and the temperature of the particle (*solid line*). Unit is °C. Note that the slope of the *long-dashed line* gives the sign of $\frac{DT_s}{Dt}$. **c** MLD (*short-dashed line*) and the depth of the particle (*long-dashed line*)

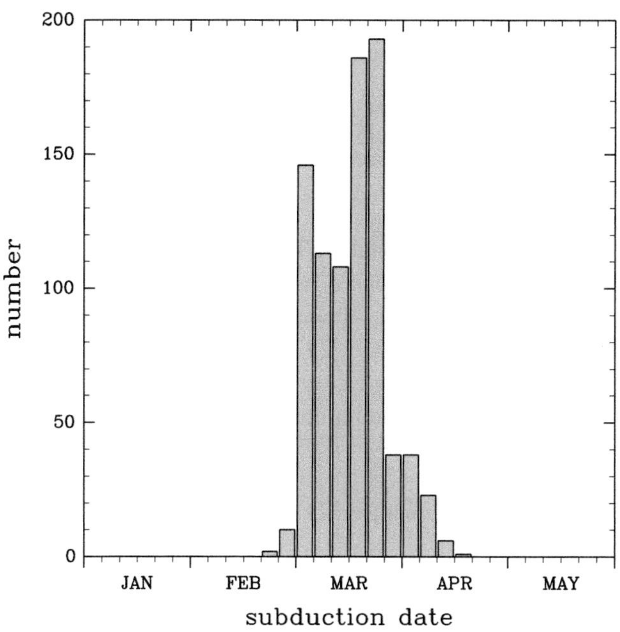

Fig. 7 Number of particles subducted at the MLD front binned according to their subduction dates; the bin width is 5 days. The total number of samples is 864 and the 12 release points are shown in Fig. 6a (72 samples per point)

h is the MLD, $\boldsymbol{u}$ is the horizontal velocity, and w is the vertical velocity. Volumes of the subducted water at the MLD front are calculated by integrating S temporally and regionally (Fig. 8a). From the above result of the subduction period, we take the integration period as 35 days from February 28 to April 3. The integration region (Fig. 8b) covers the migration of the MLD front in the period. A similar calculation is also conducted with the steady-forcing model using the subduction period of 1 year (Fig. 8c, d). For both models, the density range of subduction (Fig. 8a, c) corresponds well to that in which the MLD front intersects the outcrops (Fig. 9). The comparison of Figs. 8a and c indicates that the subduction volume at the MLD front during the subduction period (about 1 month) in the seasonal-forcing case is comparable to that during 1 year in the steady-forcing case. This result together with the above result of the subduction period (Fig. 7) gives a confirmation of the Stommel demon mechanism (Stommel 1979; Williams et al. 1995). In the present study, it is connected with the existence of the MLD front. Since the intersections of the outcrops and the MLD front vary during the subduction period, the volume distribution (Fig. 8a) is smoother and the density range is somewhat broader than in the steady-forcing case (Fig. 8c). The total volume of the subducted water at the MLD front is about 29×10^{13} m^3 (Fig. 8a), which is much larger than the annual volume by Ekman pumping in the north of the MLD front (about 4×10^{13} m^3), indicating that most of the subducted water at the MLD front is a part of the upper-layer water which circulates horizontally in the subtropical gyre. In the real oceans, it is known that total subduction rate integrated over the subtropical gyre is larger than area-integrated annual-mean Ekman pumping (e.g., Qiu and Huang 1995; Suga et al. 2008). It is implied that the difference would be similarly explained by the horizontally circulating water through the MLD front.

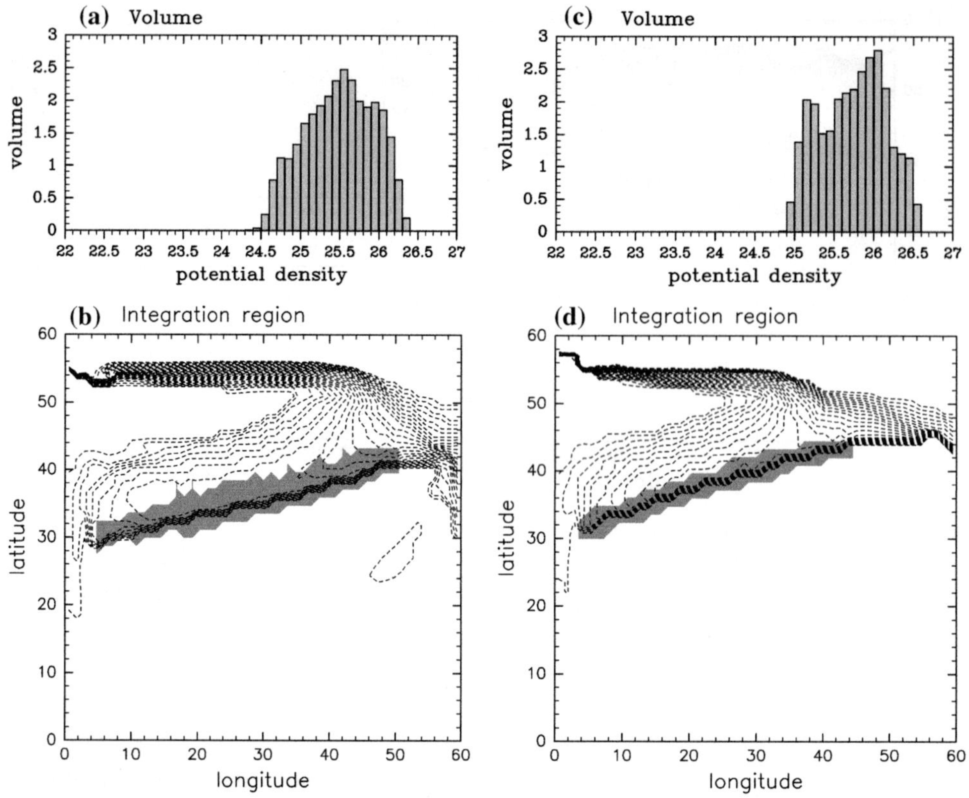

Fig. 8 **a** Volumes $[10^{13}\ m^3]$ of subduction at the MLD front binned at an interval of 0.1 σ_θ during 35 days from February 28 to April 3. **b** The integration region for the calculation of **a** is denoted by *shading*. *Dashed contours* denote MLD on March 6. **c** Same as **a** but during 1 year in the steady-forcing model. **d** The integration region for the calculation of **c** is denoted by *shading*. *Dashed contours* denote MLD in the steady-forcing model. **a** and **c** are smoothed using 1-2-1 smoothing

4.3 Differences in the MLD front between the steady- and seasonal-forcing models

Although the restoring temperature T_a at the coldest time of the year in the seasonal-forcing model coincides with that used in the steady-forcing model of NK07, differences arise in the MLD front structure between these two models (Fig. 9). First, the densities of the subducted water are a little lower in the seasonal-forcing model than those in the steady-forcing model (Fig. 8a, c). Second, the position of the MLD front is shifted a little to the east in the seasonal-forcing model (Fig. 9a, b). These differences are attributed to the difference in the process of cooling deep mixed-layer water in the northwestern subtropical gyre. The deep mixed-layer water in the steady-forcing model is always cooled over 1 year. On the other hand, that in the seasonal-forcing model is intermittently cooled in winter (see Fig. 6b). Because of the large heat capacity of the deep mixed layer, the damping toward T_a of SST is so slow that SST cannot catch up with seasonally varying T_a, and SST cannot reach the lowest T_a in midwinter. This effect reduces the cooling rate of SST, and reduces the density of the subducted water. The reduction of the cooling rate affects the zonal gradient of SST in the northwestern subtropical gyre as well; the zonal SST gradient near the western boundary becomes weaker, while that in the downstream slightly stronger. This alteration in the SST distribution pushes the position of $\boldsymbol{u}_g \cdot \nabla T_s = 0$ eastward in comparison with the steady-forcing case, and as a result, the position of the MLD front is shifted a little to the east (Fig. 9).

4.4 Effect of seasonality in winds

In this study, we also include a seasonal variation in the surface wind stress (Fig. 2a). It is found that the effect of the seasonal variation of the surface wind stress on the formation and variation of the MLD front is very little. This is likely because the response of the velocity field to the seasonality in winds is confined in the Ekman layer and $\boldsymbol{u}_g$ does not change much. In fact, there is no significant difference in the structure of mixed layer or subduction between the standard run and a test run with an annual mean wind stress (not shown). Thus, the formation of the MLD front is determined by annual mean wind stress. Note that the critical latitude of annual baroclinic Rossby waves is near 40° (Fu 2004), so that lower-frequency variability in winds may affect the formation of the MLD front.

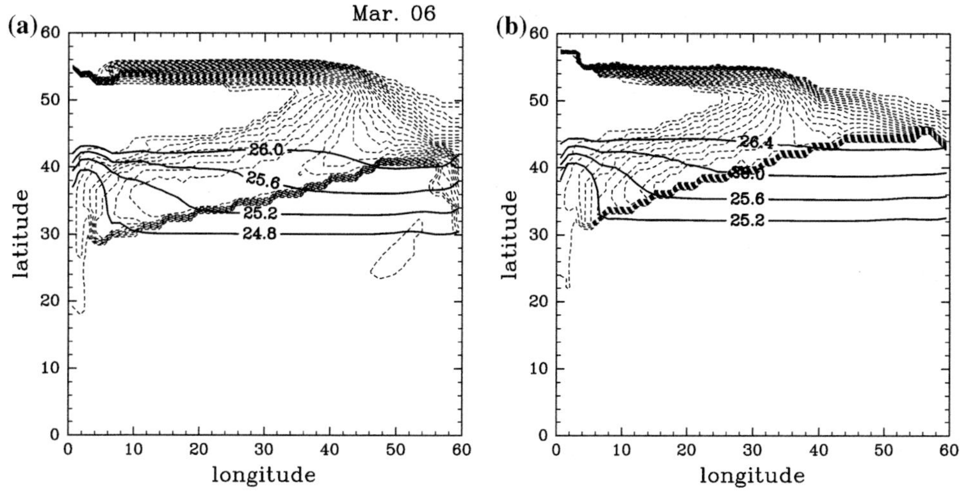

Fig. 9 **a** Selected sea-surface density contours (*solid*) that cross the MLD front on March 6. *Dashed contours* denote MLD at that time. **b** Same as **a** but in the steady-forcing model

5 Summary

NK07 has demonstrated a simple structure of the MLD front and subduction in an idealized steady-forcing model. As an extension, this paper introduces idealized seasonal forcings to NK07's model and investigates the MLD front and subduction at the front under seasonal variability.

In this model, a sharp MLD front develops and subduction occurs at the MLD front from late winter to early spring. The position of the MLD front agrees with the curve $\frac{DT_s}{Dt} = 0$, which gives the boundary of stably stratified region ($\frac{DT_s}{Dt} > 0$) and vertically mixed region ($\frac{DT_s}{Dt} < 0$) in the subtropical gyre. This result means that deep mixed-layer water is subducted parallel to the SST contour even in a seasonally varying field, which is a generalization of NK07's result. This subduction process results in the formation of the sharp MLD front structure in that season. Irreversible subduction at the MLD front is limited to about 1 month, estimated from the statistics using particles. The beginning of the subduction period agrees with the first coincidence of the MLD front and $\frac{DT_s}{Dt} = 0$ in late winter when the MLD front comes to the southernmost position, and the end of the subduction period roughly corresponds to the disappearance of the MLD front in early spring. Subduction volume at the MLD front during this period is similar to that during 1 year in the steady-forcing model. Since the cooling of the deep mixed-layer water occurs only in winter and the damping toward T_a is so slow that SST cannot catch up with seasonally varying T_a, the cooling rate of SST is reduced and the zonal gradient of the SST in the northwestern subtropical gyre is altered to push the position of $\boldsymbol{u}_g \cdot \nabla T_s = 0$ eastward in comparison with the steady-forcing case (Fig. 9). From these effects, the densities of subducted water are a little lower (Fig. 8a, c) and the MLD front is shifted a little to the east (Fig. 9).

It was suggested that the reentrainment is important in estimating the annual subduction rate at the MLD front. The reentrainment will be important further in considering interannual variability such as between cold and warm winters, because the reentrained water brings the properties of the previous winter's water into the mixed layer and will affect the formation of the MLD front. Interannual variability of the MLD front and subduction is an important next topic. For application of the present results to real oceans, it seems that the spatial and temporal resolution of observations is not sufficient so far to reconstruct the seasonal evolution of the MLD front or to compute $\frac{DT_s}{Dt}$ to test our results. Therefore, comparison with realistic GCM simulations would be interesting. Since a sharp MLD front is also seen in eddy-resolving models (e.g., Qu et al. 2002; Nishikawa et al. 2010), effects of eddies on the MLD front structure and subduction at the front under seasonal variability would also be an interesting subject.

Acknowledgments The authors thank Prof. Hiroyasu Hasumi for offering COCO. Many comments and suggestions from the reviewers were very helpful in improving the manuscript. The numerical experiments were performed on the supercomputer, HITACHI SR8000, at the Information Initiative Center, Hokkaido University. Most of the figures were produced with the GFD-DENNOU Library.

References

Fu L-L (2004) Latitudinal and frequency characteristics of the westward propagation of large-scale oceanic variability. J Phys Oceanogr 13:1907–1921

Hasumi H (2000) CCSR report no 13. Center for Climate System Research (CCSR), Chiba

Hellerman S, Rosenstein M (1983) Normal monthly wind stress over the world ocean with error estimates. J Phys Oceanogr 13:1093–1104

Huang RX, Qiu B (1994) Three-dimensional structure of the wind-driven circulation in the subtropical North Pacific. J Phys Oceanogr 24:1608–1622

Kubokawa A (1999) Ventilated thermocline strongly affected by a deep mixed layer: a theory for subtropical countercurrent. J Phys Oceanogr 29:1314–1333

Kubokawa A, Inui T (1999) Subtropical countercurrent in an idealized ocean GCM. J Phys Oceanogr 29:1303–1313

Levitus S, Boyer TP (1994) World Ocean Atlas 1994. NOAA Atlas NESDIS 4, NODC, Washington, DC

Liu Z, Pedlosky J (1994) Thermocline forced by annual and decadal surface temperature variation. J Phys Oceanogr 24:587–608

Luyten JR, Pedlosky J, Stommel H (1983) The ventilated thermocline. J Phys Oceanogr 13:292–309

Marshall JC, Jamous D, Nilsson J (2001) Entry, flux, and exit of potential vorticity in ocean circulation. J Phys Oceanogr 31:777–789

Marshall JC, Nurser AJG (1992) Fluid dynamics of thermocline ventilation. J Phys Oceanogr 22:583–595

Nishikawa S, Kubokawa A (2007) Mixed layer depth front and subduction of low potential vorticity water in an idealized ocean GCM. J Oceanogr 63:125–134

Nishikawa S, Tsujino H, Sakamoto K, Nakano H (2010) Effects of mesoscale eddies on subduction and distribution of Subtropical Mode Water in an eddy-resolving OGCM of the Western North Pacific. J Phys Oceanogr 40:1748–1765

Oka E, Suga T (2003) Formation region of North Pacific subtropical mode water in the late winter of 2003. Geophys Res Lett 30:2205. doi:10.1029/2003GL018581

Qiu B, Huang RX (1995) Ventilation of the North Atlantic and North Pacific: subduction versus obduction. J Phys Oceanogr 25: 2374–2390

Qu T, Xie S-P, Mitsudera H, Ishida A (2002) Subduction of the north pacific mode water in a global high-resolution GCM. J Phys Oceanogr 32:746–763

Stommel H (1979) Determination of water mass properties of water pumped down from the Ekman layer to the geostrophic flow below. Proc Natl Acad Sci USA 76:3051–3055

Suga T, Aoki Y, Saito H, Hanawa K (2008) Ventilation of the North Pacific subtropical pycnocline and mode water formation. Prog Oceanogr 77:285–297

Suga T, Motoki K, Aoki Y, Macdonald AM (2004) The North Pacific climatology of winter mixed layer and mode waters. J Phys Oceanogr 34:3–22

Tsujino H, Yasuda T (2004) Formation and circulation of mode water of the North Pacific in a high-resolution GCM. J Phys Oceanogr 34:399–415

Williams RG (1991) The role of the mixed layer in setting the potential vorticity. J Phys Oceanogr 21:1803–1814

Williams RG, Spall MA, Marshall JC (1995) Does Stommel's mixed layer "demon" work? J Phys Oceanogr 25:3089–3102

Xie S-P, Kunitani T, Kubokawa A, Nonaka M, Hosoda S (2000) Interdecadal thermocline variability in the North Pacific for 1958–1997: a GCM simulation. J Phys Oceanogr 30:2798–2813

J Oceanogr (2012) 68:63–77
DOI 10.1007/s10872-011-0049-9

SPECIAL SECTION: ORIGINAL ARTICLE

New developments in mode-water research: Dynamic and climatic effects

The role of meso-scale eddies in mixed layer deepening and mode water formation in the western North Pacific

Shinya Kouketsu · Hiroyuki Tomita · Eitarou Oka · Shigeki Hosoda · Taiyo Kobayashi · Kanako Sato

Received: 25 January 2011 / Revised: 31 May 2011 / Accepted: 6 June 2011 / Published online: 2 July 2011

Abstract Distributions of mixed layer depths around the centers of anti-cyclonic and cyclonic eddies in the North Pacific Ocean were composited by using satellite-derived sea surface height anomaly data and Argo profiling float data. The composite distributions showed that in late winter, deeper mixed layers were more (less) frequently observed inside the cores of the anti-cyclonic (cyclonic) eddies than outside. This relationship was the clearest in the region of 140°E–160°W and 35°N–40°N, where the temperature and salinity of the deep mixed layers were similar to those of the lighter variety of central mode water (L-CMW). A simple one-dimensional bulk mixed layer model showed that both strong sea-surface heat and momentum fluxes and weak preexisting stratification contributed to formation of the deep mixed layer. These conditions were associated with the anti-cyclonic eddies, suggesting that these eddies are important in the formation of mode waters, particularly L-CMW.

Keywords Mixed layer · Mode water · Meso-scale eddy · Sea-surface flux

S. Kouketsu (✉) · H. Tomita · S. Hosoda · T. Kobayashi · K. Sato
Research Institute for Global Change,
Japan Agency for Marine-Earth Science and Technology,
2-15 Natsushima-cho, Yokosuka 237-0061, Japan
e-mail: skouketsu@jamstec.go.jp

H. Tomita
e-mail: tomitah@jamstec.go.jp

S. Hosoda
e-mail: hosodas@jamstec.go.jp

T. Kobayashi
e-mail: taiyok@jamstec.go.jp

K. Sato
e-mail: k_sato@jamstec.go.jp

E. Oka
Atmosphere and Ocean Research Institute,
The University of Tokyo, Kashiwa, Japan
e-mail: eoka@aori.u-tokyo.ac.jp

1 Introduction

Ocean surface mixed layers, which are the layers at the ocean surface with vertically uniform water properties, develop as a result of momentum and buoyancy forcing, especially in winter. Because heat, freshwater and chemical components (e.g., CO_2) are exchanged between the ocean and the atmosphere through the mixed layers (Takahashi et al. 2009; Hosoda et al. 2010), mixed layer depths (MLDs) and properties are strongly related to climate and environmental variability. Furthermore, deep mixed layers that develop in winter are separated from surface forcing by surface re-stratification, mainly because of heating in spring, and remain in the subsurface as the frequently observed thick, vertically homogeneous layers called "mode waters" (Hanawa and Talley 2001). In association with mode water formation and subduction, the heat, freshwater, and chemical components that are exchanged with the atmosphere and the materials produced by the biological processes near the sea surface are transported to the subsurface layers in the ocean. The part of mode waters remains under the seasonal thermocline, and is entrained again in the deep mixed layer during the next winter. This process affects the late-winter sea surface temperature anomaly in some regions (Hanawa and Sugimoto 2004; Sugimoto and Hanawa 2005). The other part is subducted to the permanent pycnocline. The mode water formation and subduction are crucially involved in long-term subsurface temperature or other environmental changes (Talley and Raymer 1982; Deser et al. 1996; Emerson et al. 2004), and might have an important effect on decadal climate variability (Latif and Barnett 1994; Gu and Philander 1997).

On the western side of the North Pacific Ocean, subtropical mode water (STMW; Masuzawa 1969), central mode water (CMW; Nakamura 1996; Suga et al. 1997), and transition region mode water (TRMW; Saito et al. 2007) are frequently observed as the subsurface vertically homogeneous layers. STMW and CMW are formed as deep winter mixed layers in the region south and north, respectively, of the Kuroshio Extension (Suga et al. 2004), and TRMW is formed south of the subarctic front (Saito et al. 2007). Investigations into the formation and seasonal evolution of the deep mixed layers in the mode-water formation regions have mainly used climatology or spatially sparse observations (Suga et al. 2004; Ohno et al. 2009). Recently, profiling float data became available for the global ocean (Roemmich 2001), and the distribution and structure of the deep mixed layers and mode waters have been investigated in more detail (Oka et al. 2011; Toyama and Suga 2010).

Oka et al. (2011) showed the detailed distribution of the late-winter deep mixed layer in the North Pacific and its relationships with mode waters. They divided the mode waters into four types—STMW, the lighter and denser varieties of CMW (L-CMW and D-CMW; Tsujino and Yasuda 2004; Oka and Suga 2005), and TRMW—by using the temperature and salinity relationships of deep mixed layers as observed by profiling floats. They found that L-CMW, D-CMW, and TRMW become lighter to the east and subduct into the permanent thermocline from the eastern part of the formation region where the deep mixed layers were observed.

In the western North Pacific, the warm (cold) and saline (fresh) anti-cyclonic (cyclonic) eddies known as warm-core (cold-core) rings, which originate from the Kuroshio Extension, are frequently observed (Mizuno and White 1983; Kawamura et al. 1986; Itoh and Yasuda 2010), and their relatively weak (strong) preexisting stratification accompanied by deep (shallow) thermocline can create favorable (unfavorable) conditions for mixed-layer deepening. The spatial changes in preexisting stratification structure associated with the eddies may cause small-scale spatial changes in MLDs. Indeed, there is small-scale spatial variability in the mixed-layer distribution map of Oka et al. (2011), and deep mixed layers have frequently been observed in the ridges and anti-cyclonic eddies along the Kuroshio Extension (Uehara et al. 2003; Qiu et al. 2006; Oka 2009).

Furthermore, small-scale spatial changes in wind and buoyancy corresponding to ocean fronts or eddies have been detected (Nonaka and Xie 2003; White and Annis 2003; Tokinaga et al. 2006), and these changes are important in the distribution of rainfall near oceanic fronts (Minobe et al. 2008; Tokinaga et al. 2010). As the mixed layer develops as a result of momentum and buoyancy fluxes from the atmosphere, spatial changes in fluxes can also contribute to changes in MLDs with the distances from the eddy core. The earlier studies did not describe in detail the effects of small-scale flux variability on the ocean mixed layer; however, more recent studies (Tomita et al. in preparation) have reported that small-scale spatial changes in the forcings across the subarctic front can contribute to those in MLDs.

Recent high-resolution numerical experiments have shown the importance of eddies in the formation and transport of mode waters (especially for STMW) (Tsujino and Yasuda 2004; Rainville et al. 2007; Nishikawa et al. 2010). There have also been many observational descriptions of deep mixed layers and mode water formation in the cores of eddies for the STMW formation area (Uehara et al. 2003; Qiu et al. 2006; Oka 2009), but there were few for the whole western North Pacific. Here we describe the relationships between the late-winter deep mixed layer and meso-scale eddies in the whole western North Pacific and clarify the effects of changes in ocean stratification, accompanied by changes in thermocline depths, and of forcing fields associated with eddies on mixed layer development and mode water formation.

2 Data and methods

Gridded sea surface height anomaly (SSHA) data were produced by Segment Sol Multimissions d'Alimétrie d'Orbitographie et de Localization Precise/Data Unification and Altimeter Combination System (Ssalto/Ducas) and distributed by Archiving, Validation and Interpretation of Satellites Oceanographic Data (AVISO) with support from Centre National d'Etudes Spatiales (CNES). The SSHA map provided was a 7-day composite with $\frac{1}{3}^\circ \times \frac{1}{3}^\circ$ grids. Meso-scale eddies were detected by use of the satellite SSHA data and essentially the same methods as in previous studies (Chelton et al. 2007; Itoh and Yasuda 2010), which defined eddies as areas where the Okubo–Weiss parameter $W\,(=4\,[(\partial u / \partial x)^2 + (\partial v / \partial x)\,(\partial u / \partial y)])$, calculated from an SSHA map was less than a critical value (Fig. 1). The center of an eddy is defined as a centroid of the SSHA. In tracking the eddies detected in each 7-day composite, we searched for the nearest eddy from the center of an eddy detected on the previous map (7 days earlier) within a set search range. In this study, we set the critical value of the Okubo–Weiss parameter to $-0.2 \times \sigma_W$ to allow for the spatial variability of eddy strength, where σ_W is the

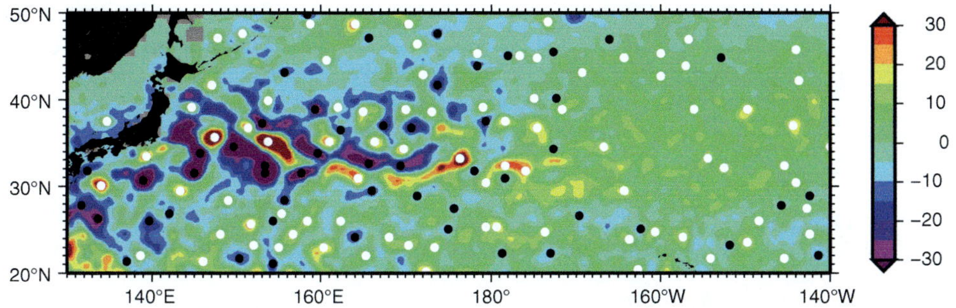

Fig. 1 Sea surface height anomaly (cm) and centers of detected anti-cyclonic (*white circles*) and cyclonic (*black circles*) eddies in the composite of the western North Pacific Ocean for Mar 12, 2008

standard deviation of the Okubo–Weiss parameter in a 5° × 2.5° box (longitude × latitude). The factor of −0.2 was chosen on the basis information presented by Isern-Fontanet et al. (2004), and the resulting critical value east of Japan was similar to that used by Itoh and Yasuda (2010) who investigated meso-scale eddies moving through the same region. The search range for tracking eddies was set to 100 km and we used only eddies that could be tracked for over 4 weeks, as recommended by Itoh and Yasuda (2010).

To determine the MLDs, we used temperature and salinity profiles obtained by Argo profiling floats in the North Pacific from 2002 to 2009. These floats drift at a parking pressure (typically 1000 dbar) and take measurements from an intermediate depth (2000 dbar) to near the sea surface at a preset time interval (10 days). In a typical profile, temperature and salinity are measured at intervals of 5 dbar over the upper 200 dbar, 10–25 dbar from 200 to 1000 dbar, and 50–100 dbar deeper than 1000 dbar. We used real-time quality-controlled float data (Wong et al. 2010) and eliminated questionable profiles, for example those with flagged measurements, by using the same method as Oka et al. (2007). We then interpolated the profile data to 1-dbar intervals by using the Akima spline method (Akima 1970).

MLDs were defined as the depths at which the density was 0.03 kg m^{-3} greater than that at the 10-m depth for each profile (Weller and Plueddemann 1996). Other studies have used vertical temperature differences as the criterion for defining MLDs (de Boyer Montégut et al. 2004), however the difference between MLDs estimated with and without the temperature criterion is not large in the North Pacific (Hosoda et al. 2010). Because large density difference criteria used in previous studies (typically 0.125 kg m^{-3}) are sometimes too large to detect MLDs using Argo float data (Hosoda et al. 2010), we used the small vertical density difference criterion. The mixed layer temperature and salinity were defined as the vertical mean over the obtained MLDs. To make composites, we set four sub regions in the region of 140°E–160°W and 25°N–45°N where almost all of deep mixed layers were observed in the western North Pacific (Oka et al. 2011). The southernmost region in the four sub regions is 25°N–30°N. The regions 30°N–35°N and 35°N–40°N approximately correspond to the main formation regions of STMW and L-CMW, respectively. In the western and eastern parts of the northernmost region of 40°N–45°N, TRMW and D-CMW are formed (Oka et al. 2011). All MLDs observed in the North Pacific region were arranged in order of their distance from the center of the eddy nearest to each profile. We then made composites of the observed MLDs along the radial direction of the eddies in the sub regions in late winter (February–April), when the mixed layers were the deepest of the year (Oka et al. 2007). In the composites, we show 50-km bin averages along the radial direction with their bootstrapped confidence intervals calculated with the bootstrap sample number of 3000 (Von Storch and Zwiers 1999). Note that all the profiles are divided into two groups: near the anti-cyclonic and cyclonic eddy centers, in this method. In this study, we especially focus on "deep mixed layers" deeper than 150 m in the composites, because the deep mixed layers contribute strongly to the formation of mode waters (Suga et al. 2004; Oka et al. 2011).

To investigate the relationship of the distribution of surface momentum and heat fluxes to the meso-scale eddies in the regions, we used a high-resolution satellite-derived air-sea heat flux dataset that was developed as version 2 of the Japanese ocean flux data sets with use of remote sensing observations (J-OFURO2; Tomita et al. 2010). This dataset provided the daily turbulent heat fluxes at the sea surface on a 0.25° × 0.25° grid from 2002 to 2007. We also used the sea surface temperature data that were used to calculate the turbulent heat fluxes in J-OFURO2.

To quantify the contributions of ocean stratification and the sea-surface fluxes to variations in MLD, we used a bulk one-dimensional mixed layer model, assuming that horizontal advection and diffusion over the mixed-layer thickness is small in the core of meso-scale eddies. In the model, we used the following equations to determine MLD (h_m):

$$
\begin{aligned}
\frac{\partial h_m}{\partial t} &= w_e, \\
\frac{\partial T_m}{\partial t} &= \frac{Q_{net} - q_d}{\rho_o c h_m} - \frac{\Delta T}{h_m} w_e, \\
w_e &= \frac{C_1 u_*^3 - C_2 B h_m}{\alpha g h \Delta T}, \\
B &= \alpha g \frac{Q_{net} + q_d}{\rho_o c}, \\
\Delta T &\equiv T_m - T(z = -h_m - 20).
\end{aligned}
\tag{1}
$$

The coefficients and variables in the equations are summarized in Table 1. The coefficients were set to values similar to those in previous studies (Yasuda et al. 2000; Kako and Kubota 2009). Q_{net} was the sum of the turbulent heat fluxes from J-OFURO2 and the radiative heat fluxes from the National Centers for Environment Prediction (NCEP (NOAA, USA); Kalnay et al. 1996), and u_* was calculated from daily mean momentum fluxes in J-OFURO2. ΔT was defined as the temperature difference between the mixed layer and 20 m below the base of the mixed layer. To create a composite view near the center of meso-scale eddies, we set the initial stratification structures of 50-km-bin average temperature profiles around the center of eddies to the averages of profiles observed from October to December in 2002–2007. The winter mean (from November to March) momentum and net heat fluxes were also estimated for each 50-km bin along the radial direction of the eddies. Because the winter mean fluxes met the requirements for mixed layer deepening, only the equations for mixed layer deepening (Eq. 1) are used in this study. Using the mean initial stratification and surface fluxes, we integrated the model from November 15 to March 15 to obtain estimated MLDs.

Table 1 Summary of variables and coefficients for a bulk ocean mixed layer model (Eq. 1)

Variable	Description
h_m	Mixed layer depth
T_m	Mixed layer temperature
w_e	Entrainment velocity
u_*	Frictional velocity
B	Net buoyancy flux
C_1	Tuning coefficients (=1.0)
C_2	Tuning coefficients (=0.2)
α	The thermal expansion coefficient of the seawater (=2.5 × 10^{-4} $^{\circ -1}$)
g	Gravitational acceleration
ΔT	The temperature difference between the mixed layer and the layer below
Q_{net}	Net surface heat flux
q_d	Downward radiative flux at the base of the mixed layer
ρ_o	The reference density
c	The specific heat of seawater (=3.99 × 10^3 kg^{-1} J^{-1})

3 Results

3.1 Relationship between deep mixed layers and meso-scale eddies

The composite analysis revealed that deep mixed layers were more frequently observed in the anti-cyclonic eddies than in the cyclonic eddies in the late winter from 2002 to 2009 (Fig. 2). In the region from 25°N to 30°N, 50-km bin averaged MLDs were approximately 50 m in most of the bins near both the anti-cyclonic and cyclonic eddies, and there was no clear relationship between MLDs and the distance from eddies of either type (Fig. 2a).

From 30°N to 35°N, the MLDs in the anti-cyclonic eddies were deeper than those in the cyclonic eddies, and were deeper (shallower) closer to the core of the anti-cyclonic (cyclonic) eddies (Fig. 2b). The differences between MLDs in the anti-cyclonic and cyclonic eddies were mainly associated with the ridges and troughs of a quasi-stationary meander along the Kuroshio Extension (Mizuno and White 1983; Tatebe and Yasuda 2001), which can be detected as eddies by this method. The deep mixed layers in the anti-cyclonic eddies or the ridges of the meander have been reported previously, and the contribution of the deep mixed layers to the formation of STMW has been shown (Uehara et al. 2003; Qiu et al. 2006; Oka 2009). Although there were clear differences between MLD distributions in the anti-cyclonic and cyclonic eddies, and the bin-averaged MLDs were significantly smaller inside 150 km from the cyclonic eddy center than outside, the differences between MLDs inside and outside of 150 km from the anti-cyclonic eddy center was insignificant in comparison with the 95% confidence intervals for the bin average (Fig. 2b). This region is mostly located south of the Kuroshio Extension, where the preexisting stratification under the seasonal thermocline was relatively weak because of STMW in the subsurface layer. Such large-horizontal-scale distribution of weak preexisting stratification resulted in a deepened mixed layer even away from the core of the anti-cyclonic eddies.

From 35°N to 40°N, differences were evident in the distributions of MLD between inside and outside 100 km from the anti-cyclonic eddy center and between inside 100 km from the anti-cyclonic and cyclonic eddy centers, whereas the difference between inside and outside the cyclonic eddies was not clear in comparison with the 95% confidence intervals (Fig. 2c). Over 70% of all observed MLDs within 50 km of the anti-cyclonic cores were deep MLDs (>150 m), and the bin-averaged MLD was approximately 200 m, whereas the bin-averaged MLDs

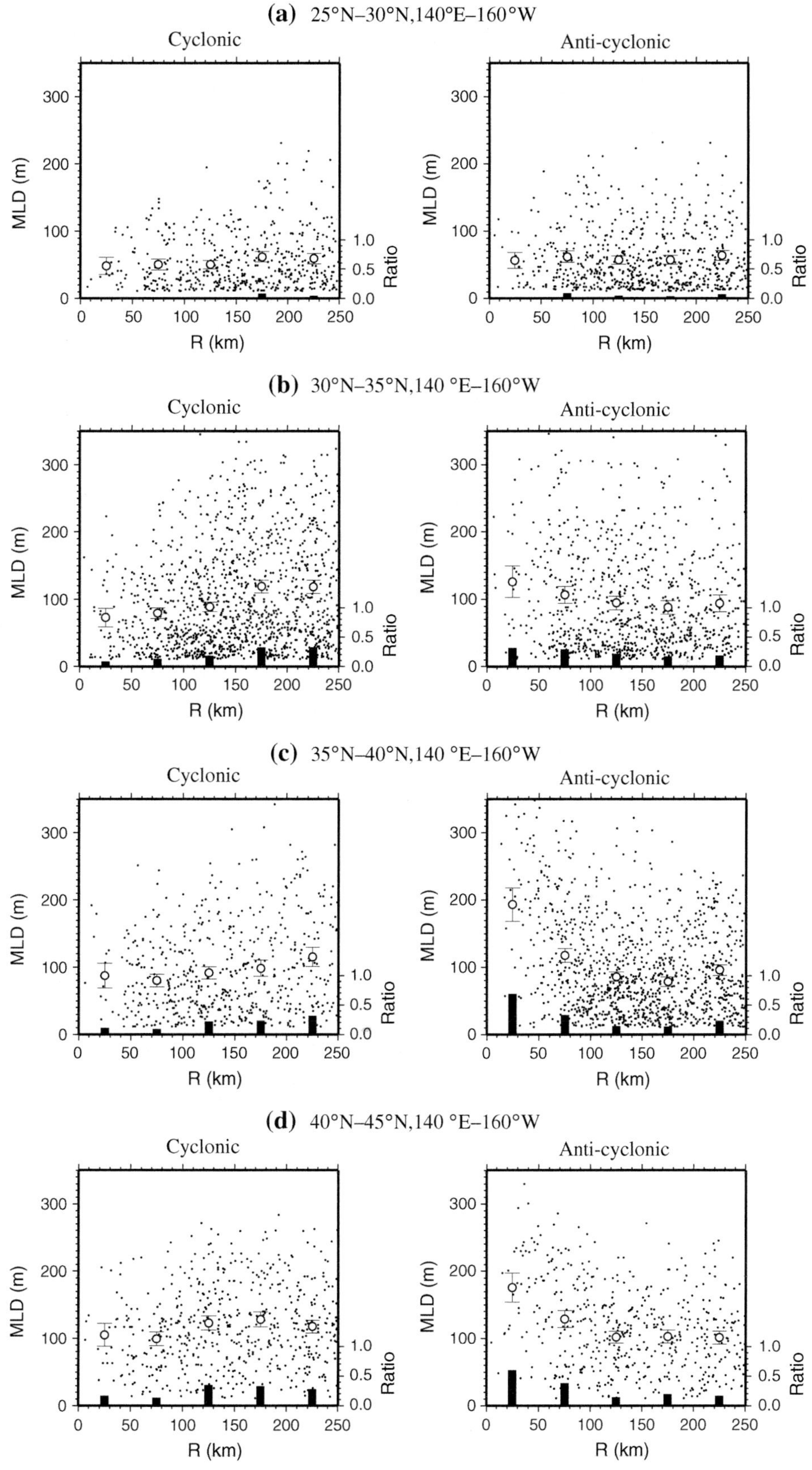

Fig. 2 Distributions of mixed layer depths as a function of distance (*R*) from the center of meso-scale eddies in late winter (from February to April). *Dots* denote observed mixed layers. *Open circles* and *bars* denote the 50-km bin averages and bootstrapped 95% confidence intervals, respectively, of the mixed layer depth. *Solid black bars* denote the ratio of deep mixed layers (>150 m) to all mixed layer observations in the 50-km bins from 2002 to 2009

more than 50 km from the core were less than 120 m. In this region, warm and saline anti-cyclonic eddies known as warm-core rings, which originate from the Kuroshio Extension, are frequently observed (Mizuno and White, 1983; Kawamura et al. 1986; Itoh and Yasuda 2010), and their relatively weak preexisting stratification create favorable conditions for mixed-layer deepening. The presence of many eddies with clearly different water properties in this region (Itoh and Yasuda 2010) could explain the clear MLD differences (Fig. 2c). From 40°N to 45°N, clear differences were also evident between the distribution of MLDs in the anti-cyclonic and cyclonic eddies and between the MLDs inside and outside 100 km from anti-cyclonic eddy center (Fig. 2d), although there were not many available observations in this region.

In the composites (Fig. 2), deep MLDs were frequently observed within 125 km from the eddy centers. Assuming that 125 km was the scale within which eddies affected mixed-layer deepening, we calculated the proportions of the number of deep MLDs (>150 m) inside of the eddies (i.e., an observation within 125 km of an eddy core) to all observed deep MLDs in each grid box in the horizontal map (Fig. 3a) to estimate the contribution of the anti-cyclonic eddies to mixed-layer deepening. Furthermore, to take into account different frequencies of observed anti-cyclonic eddies in each box, we calculate temporal mean areas of the anti-cyclonic eddy within each box, and show the spatial distribution of the ratio of the number of observed deep MLDs per unit area within the anti-cyclonic eddies to that in a whole box area (Fig. 3b). Note that 125 km is approximately double the typical eddy radius as estimated from the SSHA data (Itoh and Yasuda 2010). The typical radius of an eddy as defined using the Okubo–Weiss parameter encompasses the area where vorticity predominates over strain (i.e., only the inner part of the eddy core), and this radius is about half the radius of the entire vortex (Isern-Fontanet et al. 2006). The distance of 125 km used in this study probably corresponds to the radius of the entire vortex, which includes the areas with strong currents associated with the eddies. In addition, in this analysis, we also assume the frequencies of Argo float observations along the radial directions of the eddies were uniform, because the ratio of Argo float observations inside the anti-cyclonic eddies to all the observations was similar to that of the areas inside the eddies to all the areas (approx. 0.15).

Deep mixed layers were frequently observed from 140°E to 160°W (similar to the observations of Oka et al. 2011), and the contribution of anti-cyclonic eddies to the mixed-layer deepening, and thus to mode water formation, varied spatially (Fig. 3). In almost all the high-proportion areas in Fig. 3a, the ratios of the number of observed deep MLD per unit area were greater than unity (Fig. 3b). Thus, many deep MLDs were observed within anti-cyclonic eddies in the high-proportion areas (Fig. 3a) not just because of the presence of many eddies but also because of favorable conditions for deep MLDs associated with the anti-cyclonic eddies. As the areas with a high proportion of

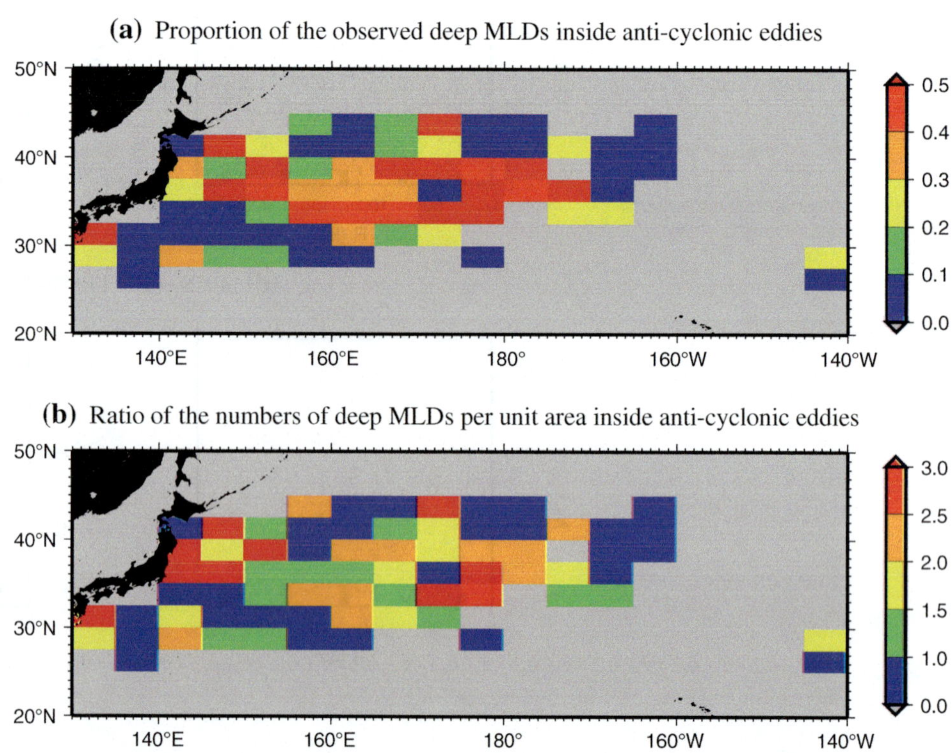

Fig. 3 The proportion of the number of the deep mixed layers (>150 m) within 125 km of the center of anti-cyclonic eddies in all the observed deep mixed layers (**a**), and the ratio of the number of the deep mixed layers per unit area inside the anti-cyclonic eddies to that in a whole box (**b**). The proportion and ratio were calculated for each 5° × 2.5° box by using profile data from 2002 to 2009, and the total areas of the anti-cyclonic eddies were calculated assuming the radius of anti-cyclonic eddies was 125 km. *Gray boxes* had fewer than five observed deep mixed layers

Table 2 Properties of mode waters around the inter-frontal zone in the North Pacific Ocean (adapted from Oka et al. 2011)

Type	Temperature (°C)	Salinity	Density (kg m^{-3})
STMW	16.0–21.5	34.65–34.95	24.2–25.6
L-CMW	10.0–16.0	34.30–34.65	25.4–26.3
D-CMW	8.0–10.5	33.80–34.20	26.1–26.5
TRMW	5.0–7.5	33.60–33.90	26.4–26.6

the number of deep MLDs within anti-cyclonic eddies were distributed in the region from 140°E to 160°W and from 35°N to 45°N (Fig. 3), the relationships evident in the composites (Fig. 2c) reflect the characteristics of a wide area with many eddies. There was also a high proportion of deep MLDs inside anti-cyclonic eddies around 35°N near the Kuroshio Extension, whereas the magnitude was relatively low around 30°N. North of 40°N, there are some regions where the proportion of the number of deep MLDs within anti-cyclonic eddies was high west of 160°E. This suggests that the relationships in the composites (Fig. 2d) reflect only a limited area mainly west of 160°E. The region with high proportion of the number of deep mixed layers inside anti-cyclonic eddies from 145°E to 160°E (Fig. 3) corresponded to the region along the quasi-stationary jet associated with the subarctic front (Isoguchi et al. 2006).

Oka et al. (2011) used the temperature and salinity relationships of mixed layers deeper than 150 m to classify water property in the deep mixed layers corresponding to mode water types, and described 4 mode water types (Table 2), because the deep mixed layers are the major source of mode waters in the North Pacific. In a similar manner, we grouped the observed deep MLDs (>150 m) by temperature and salinity characteristics and determined the proportion in each temperature-salinity bin that were distributed within anti-cyclonic eddies (Fig. 4). This grouping revealed four types of deep mixed layer waters similar to the 4 mode water types, separated by bins with fewer observations (4 rectangles surrounded white and gray boxes in Fig. 4). Note that deep MLDs were rarely observed in the cyclonic eddies (Fig. 2), and proportions of the number of deep MLDs within cyclonic eddies were less than 0.1 in almost all the temperature-salinity bins (not shown). Although deep mixed layers with the properties of STMW were not always observed in the anti-cyclonic eddies (Fig. 2b and around 30°N in Fig. 3), the relatively cold (<18°C) and fresh (<34.8) deep mixed layers in STMW (Fig. 4) were frequently found inside the anti-cyclonic eddies. Water similar to the colder and fresher range of STMW was often observed in the eastern part of the STMW formation region south of the Kuroshio Extension (Oka and Suga 2003), which suggests that the anti-cyclonic eddies contributed to the formation of

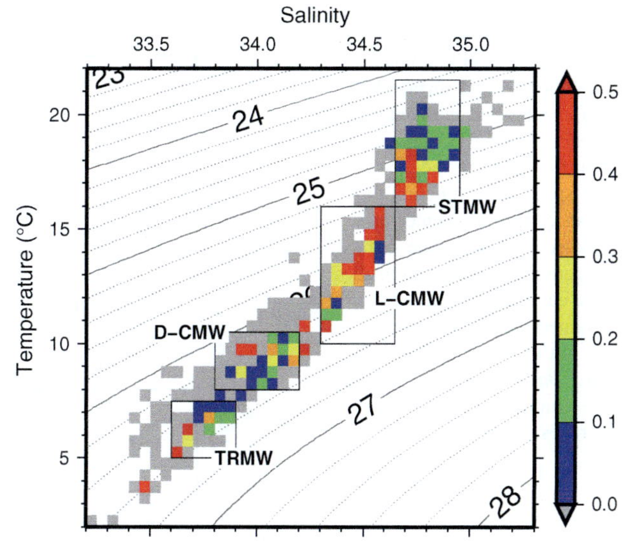

Fig. 4 Proportion of the number of deep mixed layers (>150 m) in the cores of anti-cyclonic eddies (<125 km) as a function of temperature and salinity. The *boxes* used for the summation were in increments of 0.5° for temperature and 0.05 for salinity. *Gray boxes* had fewer than five observed deep mixed layer. The four mode water types defined by Oka et al. (2011) (Table 2) are shown by *rectangles*

STMW, especially in the downstream region of the Kuroshio Extension, which is reflected in the high-proportion regions around 170°E and 35°N in Fig. 3. In contrast, the high proportions of the number of the deep mixed layers with the properties of L-CMW in the anti-cyclonic eddies (Fig. 4) reflected that the strong contribution of the anti-cyclonic eddies to deep mixed layers were detected broadly in the L-CMW formation area (Fig. 2c and from 35°N to 40°N in Fig. 3). As clearly defined large-scale fronts such as the Kuroshio Extension were not observed in the L-CMW formation region, the relatively weak preexisting stratification associated with the anti-cyclonic eddies could be more important there for mixed-layer deepening. There were smaller contributions from anti-cyclonic eddies to deep mixed layer formation in the D-CMW region, especially in its cold and fresh part, and relatively large contributions in the TRMW region (Fig. 4).

3.2 Contribution of sea-surface flux and preexisting stratification to mixed layer deepening

Because warmer (colder) water has frequently been observed in the cores of anti-cyclonic (cyclonic) eddies (Itoh and Yasuda 2010), we detected positive (negative) anomalies of bin-averaged sea surface temperature (SST), which were calculated from area-weighted averages within 350 km of the center of each eddy (Fig. 5). The differences between SST inside and outside eddies were larger north of the Kuroshio Extension (Fig. 5c, d) than south (Fig. 5a, b). In fact strong and large warm-core rings were reported north of

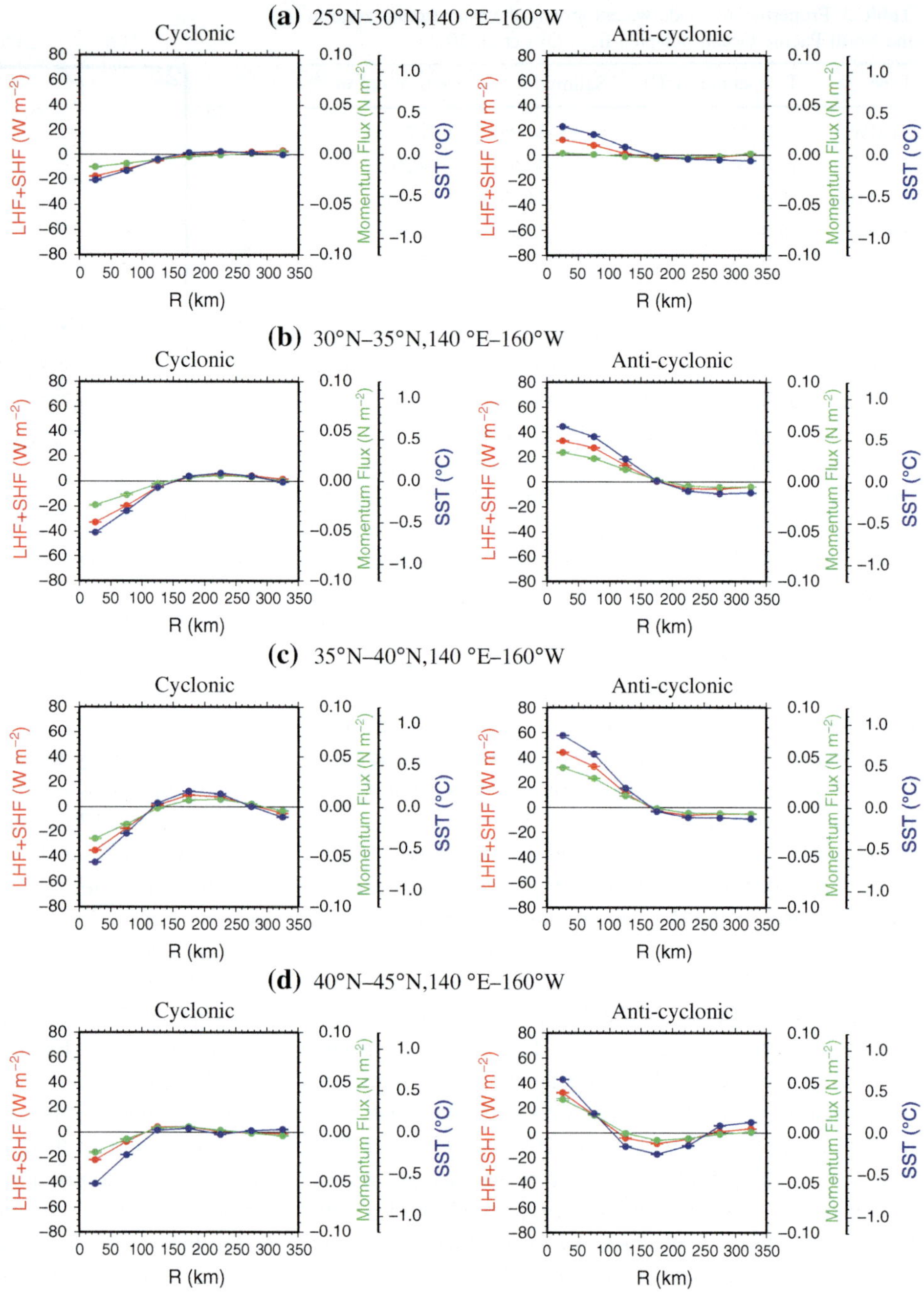

Fig. 5 Distribution of 50-km bin averages of sea surface temperature (SST) anomalies (*blue curves*) and surface heat (*red curves*) and momentum (*green curves*) flux anomalies from the ocean to the atmosphere as a function of distance from the center of meso-scale eddies. The surface temperature and flux anomalies were calculated from area-weighted averages within 350 km of the center of each eddy. *Bars* denote bootstrapped 95% confidence intervals for bin averages

the Kuroshio Extension in previous studies, whereas such clear SST anomalies associated with meso-scale eddies were not frequently observed to the south (Mizuno and White 1983; Kawamura et al. 1986). This difference between the areas south and north of the Kuroshio Extension is reflected in the amplitude of composite SST anomalies in eddies.

The anomalies of both turbulent surface heat and momentum fluxes to the ocean, which were calculated from area-weighted averages within 350 km of the center of each eddy, associated with the SST anomalies in eddies were positive (negative) in the core of anti-cyclonic (cyclonic) eddies (Fig. 5). The momentum flux anomalies reflect the tendency for wind near the ocean surface to be stronger over warmer water, as reported in previous studies (Nonaka and Xie 2003), and this relationship is also evident in the composite of the eddies detected from the SSHA (Fig. 5). Note that the 95% confidence intervals for bin-averaged anomalies of SST and surface fluxes were generally small, because of the large number of samples.

To evaluate the effects of the anomalies of turbulent heat and momentum fluxes (Fig. 5) and the preexisting stratification along a radial eddy section (Fig. 6) on the MLD distributions along a radial direction, we used the bulk one-dimensional mixed layer model (Eq. 1). From 30°N to 35°N, the differences between observed MLDs inside and outside anti-cyclonic eddies were smaller than those for the cyclonic eddies (Fig. 7b). Because changes in

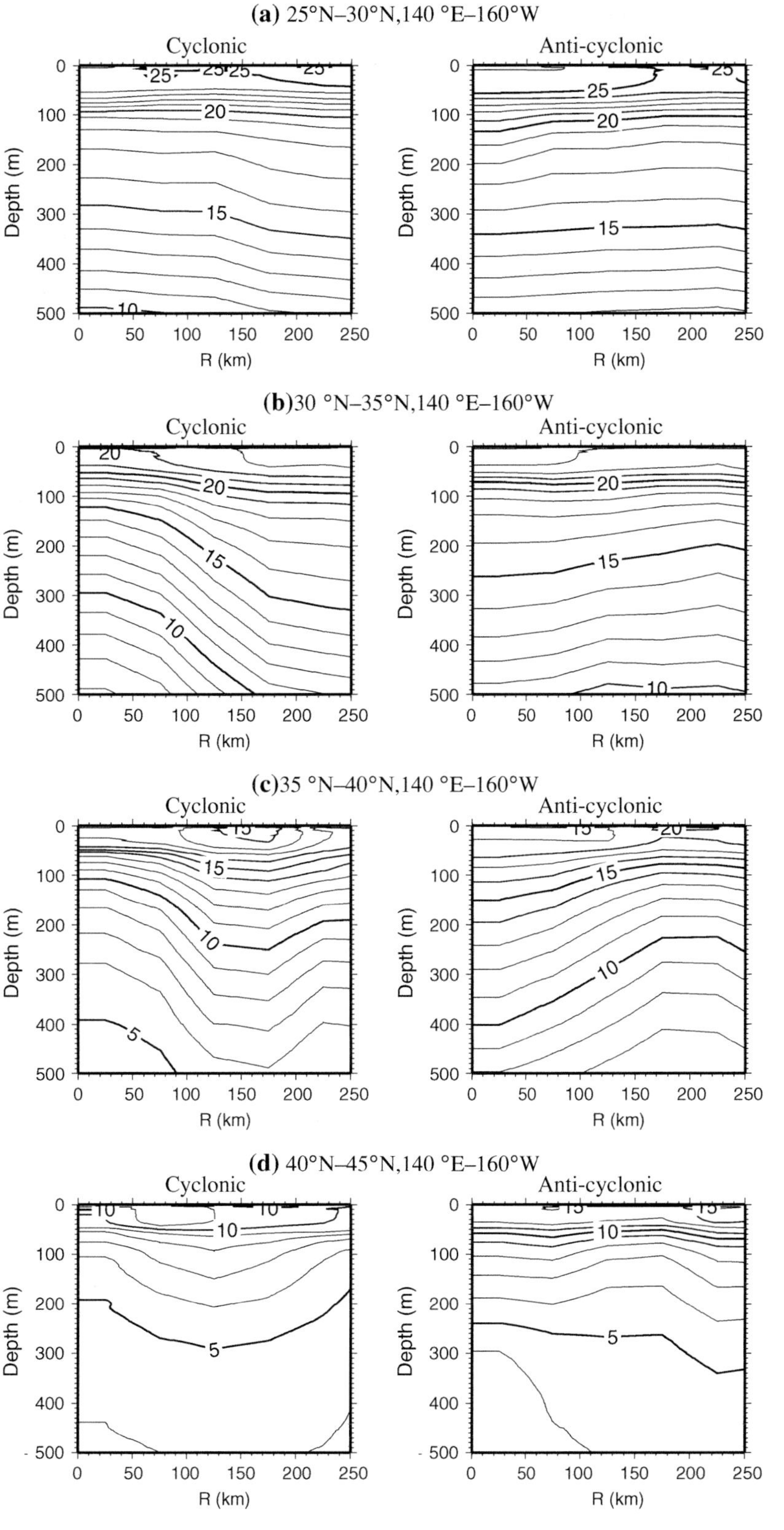

Fig. 6 Vertical temperature sections along the radial section of eddies averaged from October to December from 2002 to 2007 (°C). The contour intervals are 1°C

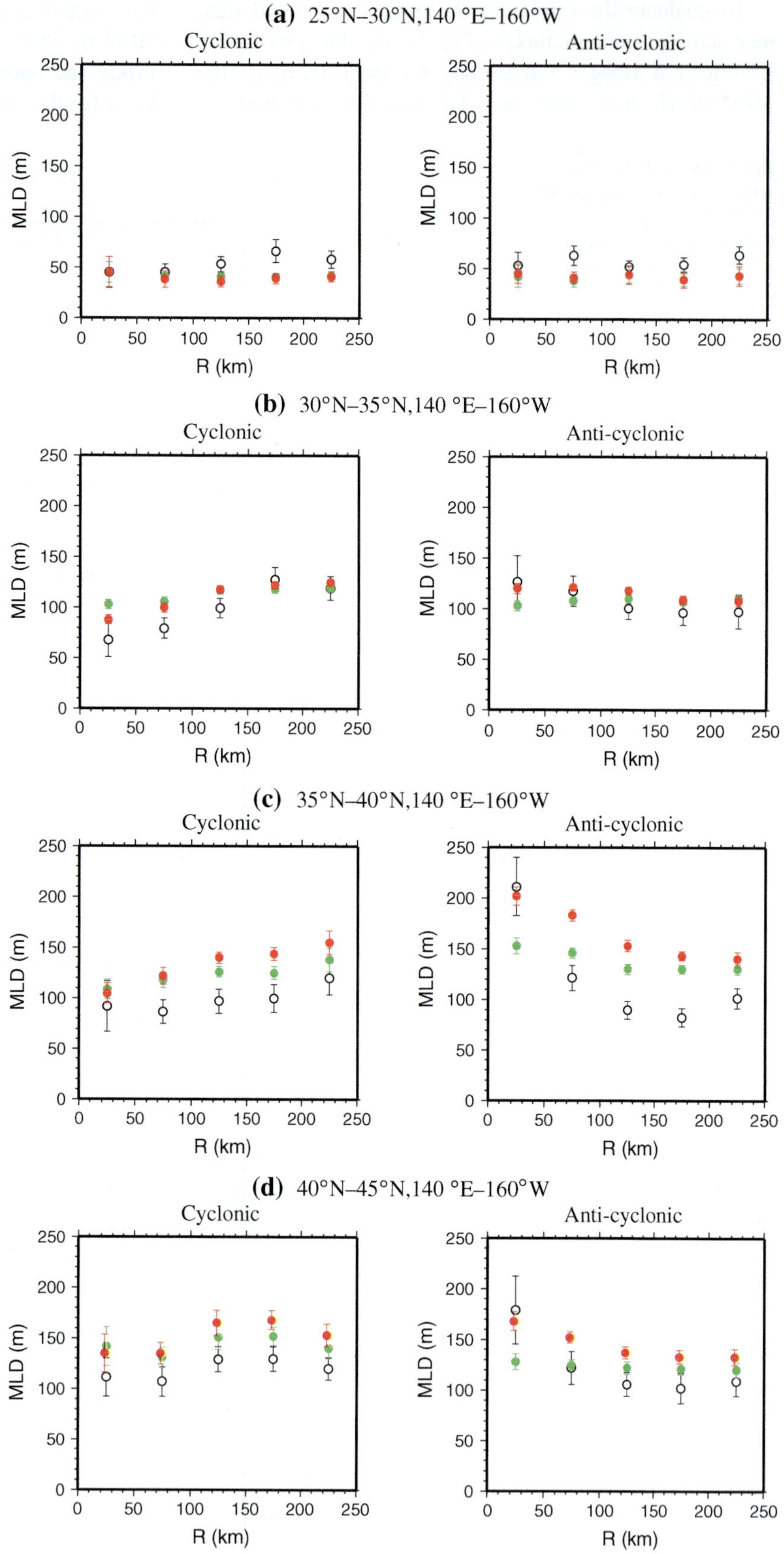

Fig. 7 Observed mixed layer depths (MLDs) and those estimated using the bulk mixed layer model as a function of distance along the radial direction of eddy. *White circles* denote the 50-km bin averages of observed MLDs similar to Fig. 2, but calculated using data from 2002 to 2007. *Green dots* denote MLDs estimated using the preexisting stratification associated with the eddies (Fig. 6) and a constant surface flux over the radial section, whereas *red dots* denote MLDs estimated using both the preexisting stratification and the surface flux associated with the eddies (Figs. 5, 6). *Bars* denote bootstrapped 95% confidence intervals for bin averages

preexisting stratification along the radial section were larger in the cyclonic eddy than in the anti-cyclonic eddy (Fig. 6b), there were large differences between MLDs inside and outside the cyclonic eddies, estimated both with and without the surface flux distribution along the radial direction (green and red circles in Fig. 7b). On the other hand, MLDs of the anti-cyclonic eddies estimated without the surface flux anomalies change little along the radial direction (green circles in Fig. 7b). This difference is mainly because clearly defined cyclonic eddies are more frequently observed than anti-cyclonic eddies in this region (Itoh and Yasuda 2010).

In the region from 35°N to 40°N, both the flux anomaly and preexisting stratification changes along the radial directions (Figs. 5c, 6c) were obvious, and there were large changes of MLDs along the radial direction, which were estimated both with and without the surface flux distribution (Fig. 7c). Thus, both the flux anomaly and the preexisting stratification changes along the radial section strongly contributed to reproduction of the pattern changes in MLDs, although the difference of the estimated and observed MLDs was significantly large except within 50 km of the eddy centers (Fig. 7c). A large portion of the deep mixed layers observed in this region was located in the anti-cyclonic eddies (as discussed in the Sect. 3.1). Thus, the strong cooling and weak preexisting stratification inside the anti-cyclonic eddies are important in the formation of the deep mixed layers in this region, which contributed to the L-CMW source region.

The flux anomalies associated with the anti-cyclonic eddies were more important for forming the MLD changes in the radial direction than the preexisting stratification changes along the radial section in the region from 40°N to 45°N, because the MLD changes estimated using the surface flux distribution along the radial direction were clear, and those without the surface flux distribution were insignificant (Fig. 7d). As shown in the Sect. 3.1, deep mixed layers in anti-cyclonic eddies have frequently been observed near the east coast of Japan, where TRMW is formed; the flux anomalies associated with the eddies might be important for the formation of TRMW (Fig. 7d). As both the flux (Fig. 5a) and the preexisting stratification changes along the radial section (Fig. 6a) are small in the region of 25°N–30°N, the differences between MLDs inside and outside the eddies were small, and there were no clear differences between the estimated MLD distributions in the anti-cyclonic and cyclonic eddies (Fig. 7a).

4 Discussion

In this study, we have shown the relationships between deep mixed layers and spatial changes of ocean preexisting stratification and surface fluxes around the centers of meso-scale eddies detected by using satellite SSHA data in the western North Pacific Ocean, wherein the weak (strong) preexisting stratification and strong (weak) forcing inside the core of anti-cyclonic (cyclonic) eddies tend to foster the development of deep (shallow) mixed layers in the winter. These relationships were clear, especially in the inter-frontal zone between the Kuroshio Extension and subarctic front. This suggests that meso-scale eddies are important in forming mode waters in this area. The small-scale distribution of MLDs associated with the meso-scale eddies could affect the temporal variability of large-scale MLD through eddy activity variability, because decadal changes of eddy activity around the Kuroshio Extension were reported after use of SSHA observations and high-resolution numerical models (Qiu and Chen 2005; Taguchi et al. 2010). Because the heat and chemical constituents that are exchanged between the atmosphere and the ocean at the sea surface can be transported to the sub-surface layer in association with CMW subduction (Deser et al. 1996; Schneider et al. 1999; Emerson et al. 2004), the eddy activity around the inter-frontal zone might be involved in inter-annual climate variability. Furthermore, many studies have reported that the distribution of eddies in the inter-frontal zone corresponds to the sea surface color pattern (Saitoh et al. 1998); therefore the distribution of MLDs observed in this study could also affect the biological activity in the eddy cores.

The contributions of the anti-cyclonic eddies to form deep MLDs, which are important for mode water formation, were different spatially (Figs. 2, 3). In the STMW formation region, the changes in MLDs along the radial direction of anti-cyclonic eddies were relatively small (Fig. 2b), and there was a low proportion of the number of deep mixed layers inside the anti-cyclonic eddies around 30°N (Fig. 3). This is because both the relatively weak preexisting stratifications associated with anti-cyclonic eddies (near 35°N in Fig. 3) and those observed broadly south of the Kuroshio Extension (in the region of 30°N), because of the existence of STMW in the subsurface layer formed in previous winters, contribute to the formation of the deep mixed layers there. Furthermore, the small MLD changes along the radial direction of the anti-cyclonic eddies also reflected that the changes in surface flux and preexisting stratification along the radial directions were relatively small (Figs. 5b, 6b, 7b), mainly because of the presence of fewer clearly defined warm-core rings (Itoh and Yasuda 2010). In the L-CMW formation region that is not accompanied by clearly-defined large-scale fronts, for example the Kuroshio Extension, the relatively weak preexisting stratification associated with the anti-cyclonic eddies could be more important there for mixed-layer deepening. Thus, there was a clear MLD difference along the radial direction of the anti-cyclonic eddy (Fig. 2c), because of the substantially different preexisting stratifications and surface fluxes along the radial direction (Figs. 5c, 6c, 7c), and large contribution of the anti-cyclonic eddies to deep MLDs were broadly detected spatially (Fig. 3) and in the temperature–salinity relationships (Fig. 4). The relatively strong contribution of anti-cyclonic eddies to deep mixed layers with the properties of TRMW (Fig. 4) might reflect that the anti-cyclonic eddies along the front (or ridges of the meander) were important for forming deep mixed layers with TRMW properties, as TRMW forms in the narrow area south of the subarctic front (Saito et al. 2007). Formation of TRMW west of 170°E and its subsequent eastward transportation are thought to set the precondition for formation and subduction of D-CMW east of 170°E (Oka et al. 2011). This might explain why there were fewer high-proportion bins with D-CMW properties in Fig. 4.

Some parts of thick homogeneous layers originating from the deep mixed layers in the anti-cyclonic eddies can be transported outside the eddies by eddy mixing, and be subducted into the permanent pycnocline by transportation along with isopycnal surfaces (Nishikawa et al. 2010). This process could contribute the broad distribution of mode waters in the permanent pycnocline. Other parts of the homogeneous layers seem to remain in the cores of mesoscale eddies during the sequential seasons. Furthermore, whereas some of these can also be subducted into the permanent pycnocline with the eddy moving (Nishikawa et al. 2010), the others remain under the seasonal pycnocline until the next winter, and can affect the deepening mixed layer. In the L-CMW formation area, while the many anti-cyclonic eddies are moving westward (Itoh and Yasuda 2010), the broad distribution of L-CMW in the main thermocline is observed in the downstream region of the interfrontal zone (Oka et al. 2011). The former process of eddy mixing and advections in the permanent pycnocline may contribute to the broad distribution of L-CMW in the permanent pycnocline near the downstream region, and the latter process of the thick homogeneous layer trapped in the cores may be dominant in the upstream regions, where the thick homogeneous layers were not broadly observed in the subsurface layer (Oka et al. 2011). However, these subduction processes around eddies were not directly shown in this study. To investigate the processes, we plan to investigate seasonal changes in the subsurface layers around eddies from future observations.

4.1 Dependency of criteria and large variances in composites

The results in this study were not strongly dependent on choices of criteria to determine MLDs (vertical density difference of 0.03 kg m^{-3}) and deep MLD (>150 m) with strong contributions to the formation of mode waters. Using the shallow (deep) criterion of 100 m (200 m) for deep MLDs, proportions of the number of deep MLDs in the anti-cyclonic eddies became only approximately 0.1 smaller (larger) than shown in this study. These changes were not very large, and the difference among the contributions of the anti-cyclonic eddies to mode waters shown in the temperature and salinity relationship were similar to Fig. 4 (not shown). Although MLDs defined with a large density difference criterion of (for example) 0.125 kg m^{-3} were approximately 40-m deeper than those shown in Fig. 1 (not shown), the proportions of the number of deep MLDs in the anti-cyclonic eddies became only about 0.1 smaller. This criterion choice for deep MLDs also did not strongly affect our results. Note that using such a large density-difference criterion, we might be able to avoid detecting many shallow mixed layers because of short-term warming in the early spring (e.g, approximately 10-m MLDs in Fig. 2), which might persist in the short-term and may cause underestimation of deep MLDs contributing to form mode waters, but the MLDs determined by this large density difference criterion tend to be overestimated because they include thermoclines in which water properties are not vertically homogeneous.

While we showed that the different MLDs and fluxes between in the anti-cyclonic and cyclonic eddies, and the changes along the radial directions of the eddies, were detectable in the 50-km bin averages with the bootstrapped confidence intervals, there were large variances in MLDs and fluxes (see dots in Fig. 2 for MLDs). Variance of eddy shapes (assuming axially symmetric shapes in this study) and uncertainty in the locations of eddy centers can cause such variances, and the large variances might be mainly due to short-term or more small-scale variability. Although warmer waters are much more frequently observed than colder waters in the anti-cyclonic eddies, anti-cyclonic eddies with cold waters, which are sometimes observed (Hosoda and Hanawa 2004; Itoh and Yasuda 2010), can also affect large variances of MLDs and SSTs. In the future, the phenomena causing such large variances will be described in detail by use of higher resolution SSH and long-term satellite SST observations.

In this study, the distribution of the MLDs along the radial direction of the eddies was reproduced by using the initial stratification and the surface flux distributions associated with the eddies through the bulk formula for the MLDs, although the variance of observed MLDs for each bin was large; this might be because of more small-scale or short-term variability. The effects of horizontal advection and diffusion, which were ignored in this study, are presumed to be small, because the model was integrated over winter only (for 4 months), and strong currents were unlikely to be

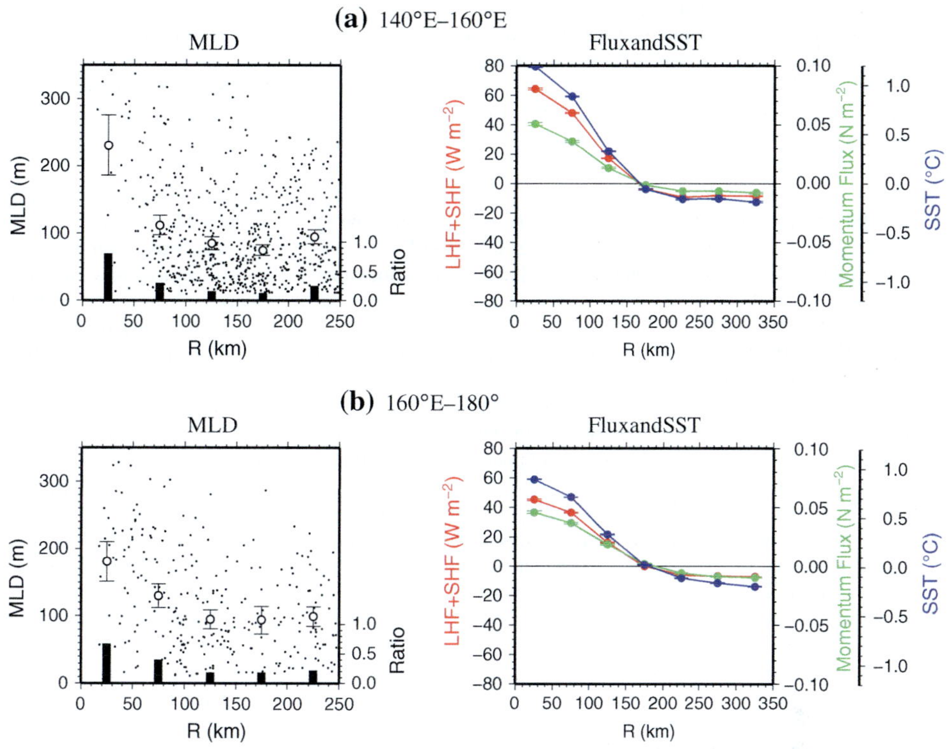

Fig. 8 Distribution of mixed layer depths (*left*) and anomalies of sea surface temperature and fluxes (*right*) along the radial direction of anti-cyclonic eddies in the upstream of 140°E–160°E (**a**) and the downstream of 160°E–180° (**b**) regions of the main L-CMW formation area (35°N–40°N). The *left* and *right figures* are same as Figs. 2 and 5 but for zonally narrow sub regions

observed in the core of the eddies. However, there could be errors in the estimates because of the neglected effects, and the effect of horizontal advection might cause the difference between observed and estimated MLDs in Fig. 7.

The composites in this study were produced by using Argo float data, which provide intensive and uniform observations of ocean subsurface structure. However, more detailed observations (e.g., observations focusing on a eddy) are needed to clarify the processes causing deepening of mixed layers and formation of mode waters in eddies, taking into account horizontal advection. Furthermore, although we showed there were some spatial changes in relationships between meso-scale eddies and deep mixed layers, by showing the proportions of the number of deep MLDs inside the anti-cyclonic eddies in the 5° × 2.5° boxes (Fig. 3), available observations are not yet sufficient to enable description of detailed spatial changes in the relationships. For example, along the radial directions of the anti-cyclonic eddy, there were the differences of MLDs and flux anomalies between in the upstream (western) and downstream (eastern) regions of the L-CMW formation area (Fig. 8). The larger flux anomalies in the anti-cyclonic eddies might cause the deeper MLDs in the upstream region than in the downstream region, and this suggested that flux anomalies corresponding to the meso-scale eddy were more important in the upstream regions. However, it was difficult to determine preexisting stratification structures along the radial sections (similar to Fig. 6) because of the small number of available observations in these small sub-regions. It is also difficult to describe relationships among the inter-annual subsurface temperature changes, eddy activity, and the process of mode waters subduction, because intense observations by Argo floats have only been carried out for approximately 5 years in the North Pacific. Continued and more intense observations by use of Argo floats, fine scale observations by ships and gliders, and analysis with high-resolution numerical models should help clarify the temporal and spatial changes in detail.

Acknowledgments The Argo float data used in this study were collected and made freely available by the International Argo Project and the national programs that contribute to it (http://www.argo.ucsd.edu, http://argo.jcommops.org). The comments from anonymous reviewers were useful for improving the manuscript. This study was supported by the Japan Society for Promotion of Science (KAKENHI, Grant-in-Aid for Young Scientists (B), no. 20740279). E. Oka is supported by the Japan Society for Promotion of Science (KAKENHI, Grant-in-Aid for Scientific Research (B), no. 21340133) and the Ministry of Education, Culture, Sports, Science and Technology, Japan (MEXT; Grant-in-Aid for Scientific Research of Innovative Areas under grant no. 22106007).

References

Akima H (1970) A new method of interpolation and smooth curve fitting based on local procedures. J Assoc Comput Mach 17:589–603

Chelton DB, Schlax MC, Samelson RM, de Szoeke RA (2007) Global observations of large oceanic eddies. Geophys Res Lett 34:L15606. doi:10.1029/2007GL030812 de Boyer Montégut C, Madec G, Fischer AS, Lazar A, Iudicone D (2004) Mixed layer depth over the global ocean: an examination of profile data and a profile-based climatology. J Geophys Res 109:C12,003

Deser C, Alexander MA, Timlin MS (1996) Upper-ocean thermal variations in the North Pacific during 1970–1991. J Climate 9:1840–1855

Emerson S, Watanabe YW, Ono T, Mecking S (2004) Temporal trends in apparent oxygen utilization in the upper pycnocline of the North Pacific: 1980–2000. J Oceanogr 60:139–147

Gu D, Philander SGH (1997) Interdecadal climate fluctuations that depend on exchanges between the tropics and extratropics. Sci Agric 275(5301):805

Hanawa K, Sugimoto S (2004) 'Reemergence' areas of winter sea surface temperature anomalies in the world's oceans. Geophys Res Lett 31:l10303. doi:10.1029/2004GL019904

Hanawa K, Talley LD (2001) Mode waters. In: Siedler G, Church J, Gould J (eds) Ocean circulation and climate. Academic Press, San Diego, pp 373–386

Hosoda K, Hanawa K (2004) Anticyclonic eddy revealing low sea surface temperature in the sea south of Japan: case study of the eddy observed in 1999–2000. J Oceanogr 60(4):663–671

Hosoda S, Ohira T, Sato K, Suga T (2010) Improved description of global mixed-layer depth using Argo profiling floats. J Oceanogr 66:773–787

Isern-Fontanet J, Font J, García-Ladona E, Emelianov M, Millot C, Taupier-Letage I (2004) Spatial structure of anticyclonic eddies in the Algerian basin (Mediterranean Sea) analyzed using the Okubo–Weiss parameter. Deep Sea Res II 51(25–26):3009–3028

Isern-Fontanet J, García-Ladona E, Font J (2006) Vortices of the Mediterranean Sea: an altimetric perspective. J Phys Oceanogr 36(1):87–103

Isoguchi O, Kawamura H, Oka E (2006) Quasi-stationary jets transporting surface warm waters across the transition zone between the subtropical and the subarctic gyres in the North Pacific. J Geophys Res 111(C10):C10,003

Itoh S, Yasuda I (2010) Characteristics of mesoscale eddies in the Kuroshio–Oyashio extension region detected from the distribution of the sea surface height anomaly. J Phys Oceanogr 40:1018–1034

Itoh S, Yasuda I (2010) Water mass structure of warm and cold anticyclonic eddies in the western boundary region of the subarctic North Pacific. J Phys Oceanogr 40:2624–2642

Kako S, Kubota M (2009) Numerical study on the variability of mixed layer temperature in the North Pacific. J Phys Oceanogr 39(3):737–752

Kalnay E, Kanamitsu M, Kistler R, Collins W, Deaven D, Gandin L, Iredell M, Saha S, White G, Woollen J, Zhu Y, Leetmaa A, Reynolds R, Chelliah M, Ebisuzaki W, Higgins W, Janowiak J, Mo KC, Ropelewski C, Wang J, Jenne R, Joseph D (1996) The NCEP/NCAR 40-year reanalysis project. Bull Am Meteor Soc 77(3):437–471

Kawamura H, Mizuno K, Toba Y (1986) Formation process of a warm-core ring in the Kuroshio–Oyashio frontal zone–December 1981–October 1982. Deep Sea Res A 33(11–12):1617–1640

Latif M, Barnett TP (1994) Causes of decadal climate variability over the North Pacific and North America. Sci Agric 266(5185):634

Masuzawa J (1969) Subtropical mode water. Deep Sea Res 16:453–472

Minobe S, Kuwano-Yoshida A, Komori N, Xie S, Small R (2008) Influence of the Gulf Stream on the troposphere. Nat Biotechnol 452(7184):206–209

Mizuno K, White W (1983) Annual and interannual variability in the Kuroshio current system. J Phys Oceanogr 13:1847–1867

Nakamura H (1996) A pycnostad on the bottom of the ventilated portion in the central subtropical North Pacific: its distribution and formation. J Oceanogr 52(2):171–188

Nishikawa S, Tsujino H, Sakamoto K, Nakano H (2010) Effects of mesoscale eddies on subduction and distribution of Subtropical Mode Water in an eddy-resolving OGCM of the western North Pacific. J Phys Oceanogr 40:1748–1765. doi: 10.1175/2010JPO4261.1

Nonaka M, Xie SP (2003) Covariations of sea surface temperature and wind over the Kuroshio and its extension: evidence for ocean-to-atmosphere feedback. J Climate 16:1404–1413

Ohno Y, Iwasaka N, Kobashi F, Sato Y (2009) Mixed layer depth climatology of the North Pacific based on Argo observations. J Oceanogr 65(1):1–16

Oka E (2009) Seasonal and interannual variation of North Pacific subtropical mode water in 2003–2006. J Oceanogr 65(2):151–164

Oka E, Suga T (2003) Formation region of North Pacific subtropical mode water in the late winter of 2003. Geophys Res Lett 30(23):2205. doi:10.1029/2003GL018581

Oka E, Suga T (2005) Differential formation and circulation of North Pacific central mode water. J Phys Oceanogr 35(11):1997–2011

Oka E, Talley LD, Suga T (2007) Temporal variability of winter mixed layer in the mid-to high-latitude North Pacific. J Oceanogr 63(2):293–307

Oka E, Kouketsu S, Toyama K, Uehara K, Kobayashi T, Hosoda S, Suga T (2011) Formation and subduction of central mode water based on profiling float data, 2003–2008. J Phys Oceanogr. 41:113–129. doi:10.1175/2010JPO4419.1

Qiu B, Chen S (2005) Variability of the Kuroshio extension jet, recirculation gyre, and mesoscale eddies on decadal time scales. J Phys Oceanogr 35(11):2090–2103

Qiu B, Hacker P, Chen S, Donohue KA, Watts DR, Mitsudera H, Hogg NG, Jayne SR (2006) Observations of the subtropical mode water evolution from the Kuroshio extension system study. J Phys Oceanogr 36(3):457–473

Rainville L, Jayne SR, McClean JL, Maltrud ME (2007) Formation of subtropical mode water in a high-resolution ocean simulation of the Kuroshio extension region. Ocean Model 17(4):338–356

Roemmich D (2001) Observing the Oceans in the 21st Century, Bureau of Meteorology. In: Argo: the global array of profiling floats, pp 248–258

Saito H, Suga T, Hanawa K, Watanabe T (2007) New type of pycnostad in the western subtropical-subarctic transition region of the North Pacific: transition region mode water. J Oceanogr 63(4):589–600

Saitoh SI, Inagake D, Sasaoka K, Ishizaka J, Nakame Y, Saino T (1998) Satellite and ship observations of Kuroshio warm-core ring 93A off Sanriku, northwestern North Pacific, in spring 1997. J Oceanogr 54(5):495–508

Schneider N, Miller AJ, Alexander MA, Deser C (1999) Subduction of decadal North Pacific temperature anomalies: observations and dynamics. J Phys Oceanogr 29(5):1056–1070

Suga T, Takei Y, Hanawa K (1997) Thermostad distribution in the North Pacific subtropical gyre: the central mode water and the subtropical mode water. J Phys Oceanogr 27(1):140–152

Suga T, Motoki K, Aoki Y, Macdonald A (2004) The North Pacific climatology of winter mixed layer and mode waters. J Phys Oceanogr 34(1):3–22

Sugimoto S, Hanawa K (2005) Remote reemergence areas of winter sea surface temperature anomalies in the North Pacific. Geophys Res Lett 32(1):L01606. doi:10.1029/2004GL021410

Taguchi B, Qiu B, Nonaka M, Sasaki H, Xie S, Schneider N (2010) Decadal variability of the Kuroshio extension: mesoscale eddies and recirculation. Ocean Dyn 60:673–691

Takahashi T, Sutherland SC, Wanninkhof R, Sweeney C, Feely RA, Chipman DW, Hales B, Friederich G, Chavez F, Sabine C, Watson A, Bakker DCE, Schuster U, Metzl N, Yoshikawa-Inoue H, Ishii M, Midorikawa T, Nojiri Y, Körtzinger A, Steinhoff T, Hoppema M, Olafsson J, Arnarson TS, Tilbrook B, Johannessen T, Olsen A, R B, Wong CS, Delille B, Bates NR, de Baar HJW (2009) Climatological mean and decadal change in surface ocean pCO_2, and net sea-air CO_2 flux over the global oceans. Deep Sea Res II 56(8–10):554–577

Talley LD, Raymer ME (1982) Eighteen degree water variability. J Mar Res 40:757–775

Tatebe H, Yasuda I (2001) Seasonal axis migration of the upstream Kuroshio extension associated with standing oscillations. J Geophys Res 106:16,685–16,692

Tokinaga H, Tanimoto Y, Nonaka M, Taguchi B, Fukamachi T, Xie S, Nakamura H, Watanabe T, Yasuda I (2006) Atmospheric sounding over the winter Kuroshio extension: effect of surface stability on atmospheric boundary layer structure. Geophys Res Lett 33(4)

Tokinaga H, Tanimoto Y, Xie S, Sampe T, Tomita H, Ichikawa H (2010) Ocean frontal effects on the vertical development of clouds over the western North Pacific: In situ and satellite observations. J Climate 22:4241–4260

Tomita H, Kubota M, Cronin M, Iwasaki S, Konda M, Ichikawa H (2010) An assessment of surface heat fluxes from J-OFURO2 at the KEO and JKEO sites. J Geophys Res 115(C3):C03,018

Toyama K, Suga T (2010) Vertical structure of North Pacific mode waters. Deep Sea Res II 57:1152–1160

Tsujino H, Yasuda T (2004) Formation and circulation of mode waters of the North Pacific in a high-resolution GCM. J Phys Oceanogr 34(2):399–415

Uehara H, Suga T, Hanawa K, Shikama N (2003) A role of eddies in formation and transport of North Pacific subtropical mode water. Geophys Res Lett 30(13):1705

Von Storch H, Zwiers FW (1999) Statistical Analysis in Climate Research. Cambridge University Press, Cambridge

Weller RA, Plueddemann AJ (1996) Observations of the vertical structure of the oceanic boundary layer. J Geophys Res 101(C4):8789–8806

White WB, Annis JL (2003) Coupling of extratropical mesoscale eddies in the ocean to westerly winds in the atmospheric boundary layer. J Phys Oceanogr 33(5):1095–1107

Wong A, Keeley R, Carval T, Argo Data Management Team (2010) Argo quality control manual. Argo Data Management Team, version 2.5 edn, available from web site of Argo Data Management, http://www.argodatamgt.org/

Yasuda I, Tozuka T, Noto M, Kouketsu S (2000) Heat balance and regime shifts of the mixed layer in the Kuroshio Extension. Prog Oceanogr 47:257–278

J Oceanogr (2012) 68:79–92
DOI 10.1007/s10872-011-0040-5

SPECIAL SECTION: ORIGINAL ARTICLE

New developments in mode-water research: Dynamic and climatic effects

Roles of mode waters in the formation and maintenance of central water in the North Pacific

Katsuya Toyama · Toshio Suga

Received: 28 February 2011 / Revised: 19 April 2011 / Accepted: 2 May 2011 / Published online: 1 June 2011

Abstract This study describes the three-dimensional distributions of the Turner angle (Tu) and the potential vorticity (PV) of the main pycnocline water in the subtropical North Pacific (10–50°N, 120°E–120°W) using a large in situ CTD data set taken by the Argo profiling floats during June to October of 2001–2009 to clarify the detailed distribution of the central water and the mode waters as well as the relationship between these water masses. The ventilated part of the main pycnocline water ($\sigma_\theta < 26.7$ kg m^{-3}) in the subtropical gyre generally displays a sharp peak in Tu value of 59° in the histogram. The Tu histograms for 10° × 10° geographical boxes mostly show that the mode for the Tu value is 59° too, but they also show some regional differences, suggesting some types of relations with the North Pacific mode waters. To further investigate this relationship, the appearance probability density function of the central water (defined as the main pycnocline water with Tu = 56°–63°) and those of the mode waters with PVs lower than the critical value on each isopycnal surface were analyzed. The distribution area of the central mode water (CMW) corresponds so well with that of the central water that a direct contribution of the CMW to the formation and maintenance of the central water is suggested. On the other hand, the distribution areas of subtropical mode water (STMW), Eastern STMW, and transition region mode water do not correspond to that of the central water. Nevertheless, indirect contributions of these mode waters to the formation and maintenance of the central water through salt finger type convection or diapycnal mixing are suggested.

Keywords Mode water · Central water · Argo

K. Toyama (✉) · T. Suga
Department of Geophysics, Graduate School of Science, Tohoku University, Sendai, Miyagi 980-8578, Japan
e-mail: katsuya@pol.gp.tohoku.ac.jp

T. Suga
Japan Agency for Marine-Earth Science and Technology, Yokosuka, Kanagawa 237-0061, Japan

1 Introduction

It has been noticed for a long time that the winter sea surface temperature–salinity (T–S) relation in the meridional direction resembles the vertical T–S profile of the main pycnocline water in the subtropical gyre (Iselin 1939). This is basically explained by subduction—the irreversible transfer of water from the surface mixed layer into the main pycnocline, which occurs in late winter (Stommel 1979). The subducted water circulates in the subtropical gyre on isopycnal surfaces while conserving its potential vorticity (Luyten et al. 1983).

Mode water is a prominent feature of the subtropical gyres of the world oceans (Hanawa and Talley 2001). It is formed as a consequence of the subduction of the deep late-winter mixed layer water, and is characterized by vertically and horizontally homogeneous water properties. It has a direct influence on absolute density stratification due to its vertically homogeneous density. In other words, the mode water is an important source of low potential vorticity (PV) for the subtropical main pycnocline. Because the PV is conserved along an advection path on isopycnal surfaces, it can be used as a tracer for the mode water (McCartney 1982). Therefore, the density stratification or the PV field of the subtropical gyre has been extensively documented in terms of the distribution of the mode water (Talley 1988; Suga et al. 2004; Oka et al. 2011). In the open North Pacific, there are at least four mode waters: the Subtropical Mode Water (STMW; Masuzawa 1969), the Central Mode

Water (CMW; Nakamura 1996; Suga et al. 1997), the Eastern STMW (ESTMW; Hautala and Roemmich 1998), and the Transition Region Mode Water (TRMW; Saito et al. 2007).

On the other hand, the density of the seawater is determined by both temperature and salinity. If we plot the subtropical main pycnocline water of each ocean basin on a T–S diagram, it is represented by a specific single T–S curve of its own, and is termed the "central water" by Sverdrup et al. (1942). The T–S curve of the central water shows a constant density ratio (Ingham 1966; Schmitt 1981). The density ratio (R_ρ) is a measure of the relative effects of temperature and salinity stratifications on density stratification, and is a parameter used to define the local stability. The Turner angle (Tu), which indicates the potential for double diffusive convection (Turner 1973; Ruddick 1983), can be used as an equivalent parameter to the density ratio. Central water is thought to acquire its characteristics through the subduction process, but it is not clear how it acquires the constant Tu. In the North Pacific, Sverdrup et al. (1942) reported two types of central waters: the Western North Pacific Central Water (WNPCW) and the Eastern North Pacific Central Water (ENPCW). The WNPCW and ENPCW are distributed in the western subtropical gyre west of 170°W and the eastern subtropical gyre east of 170°W, respectively, and are thought to have formation histories of their own (Tomczak and Godfrey 1994).

From a historical point of view, studies on the formation and circulation of the main pycnocline water in the subtropical gyre originally began with some insights into the T–S relation (i.e., central water). However, recent studies of these issues have paid more attention to density or the PV field (i.e., mode water). This is presumably because the conservative nature of low PV, which is characteristic of mode waters, is very useful when considering the formation and circulation of subducted water in the framework of the ventilated thermocline theory of Luyten et al. (1983). For a deeper understanding of the three-dimensional structure of the subtropical main pycnocline, however, a comprehensive understanding of all temperature, salinity, and density structures is needed.

Although both the mode water and the central water are important and useful concepts for describing the stratification of the main pycnocline, the relationship between these two water masses is yet to be studied. The mode water is a feature of density stratification; on the other hand, the central water is a feature of the T–S stratification. The relationship between the mode water and the central water implies a relationship between the density stratification and the T–S stratification of the subtropical main pycnocline. Shimada et al. (2007) investigated the spatial distribution of the density ratio of the upper North Pacific using the WOCE data and found "hot spots" of double diffusive convection in the area off eastern Hokkaido and in the Eastern North Pacific where the North Pacific Intermediate Water (NPIW) and the ESTMW are formed, respectively. They suggested that salt finger convection is important in the formation and modification of these water masses. Toyama and Suga (2010) showed that the vertical temperature and salinity structures differ among the four North Pacific mode waters. This implies that each mode water, which is characterized by low Q, could make a different contribution to the three-dimensional T–S structure of the subtropical North Pacific. These studies are suggestive, but more detailed studies of the relationship between the mode waters and the T–S structure (e.g., the central water) are warranted.

One of the possible reasons for the lack of this type of study is the lack of a sufficiently large in situ CTD data set with an even distribution. Previous studies on the central water were mainly based on ocean climatological data or sparsely distributed and limited shipboard observational data. To understand the relationship between the mode water and the central water, and to clarify the formation, maintenance, and circulation of both of these water masses, it is essential to obtain detailed distributions of them using a large observational data set that is as evenly distributed as possible.

Recent progress in the Argo Program (Freeland et al. 2010) allows us to describe the detailed spatial distributions and vertical structures of these water masses, because more than 3,000 Argo floats have been observing the world's open oceans mostly uniformly in near real time (Fig. 1). Using these Argo data, in this study, we describe the three-dimensional distribution of the central water by means of the Turner angle (Tu) and that of the mode water by means of the PV of the main pycnocline in the subtropical North Pacific. The relationship between the two water masses, including the possible roles of the mode waters in the formation and maintenance of the central water, is also discussed.

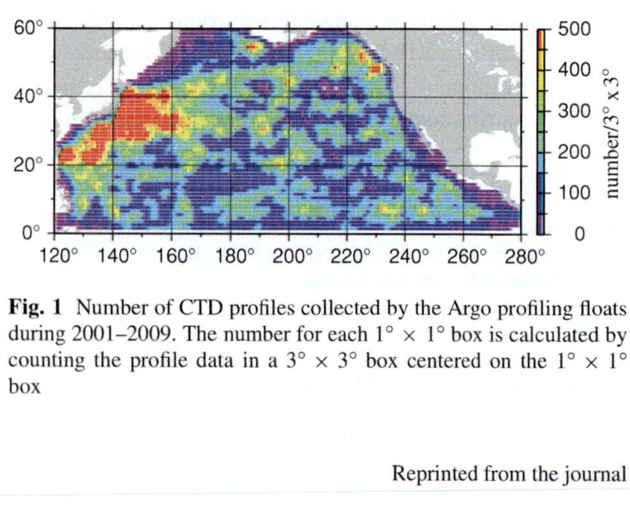

Fig. 1 Number of CTD profiles collected by the Argo profiling floats during 2001–2009. The number for each 1° × 1° box is calculated by counting the profile data in a 3° × 3° box centered on the 1° × 1° box

The rest of this paper is organized as follows. Section 2 describes the data used and the data processing procedures employed. In Sect. 3, we show the main results, including the distributions of the mode waters and the central water. A discussion and concluding remarks are presented in Sects. 4 and 5, respectively.

2 Data

CTD profile data obtained by the Argo profiling floats in the open North Pacific (defined as 10°–50°N, 120°E–120°W) during June–October of 2001–2009 were used to describe the North Pacific central waters and mode waters. The real-time quality-controlled data were downloaded from the Argo Global Data Assembly Center (GDAC). CTD profiles with very coarse vertical sampling and defective layers, such as those with measurements flagged as "bad" and those with obviously erroneous measured values, were excluded using the procedures of Oka et al. (2007). Each CTD profile was vertically interpolated onto the 10-dbar grid using the spline method described by Akima (1970). Potential temperature (θ) and potential density (σ_θ) were calculated using splined data. We then calculated vertical gradients of potential temperature, salinity, and potential density at each depth level with the data for the six closest vertical grid points (±30 dbar) to it using the least-squares fitting method. Potential vorticity (Q) was defined using the vertical gradient of potential density as follows:

$$Q = \frac{f}{\rho}\frac{d\sigma_\theta}{dz}, \quad (1)$$

where f is the Coriolis parameter, ρ is the in situ density, and z is the vertical coordinate (positive upward). The relative vorticity is disregarded in Eq. 1.

The Turner angle (Tu) was calculated using the vertical gradient of potential temperature and that of salinity after McDougall et al. (1988) as follows:

$$\mathrm{Tu}\ (^\circ) = \tan^{-1}\left(\alpha\frac{\partial\theta}{\partial z} - \beta\frac{\partial S}{\partial z},\ \ \alpha\frac{\partial\theta}{\partial z} + \beta\frac{\partial S}{\partial z}\right), \quad (2)$$

where α is the thermal expansion coefficient and β is the haline contraction coefficient. In this study, α/β was calculated by the polynomial equation proposed by McDougall (1987), which is a function of θ, S, and p. Tu can be used to examine either the fine structure of double diffusion in high-resolution CTD profiles or the mean state with climatological data (You 1998). In the latter case, You (2002) proposed that it would probably be better to term it the "water mass Turner angle." The Tu calculated in this study is thus the water mass Turner angle for interpreting large-scale water mass structure.

The density ratio (R_ρ), which is an equivalent parameter to Tu, can be calculated using the value of Tu by the following equation:

$$R_\rho = -\tan(\mathrm{Tu} + 45). \quad (3)$$

The 10-dbar gridded data are then linearly interpolated onto a 0.1 kg m^{-3} isopycnal grid. Note that the data were neither averaged nor smoothed with time. In this study, we will not consider seasonal and interannual variations.

3 Results

3.1 The Tu distribution of the subtropical main pycnocline water in the North Pacific

The formation of the North Pacific central water and mode waters is regarded as a consequence of the subduction of the winter mixed layer water. Additionally, the densest water outcrops in late winter in the open North Pacific rarely exceed 26.7 kg m^{-3} (e.g., Suga et al. 2004). Therefore, water in the analysis region (10°–50°N, 120°E–120°W) with a potential density of less than 26.7 kg m^{-3} is regarded as the ventilated part of the main pycnocline water, and is analyzed in this study.

Toyama and Suga (2010) showed that the vertical structures of temperature and salinity differ among the four mode waters in the North Pacific, despite the fact that they all have low Q. The Tu values of the STMW and CMW are relatively uniform within the low-Q layer; they are about 50°–60° and 60°–70°, respectively. The Tu values of the ESTMW and TRMW, on the other hand, are not uniform within the low-Q layer: they are greater than 70° on the upper part of the TRMW and lower part of the ESTMW. Since the mode waters occupy large volumes of water within the main pycnocline, these typical Tu values of the mode waters could affect the regional and/or basin-wide distribution of the Tu values of the main pycnocline.

The histogram of the Tu values of the main pycnocline water ($\sigma_\theta < 26.7$ kg m^{-3}) in the North Pacific shows that the mode for the Tu values is 59° (Fig. 2). This Tu value is equivalent to an R_ρ value of about 4.0 from Eq. 3, which is very close to the previous results of $R_\rho = 3.82$ (from Schmitt 1981) and $R_\rho = 3$–4 (from Shimada et al. 2007). As pointed out in the previous studies, this R_ρ value of the North Pacific pycnocline water suggests moderate salt finger activity, and is larger than those of the pycnocline waters of the other oceans. The previous results are based on limited shipboard observational data with a sparse distribution. The results shown here, based on more evenly distributed and larger Argo data set, can be regarded as more reliable.

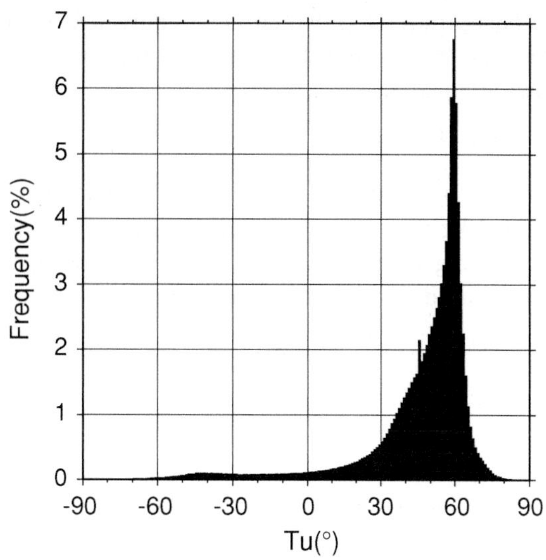

Fig. 2 Histogram of the Turner angle (Tu) of the main pycnocline water (with a potential density of less than 26.7 kg m^{-3}) in the subtropical North Pacific (10–50°N, 120°E–120°W) in June–October of 2001–2009

Taking advantage of the large size and unbiased distribution of the Argo data set, a similar analysis was carried out for each longitude and latitude 10° × 10° box (Fig. 3). Some regional differences in the Tu distribution were found in this analysis. Four characteristic regions are selected and discussed in the following several paragraphs. The representative T–S diagram for each of the selected regions is presented in Fig. 4. Note that Fig. 4 only shows the T–S observed in June 2009 as a representative sample, because we wish to focus on general features that are not largely affected by the months or years chosen for the analysis. There are some minor differences between Fig. 4 and the figures for the other months included in this analysis period due to interannual and seasonal variations, which are not considered in this study.

The Tu histograms for the box regions west of 180° at 10–20°N, west of 150°W at 20–30°N, and west of 140°W at 30–40°N mostly show clear single peaks at Tu = 59° (Fig. 3). The area occupied by these boxes mostly corresponds to the subtropical gyre region. These waters with Tu values of around 59° within the subtropical gyre correspond to the WNPCW and presumably have a certain degree of correspondence to the CMW, because the CMW distributed in the central part of the gyre has Tu values of 60°–70° (Fig. 4b; Toyama and Suga 2010).

The Tu histograms in the boxes just south of Kuroshio and its extension front (in the recirculation gyre, where the STMW is distributed) show a weak secondary peak around Tu = 40°–50° (Fig. 3). These waters are possibly related to the STMW, since it has Tu value of 50°–60° (Fig. 4a; Toyama and Suga 2010).

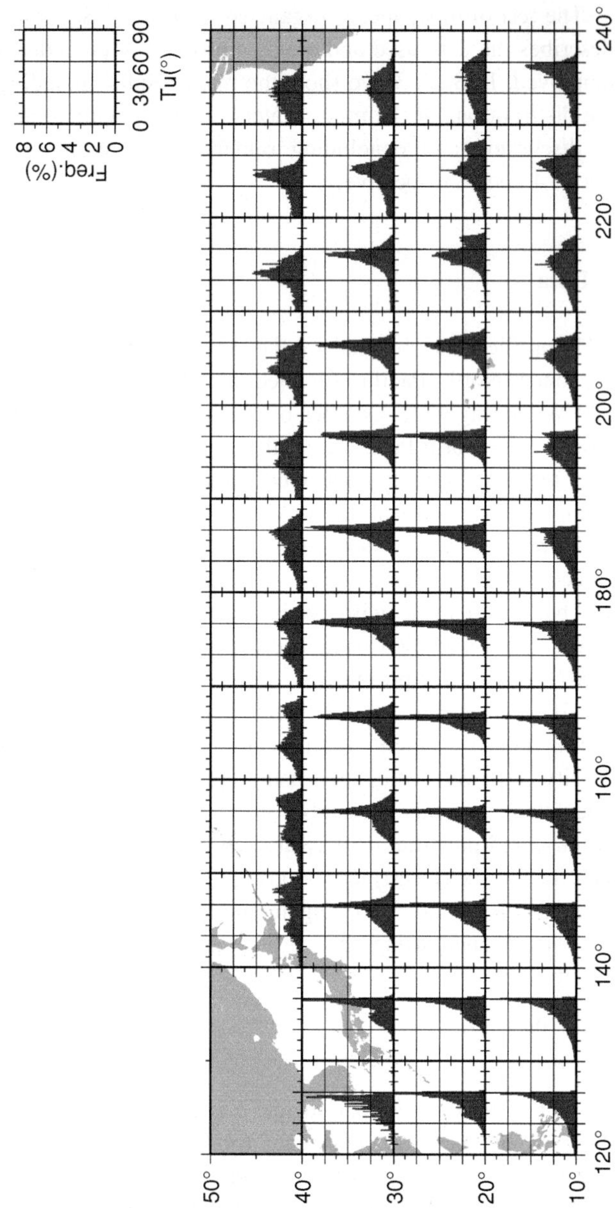

Fig. 3 Histograms of the Tu value of the main pycnocline water in the subtropical North Pacific for each longitude × latitude 10° × 10° box

In the eastern part of the basin, some histograms show two Tu peaks; one is greater than 60° and the other is less than 60° (Fig. 3). It is suggested that the ENPCW is fresher than the WNPCW in the temperature range below 17°C but saltier in the upper pycnocline (Fig. 4a–c; Tomczak and Godfrey 1994). Nevertheless, water types (T–S combination) near the salinity minimum at intermediate depth are the same between the T–S curve including the ENPCW (Fig. 4c) and the one including the WNPCW (Fig. 4a, b). This means that the slope of the T–S curve including the ENPCW is small in the upper pycnocline but increases with increasing depth. Since the Tu roughly represents the

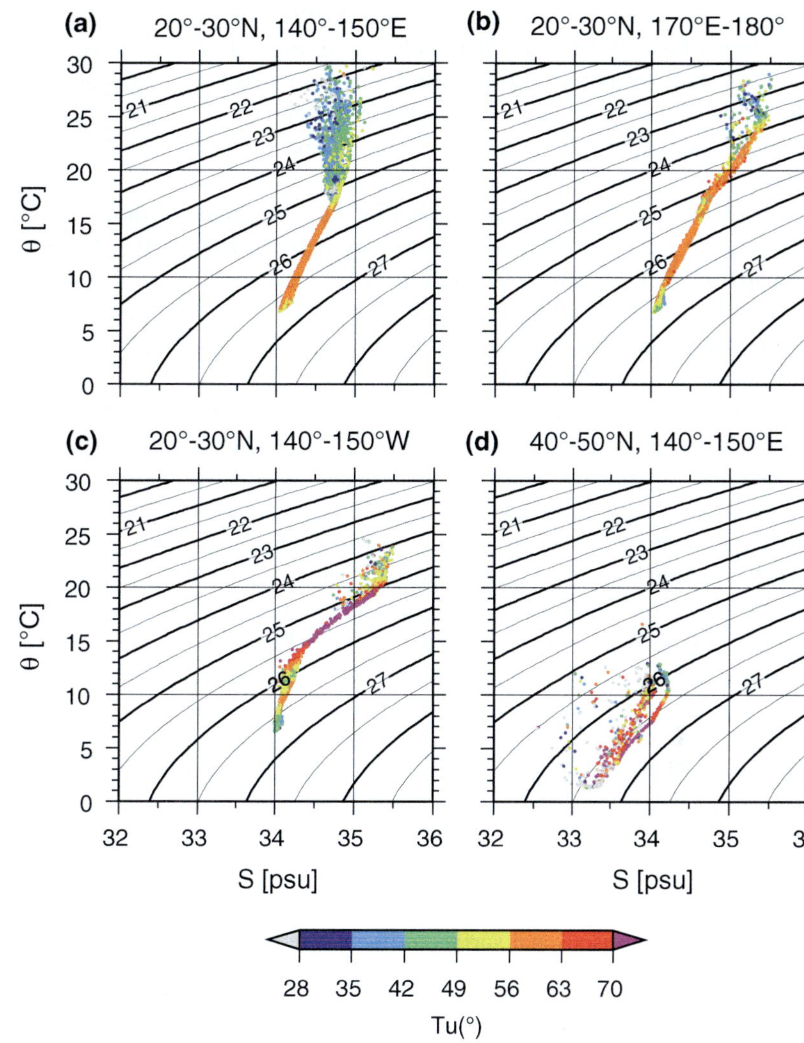

Fig. 4 T–S diagrams corresponding to the four typical regions with different characteristic Tu distributions. The observed T–S data collected by the Argo floats in June of 2009 at **a** 20–30°N, 140–150°E, **b** 20–30°N, 170–180°E, **c** 20–30°N, 140–150°W, and **d** 40–50°N, 140–150°E are plotted; the color relates to the Tu value as shown in the *scale at the bottom of the figure*

slope of the T–S curve on the T–S diagram ($\partial\theta/\partial S$), this difference in slopes at different depths represents the difference in Tu at different depths. Namely, the upper pycnocline water with the small T–S slope on the T–S diagram has a large Tu value, while the lower pycnocline water with the large T–S slope has a smaller Tu value (Fig. 4c). The water with the greater Tu value probably corresponds to the ENPCW in the upper pycnocline, while the other with the smaller Tu value corresponds to the lower pycnocline water at around intermediate levels near the salinity minimum. Since the ESTMW is the constituent water of the upper pycnocline in the eastern part of the gyre, and was shown to have large Tu values of >70° (Fig. 4c; Toyama and Suga 2010), the larger Tu peak is also related to the ESTMW. This is a result that is consistent with that of Shimada et al. (2007), who found a "hot spot" of double diffusion in the Eastern North Pacific, and suggested that it was related to ESTMW formation and modification.

Most Tu histograms in the boxes north of 40°N also show two peaks; one occurs at around 10°–40° and gradually increases to the east, while the other is at 50–70° and gradually decreases to the east (Fig. 3). If the density gradient is completely determined by the temperature gradient, the Tu value is 45°. For the positive range, the Tu is greater (less) than 45° when the salinity has a negative (positive) effect on the density stratification. Since the effect of the salinity on the density gradient is generally negative (i.e., salinity is unstably stratified) in the subtropical region and positive (i.e., salinity is stably stratified) in the subpolar region, the smaller Tu peak at less than 45° is suggested to originate in the subpolar region, while the larger Tu peak has a subtropical origin (Fig. 4d). The bimodality occurs because this latitude range corresponds to the frontal region between the subtropical and subpolar oceans. The Tu peak larger than 70° west of 170°E is possibly related to the TRMW, since the TRMW distributed in the transition domain is suggested to have a Tu value of >70° (Fig. 4d; Toyama and Suga 2010). This Tu peak larger than 70° west of 170°E probably corresponds to the "hot spot" of double diffusion in the area off eastern Hokkaido found by Shimada et al. (2007) which, in their study, is suggested to be related to the formation and modification of the NPIW.

The density ratio for the world ocean for each longitude and latitude 10° × 10° box was previously described by Figueroa (1996). Figueroa (1996) used the climatological T–S curve of Levitus (1982) averaged for each 10° × 10° square and estimated a single best-fit density ratio for each. The analysis method used in Figueroa (1996) can only provide the dominant density ratio for each 10° × 10° square, and cannot provide volume information and/or a second most dominant density ratio, which are clearly shown in this study. Since the analysis of Figueroa (1996) was based on the climatology of Levitus (1982), which was largely smoothed spatially and temporally, direct comparison with the present result is difficult. Additionally, the method used to estimate the density ratio (or the Tu) of the central water differs between the two studies, which makes the comparison even more difficult. Nevertheless, the result shown in this study, based on a large number of individual CTD observations, presumably gives a more realistic distribution of Tu for the subtropical main pycnocline water in the North Pacific.

3.2 Distribution of the mode waters and the central water on each isopycnal surface in the North Pacific

The Tu histograms of the North Pacific subtropical main pycnocline waters mostly show a constant Tu value of 59°. It is also clear that there is a relation between the Tu distribution and the mode waters. To further investigate this relation, the Tu distribution was examined in more detail on each isopycnal surface, paying particular attention to the distribution of Tu in the range 56–63° in order to include the dominant mode Tu value of 59°. This ±3° band was selected because the relative frequencies of the 56° and 63° bins are roughly the half that of the peak value at 59°. The cumulative relative frequency between 56° and 63° is about 35%. Hereafter, the "central water" in the North Pacific is defined as the main pycnocline water with Tu values ranging from 56° to 63°. This definition is somewhat different from the definition (the straight line part of the T–S curve) used by Sverdrup et al. (1942). We discuss this difference in a later section.

To consider the relation between the central water and the mode waters, the appearance frequency of the distribution of Tu = 56–63° and that of PV values lower than the critical value on each isopycnal surface between $\sigma_\theta = 24.8$ and 26.6 kg m^{-3} for the warm season (June–October) of 2001–2009 were analyzed using a probability density function (PDF). The frequency for each 1° × 1° grid box was calculated using the float data in the 3° × 3° grid box centered on the 1° × 1° grid box, but this was only done if 10 or more data existed in the 3° × 3° grid box (Oka et al. 2011).

The frequency of low Q on each isopycnal surface is shown in Fig. 5. On the isopycnal of $\sigma_\theta = 26.6$ kg m^{-3}, the frequency of low Q is high at around 40–42°N, 152–163°E, which represents the TRMW (Fig. 5a). On $\sigma_\theta = 26.4$ kg m^{-3}, the frequency of low Q extends anticyclonically from 41°N, 170°E to 18°N, 135°E while decreasing its intensity downstream, as shown by Oka et al. (2011). Part of this is considered to correspond to the TRMW, while the other part corresponds to the CMW (Fig. 5b). On the isopycnals between $\sigma_\theta = 25.8$ and 26.2 kg m^{-3}, an anticyclonic circulation pattern of low Q is apparent, and this is considered the circulation pattern of the CMW (Fig. 5c–e). On the isopycnals of $\sigma_\theta = 25.2$–25.4 kg m^{-3}, the frequency of low Q is high in the recirculation gyre of the subtropical gyre west of the dateline (Fig. 5g, h). This high-frequency region corresponds to the STMW. On $\sigma_\theta = 25.2$ kg m^{-3}, a rather weak low-Q signature appears around 30°N, 140°W, which is regarded as the ESTMW (Fig. 5h).

The appearance frequency of Tu = 56–63° (which is defined as the central water in this study) on each isopycnal surface is shown in Fig. 6. Superimposed cross symbols denote grid points with PDFs of low Q that are greater than 10% (the colored grid in Fig. 5). On the isopycnal of $\sigma_\theta = 26.6$ kg m^{-3}, the high-frequency region of the central water is limited only near the center of the subtropical gyre (Fig. 6a). On the isopycnals between $\sigma_\theta = 25.8$ and 26.4 kg m^{-3}, the high-frequency area of Tu = 56–63° appears mostly within the subtropical gyre (Fig. 6b–e). These isopycnals are ventilated by the CMW, as described above. The distribution area of the central water on these isopycnals, especially in the central to southern part of the gyre, corresponds to the distribution pattern of the CMW along anticyclonic circulation paths. Aside from this similarity, a clear dissimilarity also exists between denser isopycnals of $\sigma_\theta = 26.2$–26.4 kg m^{-3} and lighter ones of $\sigma_\theta = 25.8$–26.0 kg m^{-3}. Namely, the frequency of Tu = 56–63° on the denser isopycnals is lower than that on lighter ones in the northern part of the gyre around 30–42°N, 170°E–160°W. Instead, this area shows a high frequency of Tu = 63–70° (Fig. 7b, c). This area corresponds to the immediate downstream region of the outcrop/formation area of the CMW on the denser isopycnals, and thus corresponds to the region with a very high appearance frequency of low Q. This area also corresponds to an area where a deep mixed layer develops in winter (Fig. 8). This dissimilarity is probably related to the different types of CMW: the denser variety of CMW

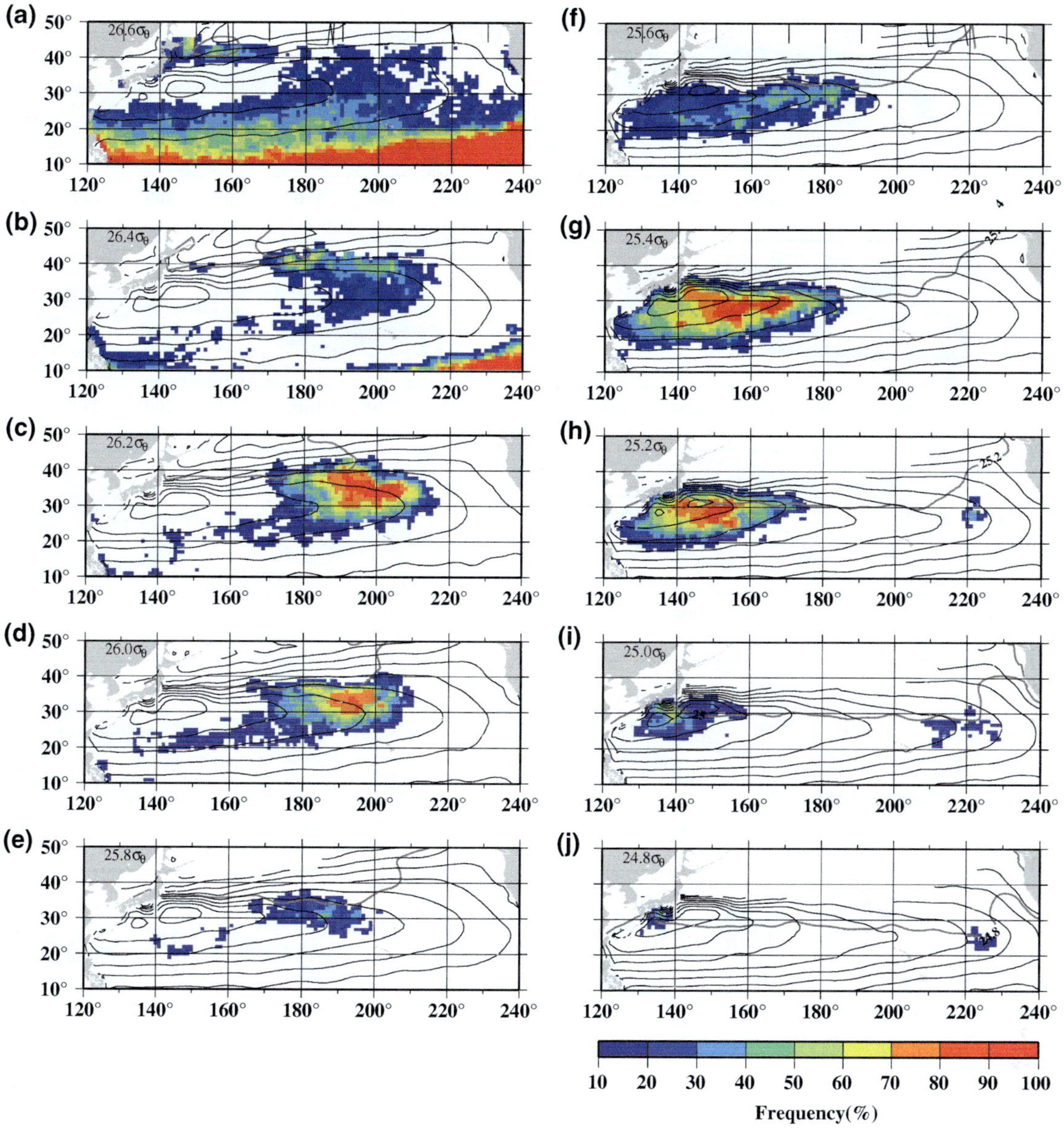

Fig. 5 The appearance frequency of Q lower than the critical value on each isopycnal (0.2 kg m^{-3} interval) between 24.8 and 26.6 kg m^{-3} in June–October of 2001–2009. The frequency for each 1° × 1° box was calculated using the Argo float data in a 3° × 3° box centered on the 1° × 1° box, but only if 10 or more data existed in the 3° × 3° box. The critical values used for the isopycnals of 24.8–25.6, 25.8–26.0, 26.2, and 26.4–26.6 kg m^{-3} were 2.25, 2.0, 1.75, and 1.5 × 10^{-10} m^{-1} s^{-1}, respectively. The *thick gray curve* denotes the outcrop of each isopycnal in March, and the *black contours* denote the annual mean acceleration potential relative to 2,000 dbar (the contour interval is 0.5 m^{-2} s^{-2}) on each isopycnal, both of which are based on the World Ocean Atlas 2001 (Boyer et al. 2002; Stephens et al. 2002)

that ventilates the isopycnals of $\sigma_\theta = 26.2$–26.4 kg m^{-3} and the lighter variety of CMW that ventilates the isopycnals of $\sigma_\theta = 25.8$–26.0 kg m^{-3} (Oka and Suga 2005).

On $\sigma_\theta = 25.2$ kg m^{-3}, which roughly represents the core density of the STMW for the analysis period, the frequency of Tu = 56–63° is very low (nearly zero) in the western part of the subtropical gyre south of the Kuroshio, and its extension where the STMW is distributed (Fig. 6h). On the other hand, on $\sigma_\theta = 25.4$–25.6 kg m^{-3}, which roughly corresponds to the bottom of the STMW or just below the STMW, the frequency of Tu = 56–63° is high in the distribution area of STMW (Fig. 6f, g).

On the lighter isopycnals between $\sigma_\theta = 24.8$ and 25.6 kg m^{-3}, the high-frequency areas of Tu = 56–63° are distributed in the southern part of the subtropical gyre except for the eastern area centered around 25°N, 145°W, which roughly corresponds to the high-frequency area of low Q of the ESTMW (Fig. 6f–j). When the frequencies of Tu >70° on the same isopycnals are calculated (Fig. 9), a high-frequency area appears on $\sigma_\theta = 25.0$–25.4 kg m^{-3} in the eastern part of the basin where the ESTMW is distributed (Fig. 9c–e).

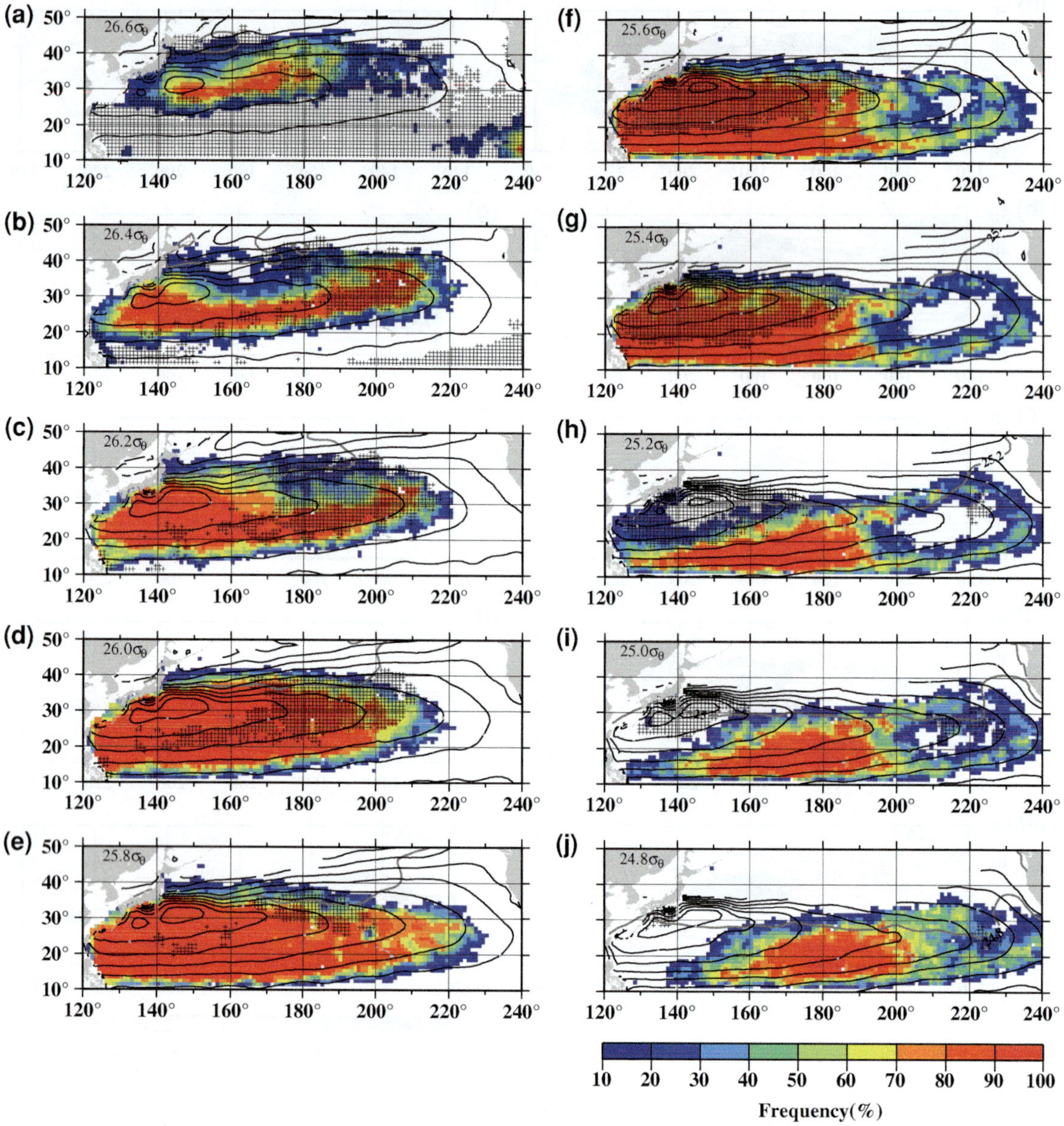

Fig. 6 Appearance frequency of Tu ranging from 56° to 63° on each isopycnal (0.2 kg m^{-3} interval) between 24.8 and 26.6 kg m^{-3} in June–October of 2001–2009. The methods used to calculate the appearance frequency, the *thick gray curve* and the *black contours* are the same as those described in the caption for Fig. 5. Each *cross symbol* denotes a grid point where the frequency of low Q on each isopycnal is greater than 10% (the *colored grid* in Fig. 5)

A high-frequency area also appears on $\sigma_\theta = 26.4$–26.6 kg m^{-3} in the subtropical–subarctic transition region where the TRMW is formed and distributed (Fig. 9a, b).

4 Discussion: possible contributions of mode waters to the formation and maintenance of central water in the North Pacific

Toyama and Suga (2010) showed that the vertical structures of temperature and salinity differ among the four mode waters in the North Pacific, despite the all having low Q. The Tu values of the STMW and CMW are relatively uniform within the low-Q layer, with Tu values of about 50°–60° and 60°–70°, respectively. The Tu values of ESTMW and TRMW, on the other hand, are not uniform within the low-Q layer (they are greater than 70° on the upper part of the TRMW and lower part of the ESTMW). They suggested that these mode waters have different formation and modification processes. Namely, it is implied that ESTMW and TRMW could be modified by salt finger type convection.

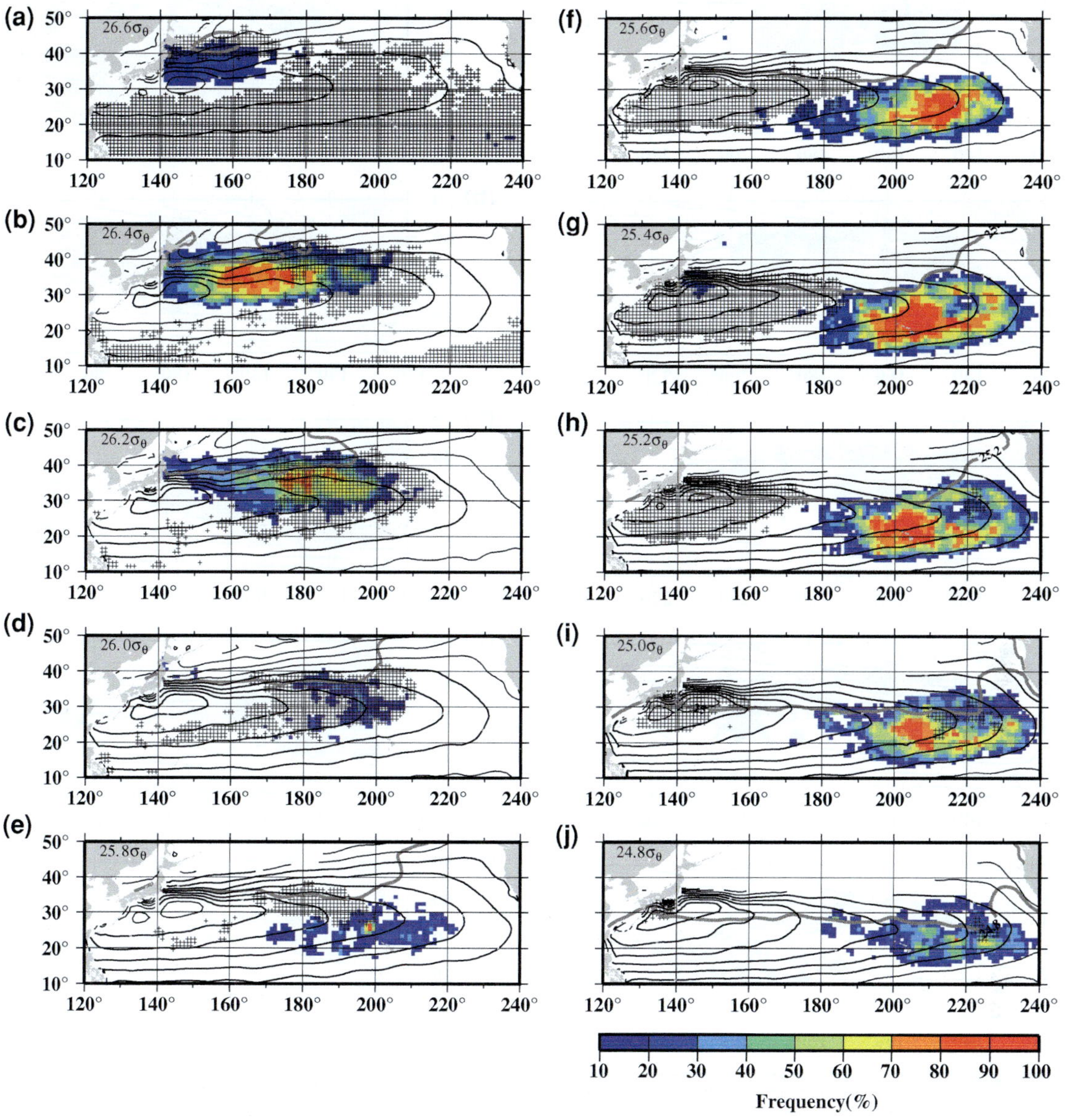

Fig. 7 Same as Fig. 6, but for the appearance frequency of Tu ranging from 63° to 70°

From the results of this study and those of Toyama and Suga (2010), the relationship between the central water (Tu = 56–63°) and the mode waters (low Q) can be interpreted as follows.

Subtropical Mode Water, which is formed as a very uniform layer in terms of temperature, salinity and density, and has a rather small Tu value (54°) around its core, is not a part of the central water defined as the main pycnocline water with Tu = 56–63°). However, the bottom part of the STMW, which has substantially larger Tu value than that at the core, possibly contributes to the central water in the recirculation gyre region of the subtropical gyre. Based on these facts, it is suggested that the central water (with a high frequency of Tu = 56°–63°) in the western part of the subtropical gyre (namely in the recirculation gyre) is possibly formed and maintained by the vertical diapycnal mixing of the STMW as an upper end member with some lower end members such as the CMW and/or the NPIW. The effect of mixing on the formation and maintenance of the T–S structure of the North Pacific pycnocline is presumably important, but it remains unclear. It is left for future study.

Central mode water is formed as relatively inhomogeneous layer in terms of temperature and salinity with Tu = 60°, and it is expected to be a subset of the central water. Since Tu is not a conservative property, the Tu of

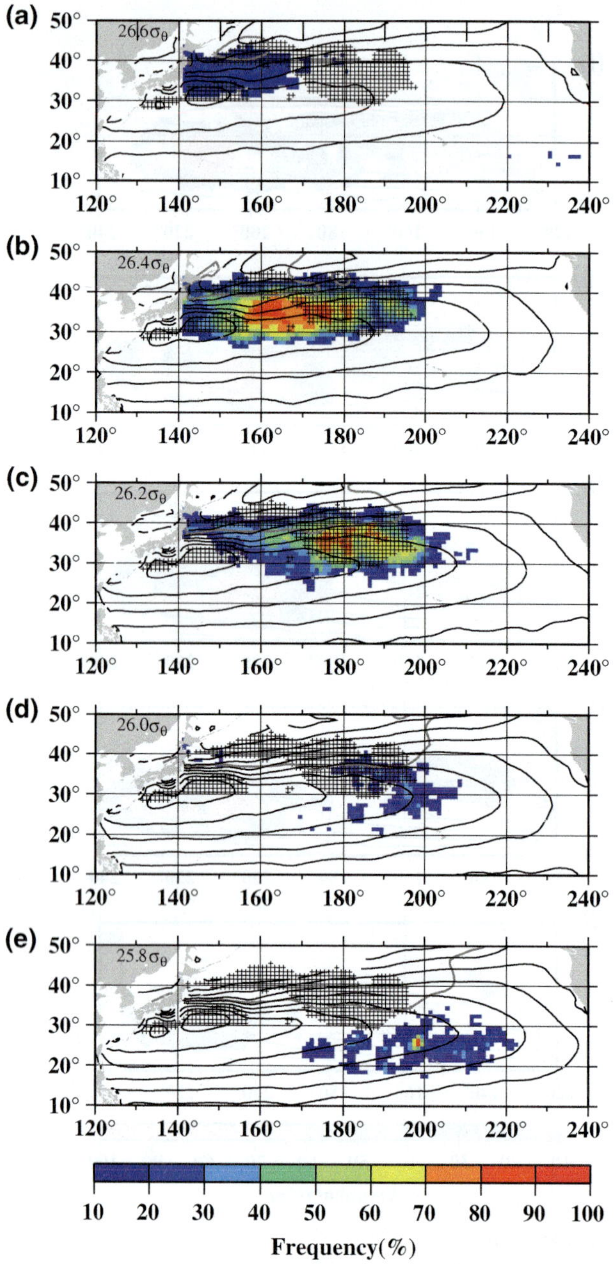

Fig. 8 Same as Fig. 7, but with the winter mixed layer depth (MLD) greater than 150 dbar (*cross symbol*) in March 2005 superimposed. The MLD was generally deep in the North Pacific in March 2005. Grid Point Values of the Monthly Objective Analysis using the Argo data (MOAA GPV; Hosoda et al. 2008) were used for the MLD calculation, and the MLD was defined as the depth at which the potential density was greater than the surface density by $\Delta\sigma_\theta = 0.125$ kg m^{-3} (Levitus 1982)

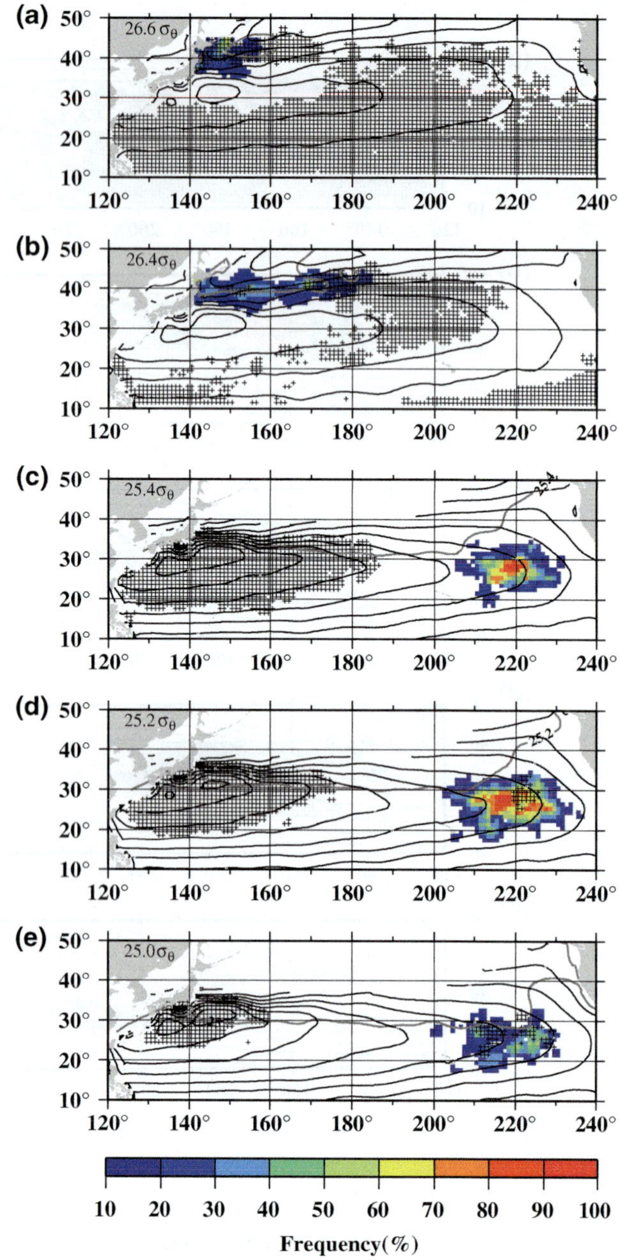

Fig. 9 Same as Fig. 6, but for the appearance frequency of Tu >70°

CMW does not need to be conserved along its advection paths. This contrasts with the low-Q signature of the CMW, which is conserved along advection paths on isopycnals (Fig. 5; Oka et al. 2011). However, the good correspondence between the anticyclonic circulation pattern of low Q (CMW) and the distribution pattern of Tu = 56°–63° (central water) suggests near-conservative advection of Tu associated with the CMW circulation. Such a situation is possible, for example, when the water circulation patterns on one isopycnal are similar to those on the isopycnals above and below and isopycnal/diapycnal diffusive mixing with water with a different T–S is small. This may be examined by analyzing the advection of spiciness (Flament 2002) or its anomaly on each isopycnal, but this is beyond the scope of this study.

Close to the formation area of the CMW on heavier isopycnals (σ_θ = 26.2–26.4 kg m^{-3}), the frequency of low Q is very high but the frequency of Tu = 56°–63° is low.

This corresponds well to the region where the mixed layer becomes very deep in late winter and the mixed-layer density ranges from 25.5 to 26.5 kg m^{-3} (Fig. 8). The density of the winter mixed layer decreases along the advection path (streamlines) of the CMW on heavier isopycnals. It is suggested that the Tu of the CMW on denser isopycnals, which is re-entrained into the deep winter mixed layer in the following years, is possibly affected by lighter water lying on top of it. This mixed layer water in the following years is lighter, warmer, and saltier (more "spicy") than the CMW on denser isopycnals, so the Tu of the denser CMW becomes larger. Away from the deep winter mixed-layer region where the seasonal cycle of near-surface layers could affect the Tu values of denser isopycnals, the Tu value of the subducted water becomes relatively stable. From this point of view, near-conservative advection of the Tu accompanied by CMW circulation is plausible.

The Tu values in the formation and distribution areas of the TRMW ($\sigma_\theta = 26.4$–26.6 kg m^{-3}) and ESTMW ($\sigma_\theta = 24.8$–25.4 kg m^{-3}) are larger (Tu > 70°) than that of the central water (Fig. 9). These large-Tu regions appear only near the formation regions of these mode waters, and do not appear in the downstream regions. In the southwestern region of the formation region of the ESTMW (the downstream region of the formation region along the gyre flow), a high-frequency region of Tu <70° appears instead (Figs. 6, 7). These features may be partly related to the modification of the mode waters. As suggested by Toyama and Suga (2010), salt finger convection is expected to be important in the modification of the TRMW and ESTMW, and modified mode waters show rather small Tu values. During this modification, the mode waters are expected to be advected by general flow on each isopycnal. That means that the larger Tu values would tend to be confined near the formation areas of these mode waters, as they appear to in Fig. 9. The modified ESTMW—which can have high Q and thus does not qualify as a mode water any more—possibly becomes a constituent of the central water, because a high-frequency area of Tu = 56°–63° appears in the downstream region of the ESTMW. On the other hand, because a high-frequency area of Tu <70° is not clearly apparent in the downstream region of the TRMW, the modified TRMW would not become a constituent of the central water. The TRMW in the mixed water region west of 170°E with a high Tu value could be re-entrained into a deeper surface mixed layer east of 170°E rather than subducted into the main pycnocline, and could lose the properties it obtained during its formation, as suggested by Oka et al. (2011).

It may be helpful to summarize the scenario stated above on the T–S diagram (Fig. 10), because Tu roughly represents the slope of the T–S curve on the T–S diagram ($\partial\theta/\partial S$). The smaller the angle between the T–S curve and the isopycnal line, the larger the Tu value. The T–S curve of the central water, with a Tu value of around 59° (gray curve in Fig. 10), is coincident with that of the CMW (orange curve in Fig. 10) in the density range of $\sigma_\theta = 25.8$–26.4 kg m^{-3}. The STMW with the smaller Tu value (the red curve in Fig. 10) is not a part of the central water defined in terms of constant Tu, although it may form and maintain the central water below it through mixing with lower pycnocline water (yellow dashed curve in Fig. 10). Because the STMW has a very uniform T–S combination compared with the other mode waters, the STMW appears as almost a single point on the T–S diagram. A small temperature gradient largely determines its Tu. In that sense, if the central water is defined as "the linear part of the T–S curve of the main pycnocline water," as in the definition of Sverdrup et al. (1942), the STMW is a part of the central water because the single T–S point of STMW is located on "the linear part of the T–S curve." The ESTMW with Tu >70° (green solid curve in Fig. 10) is possibly modified by salt fingering and decreases its Tu value. On the T–S diagram, salt fingering rotates the T–S curve (Schmitt 1981). The T–S curve of the modified ESTMW (green dashed curve in Fig. 10) could be a constituent of the central water in the lighter density ranges. It is also suggested that the TRMW (blue curve in Fig. 10) is modified by salt fingering; however, a contribution of the

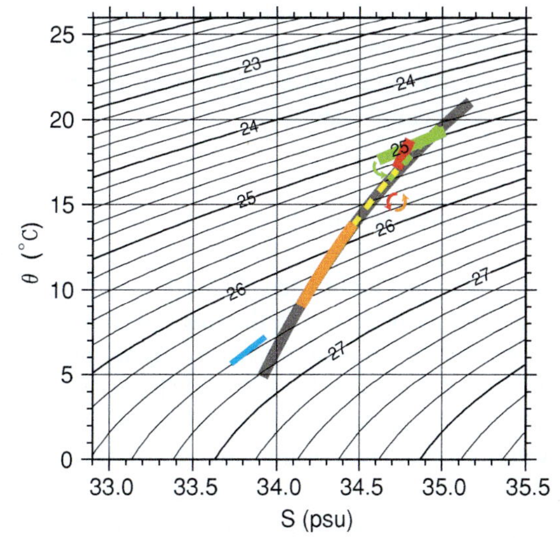

Fig. 10 Schematic T–S diagram illustrating the relations between the central water and the four mode waters in the North Pacific. *Gray curve* indicates the central water and *solid colored lines* indicate the mode waters: the STMW (*red*), CMW (*orange*), ESTMW (*green*), and TRMW (*blue*). The ESTMW (*green solid curve*) is possibly modified by salt fingering and becomes part of the central water with Tu ~ 59° in the eastern part of the gyre (*green dashed curve*). Mixing between the bottom part of the STMW and the lower pycnocline water (such as the CMW and/or the NPIW) could possibly contribute to the formation and maintenance of the central water in the western part of the gyre (*yellow dashed curve*)

modified TRMW to the central water is less expected. Part of the TRMW is possibly modified to become a denser variety of CMW (Saito et al. 2011) and it can then contribute to the central water.

5 Summary and remarks

In this work, the basin-wide distribution of the central water and the mode waters of the subtropical North Pacific was described in terms of Tu and low Q using CTD data collected by the Argo profiling floats. The main pycnocline water in the subtropical North Pacific displays a constant Tu value of 59°. The histograms of Tu for 10° × 10° geographical boxes mostly show a mode Tu value of 59°, but they also show some regional differences, suggesting some links between the distributions of Tu and the mode waters, judging from the distribution area overlap and the correspondence of the Tu with the Tu of each mode water, as suggested by Toyama and Suga (2010). The constant Tu of the main pycnocline water reported by previous studies (Ingham 1966; Schmitt 1981; Shimada et al. 2007) was confirmed based on a larger number of individual Argo CTD observations. Since the relative frequency (water volume) of the main pycnocline water is high in a band range centered around 59° (Fig. 2), the main pycnocline water with Tu ranging from 56° to 63° is defined as the "central water" in this study.

The appearance frequency of the Tu on each isopycnal surface (0.1 kg m^{-3} interval) was analyzed to examine the distribution of the central water and its relation to the mode waters. On isopycnals denser than $\sigma_\theta = 25.6$ kg m^{-3}, waters in the subtropical gyre generally show a mode Tu value of 59°, and parts of them correspond to the distribution area of CMW and the bottom part of STMW. The distribution area of the TRMW corresponds to the high-frequency area of Tu >70°. On isopycnals lighter than $\sigma_\theta = 25.6$ kg m^{-3}, on the other hand, the dominant Tu peak at around 59° is mostly distributed in the eastern to southern part of the gyre. A secondary peak with Tu >70° is distributed in the eastern part of the gyre, which corresponds well to the distribution area of the ESTMW. The recirculation gyre region, which is ventilated by the STMW on these isopycnals, has a rather small Tu value of 54°, especially near the core density of the STMW. That is, the central water is not distributed in the recirculation gyre region at the core density of STMW. Density ranges are broadly divided in two by the threshold of $\sigma_\theta = 25.6$ kg m^{-3} in this explanation; however, isopycnals of $\sigma_\theta =$ 25.4–25.8 kg m^{-3} show intermediate features or features of both the lighter and denser isopycnals.

The roles played by mode waters in the formation and maintenance of the constant Tu of the central water in the North Pacific are inferred as follows. The subduction of the CMW contributes directly to the formation and maintenance of the central water. That is, the CMW is subducted as Tu = 59° and contributes to the formation and maintenance of the central water by circulating in the subtropical gyre, mostly preserving its Tu on its own density surface corresponding to the winter outcrop. On the other hand, the STMW does not—at least directly—contribute to the central water, because the STMW is subducted and circulates with Tu = 54°. However, the central water lying below STMW in the recirculation gyre is possibly formed and maintained by the vertical diapycnal mixing of the STMW with certain water masses such as the CMW and/or the NPIW. In that sense, the subduction of the STMW is still an important factor in the formation and maintenance of the central water. The ESTMW with a relatively large Tu value (Tu > 70°) near the formation region is advected downstream (southwestward) by the general flow, decreasing its high-Tu signature through salt finger type convection. This may contribute to the central water away from its formation sites during and after its modification. The TRMW is also modified and advected downstream (mostly eastward), decreasing its Tu through salt fingering in a similar manner to the ESTMW, but the frequency of Tu <70° is not as high in the downstream region of TRMW. The TRMW is unlikely to contribute to the central water.

Very close to the CMW formation region, the Tu is higher than it is for the central water, while it is not as high as that indicating active salt fingering. Because this higher-Tu region corresponds well to the deep winter mixed-layer region, the seasonal cycle of near-surface layers should be taken into account, as suggested by Schmitt (1999). This transition of larger Tu near the formation area of CMW to the constant Tu downstream is an interesting but rather complicated issue. The lateral density ratio (R_L) distribution (Rudnick and Ferrari 1999; Rudnick and Martin 2002) and the mechanism that determines R_L in the mixed layer (Stommel 1993; Stommel and Young 1993) should be taken into consideration to address this issue. The R_L of the winter mixed layer water in the direction along its subduction path would presumably lend some insight into this transition, but this is left for future study.

The T–S structure of the North Pacific ocean was confirmed to be generally represented by a constant Tu of 59°; however, the formation and maintenance mechanisms are suggested to differ from place to place in the ocean basin. Such differences are partly explained by connecting them to the different formation and modification mechanisms of four mode waters in the North Pacific. However, the distribution area of the central water spreads far beyond the distribution areas of the four mode waters. The relations between the formation/maintenance of the central water and other water masses (including the Shallow Salinity

Minimum Water, Tropical Water, and Intermediate Water) are also suggestive, and further analysis is warranted to clarify the three-dimensional structure of the North Pacific subtropical main pycnocline. A quantitative evaluation of the contribution of each mode water to the central water is also necessary.

The mode waters should return to the surface mixed layer from the main pycnocline in the end, and affect both local and global climate variability through the redistribution of heat and freshwater. A better understanding of the roles of the mode waters in the formation and maintenance of the T–S structure of the subtropical main pycnocline should lead to a deeper understanding of climate variability and improvements in the ocean general circulation model (OGCM) and/or climate models.

Acknowledgments The authors wish to express their gratitude to members of the Physical Oceanography Group at Tohoku University and to two anonymous reviewers for their valuable comments. The first author (KT) was supported by the 21st-Century Center-Of-Excellence (COE) Program, "Advanced Science and Technology Center for the Dynamic Earth (E-ASTEC)," and the Global COE Program, "Global Education and Research Center for Earth and Planetary Dynamics" at Tohoku University. This study was partly supported by funds from a grant-in-aid for Scientific Research in Priority Areas—"Western Pacific Air–Sea Interaction Study (W-PASS)"—under grants 19030004 and 21014004 from the Ministry of Education, Culture, Sports, Science, and Technology (MEXT), Japan, and from "Studies on Prediction and Application of Fish Species Alteration (SUPRFISH)" sponsored by the Agriculture, Forestry, and Fisheries Research Council (AFFRC), Japan.

References

Akima H (1970) A new method of interpolation and smooth curve fitting based on local procedures. J Assoc Comput Mach 17:589–602

Boyer TP, Stephens C, Antonov I, Conkright ME, Locarnini RA, O'Brien TD, Garcia HE (2002) World ocean atlas 2001. In: Levitus S (ed) NOAA atlas NESDIS 50, vol 2: salinity. NOAA, Silver Spring

Figueroa HA (1996) World ocean density ratios. J Phys Oceanogr 26:267–275

Flament P (2002) A state variable for characterizing water masses and their diffusive stability: spiciness. Prog Oceanogr 54:493–501

Freeland H, Roemmich D, Garzoli S, LeTraon P, Ravichandran M, Riser S, Thierry V, Wijffels S, Belbéoch M, Gould J, Grant F, Ignazewski M, King B, Klein B, Mork K, Owens B, Pouliquen S, Sterl A, Suga T, Suk M, Sutton P, Troisi A, Vélez-Belchi P, Xu J (2010) Argo—a decade of progress. In: Hall J, Harrison DE, Stammer D (eds) Proceedings of Ocean Obs'09: sustained ocean observations and information for society, vol 2, Venice, Italy, 21–25 September 2009. ESA Publication WPP-306. doi: 10.5270/OceanObs09.cwp.32

Hanawa K, Talley LD (2001) Mode waters. In: Siedler et al (eds) Ocean circulation and climate. Academic, New York

Hautala SL, Roemmich DH (1998) Subtropical mode water in the Northeast Pacific Basin. J Geophys Res 103(C6):13055–13066

Hosoda S, Ohira T, Nakamura T (2008) A monthly mean dataset of global oceanic temperature and salinity derived from Argo float observations. JAMSTEC Rep Res Dev 8:47–59

Ingham MC (1966) The salinity extrema of the world ocean (Ph.D. dissertation). Oregon State University, Corvallis

Iselin CO'D (1939) The influence of vertical and lateral turbulence on the characteristics of the waters at mid-depths. Trans Am Geophys Un 20:414–417

Levitus S (1982) Climatological atlas of the world ocean (NOAA/ERL GFDL Prof Paper 13). GFDL, Princeton

Luyten JR, Pedlosky J, Stommel H (1983) The ventilated thermocline. J Phys Oceanogr 13:292–309

Masuzawa J (1969) Subtropical mode water. Deep-Sea Res 16:463–472

McCartney MS (1982) The subtropical recirculation of mode waters. J Mar Res 40(Suppl):427–464

McDougall TJ (1987) Neutral surfaces. J Phys Oceanogr 17:1950–1964

McDougall TJ, Thorpe SA, Gibson CH (1988) Small-scale turbulence and mixing in the ocean. In: Nihoul JCJ, Jamart BM (eds) Small-scale turbulence and mixing in the ocean. Elsevier, Amsterdam

Nakamura H (1996) A pycnostad on the bottom of the ventilated portion in the central subtropical North Pacific: its distribution and formation. J Oceanogr 52:171–188

Oka E, Suga T (2005) Differential formation and circulation of North Pacific central mode water. J Phys Oceanogr 35:1997–2011

Oka E, Talley LD, Suga T (2007) Temporal variability of winter mixed layer in the mid- to high-latitude north Pacific. J Oceanogr 63:293–307

Oka E, Kouketsu S, Toyama K, Uehara K, Kobayashi T, Hosoda S, Suga T (2011) Formation and subduction of central mode water based on profiling float data 2003–08. J Phys Oceanogr 41:113–129

Ruddick BR (1983) A practical indicator of the stability of the water column to double-diffusive activity. Deep-Sea Res 30:1105–1107

Rudnick DL, Ferrari R (1999) Compensation of horizontal temperature and salinity gradients in the ocean mixed layer. Science 283:526–529

Rudnick DL, Martin JP (2002) On the horizontal density ratio in the upper ocean. Dyn Atmos Oceans 36:3–21

Saito H, Suga T, Hanawa K, Watanabe T (2007) New type of pycnostad in the western subtropical–subarctic transition region of the North Pacific: transition region mode water. J Oceanogr 63:589–600

Saito H, Suga T, Hanawa K, Shikama N (2011) The transition region mode water of the north Pacific and its rapid modification. J Phys Oceanogr (in press). doi:10.1175/2011JPO4346.1

Schmitt RW (1981) Form of the temperature–salinity relationship in the central water: evidence for double-diffusive mixing. J Phys Oceanogr 11:1015–1026

Schmitt RW (1999) Spice and the demon. Science 283:498–499

Shimada K, Nemoto M, Yoshida J (2007) Distribution of the density ratio in the North Pacific. La Mer 45:149–158

Stephens C, Antonov I, Boyer TP, Conkright ME, Locarnini RA, O'Brien TD, Garcia HE (2002) World ocean atlas 2001. In: Levitus S (ed) NOAA Atlas NESDIS 49, vol 1: temperature. NOAA, Silver Spring

Stommel H (1979) Determination of water mass properties of water pumped down from the Ekman layer to the geostrophic flow below. Proc Natl Acad Sci USA 76:3051–3055

Stommel HM (1993) A conjectural regulating mechanism for determining the thermohaline structure of the oceanic mixed layer. J Phys Oceanogr 23:142–148

Stommel HM, Young WR (1993) The average T–S relation of a stochastically forced box model. J Phys Oceanogr 23:151–158

Suga T, Takei Y, Hanawa K (1997) Thermostad distribution in the north Pacific subtropical gyre: the central mode water and the subtropical mode water. J Phys Oceanogr 27:140–152

Suga T, Motoki K, Aoki Y, Macdonald AM (2004) The North Pacific climatology of winter mixed layer and mode waters. J Phys Oceanogr 34:3–22

Sverdrup HU, Johnson MW, Fleming RH (1942) The oceans, their physics, chemistry, and general biology. Prentice-Hall, New York

Talley LD (1988) Potential vorticity distribution in the North Pacific. J Phys Oceanogr 18:89–106

Tomczak M, Godfrey JS (1994) Regional oceanography: an introduction. Pergamon, London

Toyama K, Suga T (2010) Vertical structure of North Pacific mode waters. Deep-Sea Res II 57:1152–1160

Turner JS (1973) Buoyancy effects in fluids. Cambridge University Press, Cambridge

You Y (1998) Dianeutral mixing and transformation of Antarctic intermediate water in the Indian Ocean. J Geophys Res 103:30941–30971

You Y (2002) A global ocean climatological atlas of the Turner angle: implications for double-diffusion and water mass structure. Deep-Sea Res I 49:2075–2093

J Oceanogr (2012) 68:93–111
DOI 10.1007/s10872-011-0074-8

SPECIAL SECTION: ORIGINAL ARTICLE

New developments in mode-water research: Dynamic and climatic effects

Interannual variations of the Hawaiian Lee Countercurrent induced by potential vorticity variability in the subsurface

Hideharu Sasaki · Shang-Ping Xie · Bunmei Taguchi · Masami Nonaka · Shigeki Hosoda · Yukio Masumoto

Received: 11 January 2011 / Revised: 1 August 2011 / Accepted: 10 August 2011 / Published online: 2 October 2011

Abstract Interannual variations of the Hawaiian Lee Countercurrent (HLCC) in the 2000s were investigated using satellite and Argo profiling float observations. The satellite-observed sea surface height shows that the geostrophic eastward current was anomalously strong to the west away from Hawaii in 2003 and 2005. However, the trade winds and the orographic wind curl dipole in the lee of Hawaii that drives the climatological mean HLCC were not particularly strong in these years, suggesting that the accelerations of the HLCC were not caused by the wind stress curl forcing around Hawaii and subsequent Rossby wave propagation. Using Argo observations, we found negative potential vorticity (PV) anomalies in the subsurface north of the HLCC in these 2 years. The pycnocline is lifted northward as low PV waters of different densities stack up in the vertical, and the HLCC is then accelerated via the thermal wind. The intensification and/or southward intrusion of the eastern subtropical mode water and subtropical mode water seem to have induced negative PV anomalies in 2003 and 2005, respectively. Using high-resolution ocean simulations, we confirmed the migrations of PV anomalies and their contributions to the HLCC accelerations. Although the HLCC is located away from the cores of major mode waters, our results suggest that interannual variations of the HLCC are affected by those of mode waters.

Keywords Hawaiian Lee Countercurrent · Low potential vorticity water · Interannual variations · High-resolution ocean simulation

H. Sasaki (✉) · B. Taguchi
Earth Simulator Center, JAMSTEC, 3173-25 Showa-machi, Kanazawa-ku, Yokohama, Kanagawa 236-0001, Japan
e-mail: sasaki@jamstec.go.jp

S.-P. Xie
International Pacific Research Center and Department of Meteorology, University of Hawaii at Manoa, 2525 Correa Rd., Honolulu, HI 96822, USA

M. Nonaka · Y. Masumoto
Research Institute for Global Change, JAMSTEC, 3173-25 Showa-machi, Kanazawa-ku, Yokohama, Kanagawa 236-0001, Japan

S. Hosoda
Research Institute for Global Change, JAMSTEC, 2-15 Natsushima, Yokosuka, Kanagawa 237-0061, Japan

1 Introduction

The Hawaiian Lee Countercurrent (HLCC), a narrow eastward current extending far west of the Hawaiian Islands, has been observed from surface drifters (Qiu et al. 1997; Flament et al. 1998) and hydrography (Kobashi and Kawamura 2002). The HLCC is one of the eastward subtropical countercurrents (STCC) that exist in the North Pacific (e.g., Uda and Hasunuma 1969; Kobashi et al. 2006) and flow against both the northeasterly trade winds and the broad westward north equatorial current (NEC).

Quasi-steady, broad trade winds prevail over the Hawaiian Islands, generating a number of mechanical wakes behind individual islands (Smith and Grubišic 1993; Yang et al. 2008a, b). The individual wakes dissipate within a few hundred kilometers, and a broad wake with a meridional dipole of wind stress curl forms (Hafner and Xie 2003). Using satellite observations and an ocean general circulation model (OGCM), Xie et al. (2001) proposed a generation mechanism for the climatological mean HLCC (hereafter referred to as the wind-driven mechanism), suggesting that the HLCC is a response to the orographic wake in the lee of Hawaii that occurs via Rossby wave propagation. Using an atmosphere–ocean coupled general circulation model (GCM), Sakamoto et al. (2004) succeeded in reproducing the HLCC and confirmed that the islands trigger the current.

Several factors complicate the westward propagation of Hawaii's orographic effects in the form of Rossby waves. For example, eddy variability with a 100-day timescale west of 160°W along 19°N is likely generated by the instability of the vertically sheared NEC and HLCC (Yoshida et al. 2010). Yu et al. (2003) suggested that the westward extension of the HLCC is limited to the east of the dateline because of mesoscale eddies that extract energy from the mean flow.

The southern subtropical front (STF) along 19–21°N west of the dateline, which is distinct from the HLCC front, is one of three STFs defined using hydrographic data by Kobashi et al. (2006). The STFs are associated with large meridional gradients of isopycnal potential vorticity (PV) below the fronts. Theoretical (Kubokawa 1999) and numerical (Kubokawa and Inui 1999) studies have shown that an STCC forms with the STF as low PV waters of different density become vertically stacked north of the STF, which push up isopycnals in the lower density range. The southern STF results from the southward intrusion of subtropical mode water (STMW; e.g., Bingham 1992; Hanawa and Talley 2001) and upper central mode water (CMW; e.g., Hanawa and Talley 2001) in the subtropical gyre (Aoki et al. 2002; Kobashi et al. 2006).

Seasonal HLCC variations, which are strong from summer to winter and weak in spring, are detected in hydrographic data (Kobashi and Kawamura 2002). The seasonal HLCC variations are consistent with the wind-driven mechanism for climatological HLCC generation and with the seasonal cycle of the trade winds (Kobashi and Kawamura 2002). Using satellite observations and a high-resolution OGCM, Sasaki et al. (2010) confirmed the hypothesis of the seasonal HLCC variations.

Considerable interannual variability of the HLCC, which was stronger in 2003 and 2005 than in other years, was found in geostrophic currents estimated from satellite-observed sea surface height (SSH) by Sasaki et al. (2010). However, the trade winds were not strong in these years, and the HLCC accelerations were stronger away from Hawaii than near Hawaii, suggesting that the wind-driven mechanism forced by the wind curl dipole near Hawaii is not dominant in the HLCC accelerations. Sasaki et al. (2010) suggested from an analysis using an OGCM simulation that the HLCC acceleration in 2003 seems to be induced by the intrusion of lower PV water associated with the CMW than usual to the north of the HLCC. However, the hypothesis is still a matter of speculation. The contributions of low PV water to HLCC acceleration have not been studied quantitatively or examined in observational data.

The present study examines the interannual variations of HLCC and their mechanism near and away from Hawaii separately, using satellite and Argo profiling float observations from 2001 to 2008 with a focus on the HLCC accelerations in 2003 and 2005. We consider not only the wind-driven mechanism but also low PV water intrusions in the subsurface. To support the observations, a high-resolution OGCM is also used to examine variations of PV and those induced by a wind forcing around Hawaii. It is expected that this study will help us understand the effects of mode waters on ocean circulation and climate in the North Pacific through revealing relationships between mode waters and surface currents, including the HLCC. As in a recent research using a coupled GCM, mode waters induce variations of STCC and generate sea surface temperature anomalies along STCCs, which in turn affect wind and precipitation (Xie et al. 2011).

The contents of this paper are as follows. Section 2 describes the observational data and high-resolution simulations performed using an OGCM. In Sect. 3, interannual variations of the HLCC from 2001 to 2008 are presented, and their mechanisms are investigated using satellite and profiling float observations. Section 4 examines how the mechanism works in a high-resolution simulation. Section 5 discusses how the intrusion of low PV water masses is induced and then suggests other possible mechanisms that could interannually accelerate the HLCC. Section 6 presents our conclusions.

2 Observations and high-resolution OGCM simulations

2.1 Observational data

We examined satellite observations of SSH and surface wind stress to detect interannual variations of the HLCC and trade winds. Maps of absolute dynamic topography (MADT) products were obtained from archiving and interpretation of satellite oceanographic data (AVISO; http://www.aviso.oceanobs.com). The SSH data were constructed every 7 days from multiple satellite altimetry data since 1993 with a horizontal resolution of 0.25°. Daily mean surface wind stress data with a horizontal grid of 0.5° based on Quick Scatterometer (QuikSCAT) satellite observations from July 1999 to October 2009 included in the Japanese ocean flux data sets with the use of remote sensing observations product (J-OFURO, Kutsuwada 1998; Kubota et al. 2002) were also used.

The monthly temperature and salinity data from 2001 with a 1° horizontal resolution included in the MOAA GPV dataset (grid point value of the monthly objective analysis using Argo float data; Hosoda et al. 2008; http://www.jamstec.go.jp/ARGO/argo_web/argo/index.html) were used to detect interannual variations in subsurface PV. The grid data product was constructed from not only Argo profiling float data but also data from the Triangle Trans-Ocean Buoy Network (TRITON) and available conductivity–temperature–depth (CTD) profilers. A two-dimensional optimal interpolation (OI) method is used to produce the grid data on pressure levels from 10 to 2,000 dbar, and the first guess field is the monthly climatology data of the *World Ocean Atlas 2001* (WOA01; Boyer 2002; Stephens et al. 2002). The 10-dbar level is the first pressure level near the sea surface. We also used the potential density and dynamic height at 10 dbar referenced to 2,000 dbar in the MOAA GPV product. Interpolation errors related to the grid data for the North Pacific are not insignificant before 2002 owing to the small number of Argo profiling float observations (Hosoda et al. 2008). However, it is expected that the grid data capture the long-term mean and interannual variability of subsurface PV after 2003.

2.2 Hindcast and sensitivity simulations using an OGCM

Interannual PV variations from 2001 to 2008 were further examined using a high-resolution OGCM. We used the OGCM for the Earth Simulator (OFES; Masumoto et al. 2004) based on GFDL MOM3 (modular ocean model 3 developed by the Geophysical Fluid Dynamics Laboratory, Pacanowski and Griffies 1999). The model domain is quasi-global from 75°S to 75°N, and the horizontal resolution is 0.1°, with 54 vertical levels. A hindcast simulation using the OFES is forced at the surface by daily mean QuikSCAT wind stress with a 1° horizontal resolution in the J-OFURO product (Kutsuwada 1998; Kubota et al. 2002) and National Centers for Environmental Prediction/National Center for Atmospheric Research (NCEP/NCAR) reanalysis data (Kalnay et al. 1996) from July 1999 to 2008 (Sasaki et al. 2006). The hindcast begins from the result of a previous OFES hindcast simulation forced by NCEP/NCAR reanalysis from 1950 (Sasaki et al. 2008).

A sensitivity simulation from July 1999 to 2008 was conducted using the OFES to observe how the HLCC varied interannually, without interannual variations of local wind forcing near Hawaii. The simulation is forced by the same forcings as the hindcast simulation, but the wind stress in the region around Hawaii (170–155°W, 15–23°N) is replaced by the climatological mean fields averaged from 2001 to 2008. In a buffer zone with a width of 2° outside the region, the wind stress is gradually tapered to the interannually varying field, reducing the discontinuity at the boundary of the region.

3 Interannual variations in observations

3.1 HLCC and trade winds

Figure 1a shows the long-term mean eastward speed of the geostrophic current estimated from the AVISO SSH averaged from 2001 to 2008. A narrow eastward current along approximately 19.5°N, corresponding to the HLCC, extends westward from Hawaii across the dateline. The HLCC path tilts slightly southward toward the west. The current speed averaged between 180°W and 160°W along 19.5°N is 6.8 cm s^{-1}, which is comparable to the value of 6.3 cm s^{-1} estimated from historical hydrographic data (Kobashi and Kawamura 2002).

The annual mean HLCCs in 2003 and 2005 were markedly stronger west of 170°W compared with the multi-year average from 2001 to 2008 (Fig. 1). Sasaki et al. (2010) found the HLCC accelerations away from Hawaii and suggested that the wind-driven mechanism is not dominant in these accelerations. Here, we examine separately the interannual variations of the HLCC near (between 165°W and 160°W) and away from (between 180°W and 165°W) Hawaii.

Time–latitude plots of 13-month running means of the eastward current speed averaged between 180°W and 165°W exhibit distinct interannual variations of the HLCC, capturing the events of HLCC enhancement away from Hawaii that occurred in both 2003 and 2005 (Fig. 2a). If the HLCC accelerations can be explained by the wind-driven mechanism through trade wind variations, the winds and orographic wind curl dipole must be strong during the same periods. The meridional gradient of the wind stress curl dipole must also be large based on the Sverdrup theory for estimating mass transport. However, Fig. 3 shows that the easterly wind and meridional gradient of the wind stress curl in the lee of Hawaii were not strong in 2003 and 2005. The discrepancies between the variations of the HLCC and those of the trade winds with the wind curl dipole suggest that the HLCC accelerations are mainly induced by other mechanisms that are not associated with trade wind variability.

Near Hawaii, the HLCC was strong not only in 2003 and 2005 but also in 2001 and 2007 (Fig. 2b). The HLCC accelerations in 2003 and 2005 are similar to but weaker than those away from Hawaii (Fig. 2a), and, as discussed above, are not due to the wind-driven mechanism. However, the wind-driven mechanism associated with the strong trade winds in 2001 and 2007 (Fig. 3) seems to have accelerated the HLCC near Hawaii (Fig. 2b). The HLCC was weak in 2004 both near (Fig. 2b) and farther from Hawaii (Fig. 2a) compared with the previous and subsequent years. It is possible that the anomalous weak trade winds in 2004 induced HLCC deceleration not only near

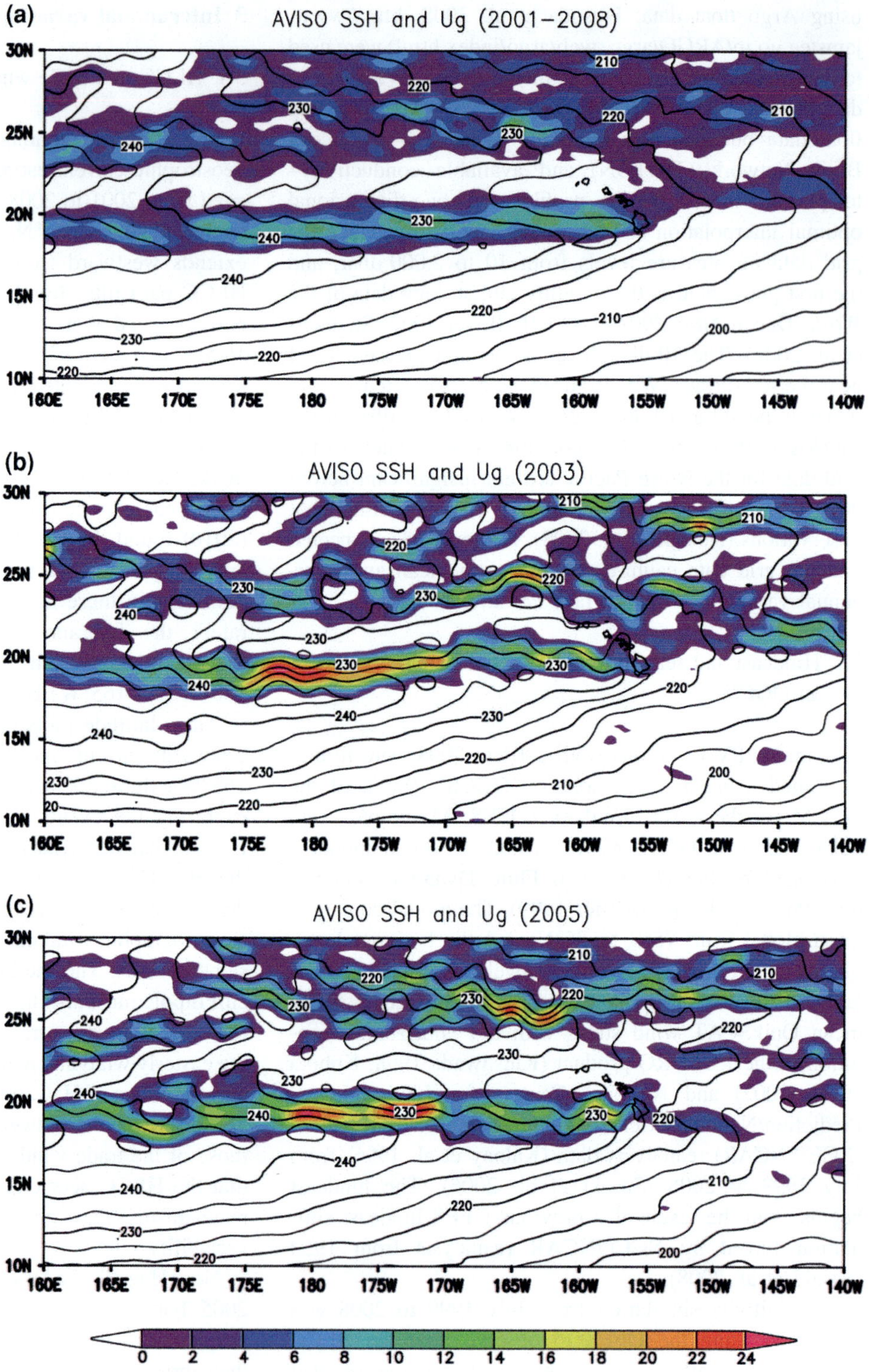

Fig. 1 SSH (*contour*, unit is cm) and geostrophic eastward current speed (*color*, unit is cm s^{-1}) based on the AVISO SSH. **a** Mean averaged from 2001 to 2008 and annual mean in **b** 2003 and **c** 2005 (color figure online)

Hawaii but also farther from Hawaii by the wind-driven mechanism. These results suggest some influence of the wind-driven mechanism on interannual variations of the HLCC both near Hawaii and farther from Hawaii, although this mechanism is not the main factor in the HLCC accelerations away from Hawaii in 2003 and 2005.

The Argo observations from the MOAA GPV dataset also feature the mean HLCC (not shown) and the accelerations away from Hawaii in 2003 and 2005 (Fig. 2c). However, there are some differences from the altimeter observations. Compared to the altimeter observations (Figs. 1a, 2a, b), the mean HLCC is relatively

Fig. 2 Time–latitude plots of 13-month running means of the geostrophic eastward current (cm s^{-1}) averaged between **a**, **c**, **e**, **g** 180°W and 165°W and **b**, **d**, **f**, **h** 165°W and 160°W based on **a**, **b** the Aviso SSH, **c**, **d** dynamic height at 10 dbar referenced to 2,000 dbar of the MOAA GPV, **e**, **f** SSH of the OFES hindcast simulation, and **g**, **h** SSH of the OFES sensitivity experiment

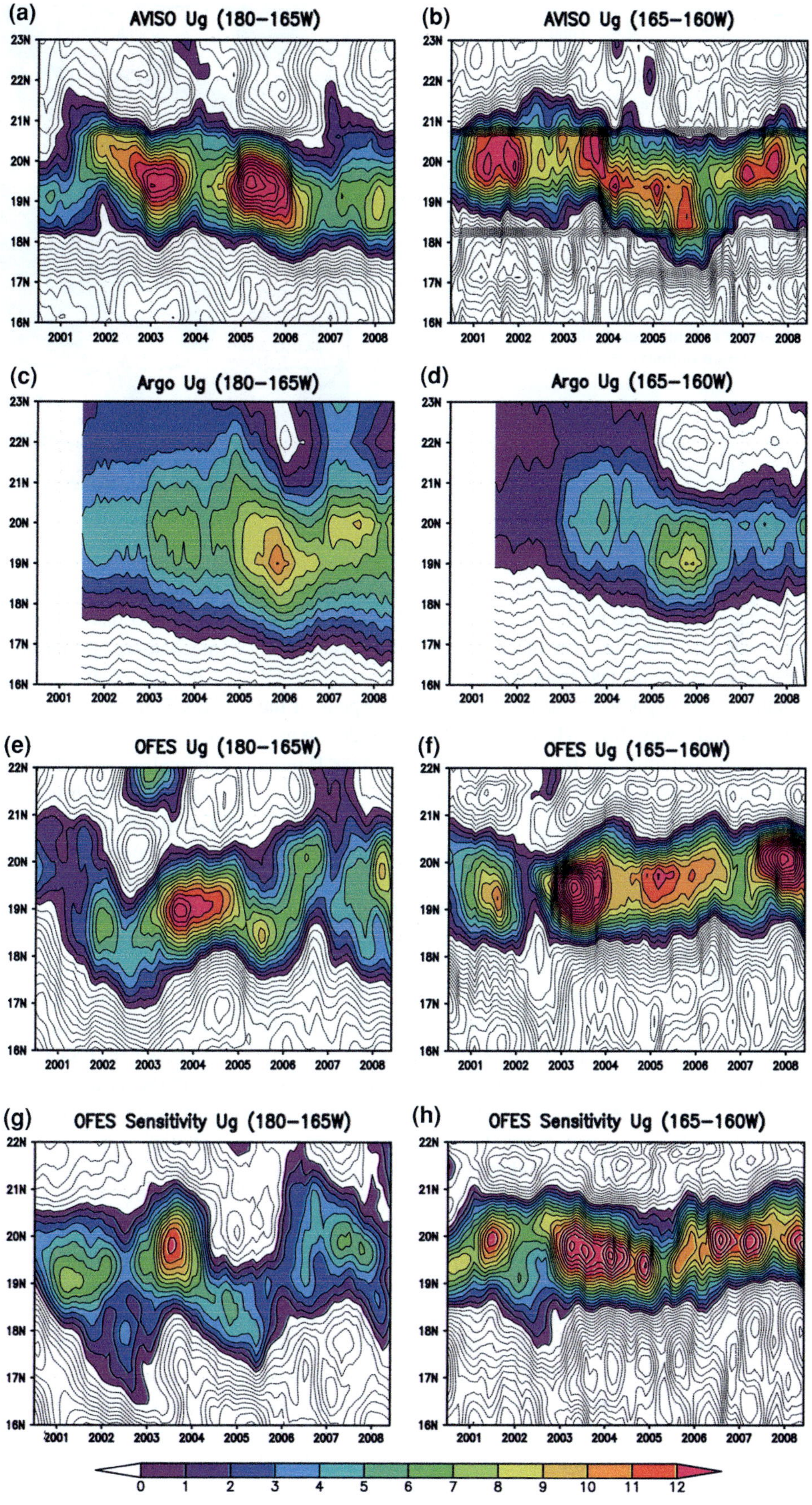

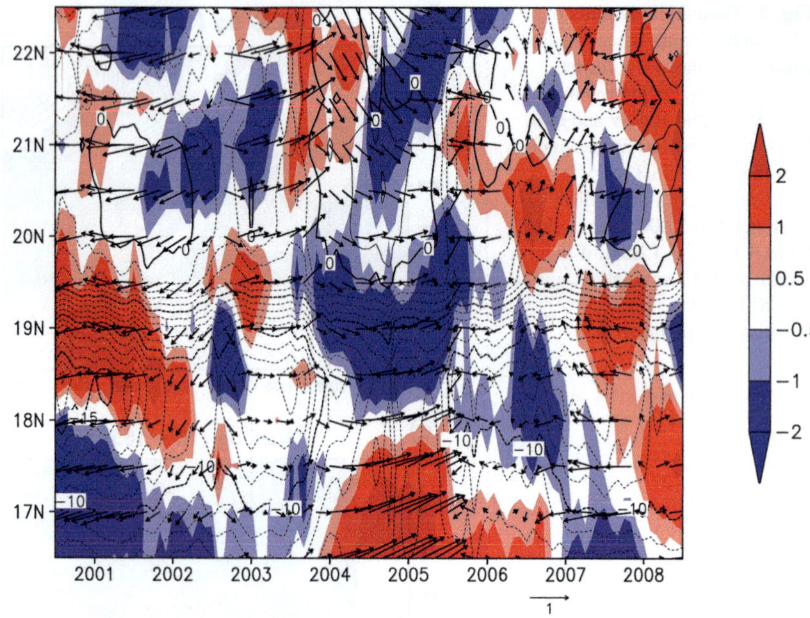

Fig. 3 Anomalies of wind stress vectors (10^{-2} N m^{-2}) and meridional gradient of wind stress curl (*color*, unit is 10^{-13} N m^{-4}) averaged between 165°W and 157°W from the mean averaged from 2001 to 2008 based on QuikSCAT satellite observations. *Contour* indicates wind stress curl (10^{-8} N m^{-3}) averaged between 165°W and 157°W, and its interval is 10^{-8} N m^{-3}. All variables are 13-month running mean values (color figure online)

broad and its speed is about half (not shown). In addition, the magnitude of the interannual variations is much smaller (Fig. 2c, d). These discrepancies are possibly because of sparse observations and smoothing of the dataset by the OI method. The Argo observations display another peak of the high speed away from Hawaii in 2007 (Fig. 2c), which could be contaminated by the strong HLCC near Hawaii forced by the strong wind (Figs. 2b, 3).

3.2 Subsurface PV

Argo observations capture the interannual accelerations in 2003 and 2005 (Fig. 2c), which encourages us to examine subsurface PV and its relationship to the HLCC variations. Figure 4 shows a latitude–depth section for PV and potential density averaged between 180°W and 165°W away from Hawaii. In the multi-year mean plot (Fig. 4a), the pycnocline around 24.2 σ_θ slopes up northward most steeply between 18°N and 21°N. A surface-intensified eastward current corresponding to the HLCC is consistently positioned at the latitudes of the steep pycnocline slope.

The annual mean main pycnocline slope in 2005 was steeper around 20°N than the multi-year mean from 2001 to 2008 (Fig. 4c), which is consistent with the highest HLCC speed away from Hawaii in that year based on the Argo dynamic height (Fig. 2c). We found negative PV anomalies in the density range of 24.6–25.8 σ_θ north of 18°N in 2005 (Fig. 4c). If the HLCC intensification was induced by the wind-driven mechanism, the PV anomalies would be localized near the HLCC in the meridional direction by Rossby wave propagations, but they actually extend north across 30°N. Therefore, the PV anomalies are likely to have been induced by some other mechanism and appear to have accelerated the HLCC on the basis of the mechanism proposed by Kubokawa (1999); however, Kobashi et al. (2006) suggested that the climatological HLCC front east of the dateline is not induced by low PV mode water intrusions. In 2003 (Fig. 4b), negative PV anomalies spread further southward than those in 2005, extending south across 20°N at the depth of 100–200 m. The pronounced negative anomalies north of the HLCC push isopycnals upward along the southern flank of the PV anomalies, intensifying the eastward HLCC. These results suggest that the negative PV anomalies north of the HLCC may contribute to interannual variations of HLCC speed.

We examine the horizontal distribution of PV anomalies. The long-term mean PV at 25.0 σ_θ in the Argo observations (Fig. 5a) captures the eastern subtropical mode water (ESTMW; Hautala and Roemmich 1998). The ESTMW is characterized by a subsurface low PV water mass with a density range of 24.0–25.4 σ_θ, which is distributed east of Hawaii in the eastern subtropical North Pacific. The STMW is another low PV water mass with lowest PV at 25.4–25.5 σ_θ. The long-term mean PV at 25.4 σ_θ (Fig. 5b) shows that the STMW is located in the northwestern subtropical gyre (e.g., Suga and Hanawa 1995; Kobashi et al. 2006). The formations of the ESTMW (Toyoda et al. 2011) and STMW (e.g., Suga and Hanawa 1995; Oka 2009) display considerable interannual variations.

Interannual variations of the isopycnal PV were examined to investigate the relationship with the HLCC variations. We observed isopycnal PV variations south of the southernmost outcrop line of the isopycnal surface among the lines from 2001 to 2008, because interannual variations

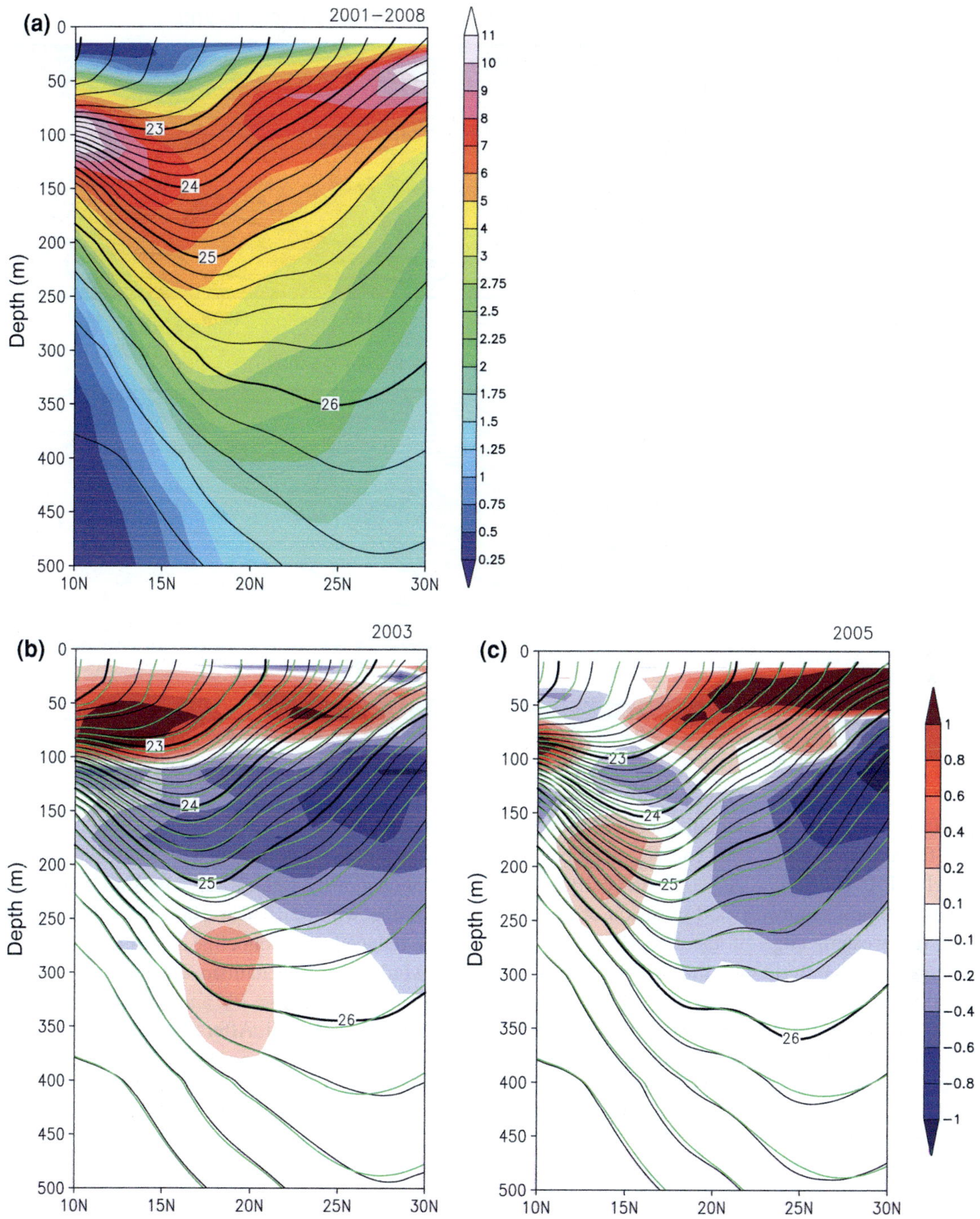

Fig. 4 Latitude–depth section of PV (*color*, unit is 10^{-10} m^{-1} s^{-1}) averaged between 180°W and 165°W based on the MOAA GPV. **a** Mean averaged from 2001 to 2008 and yearly anomalies from the mean in **b** 2003 and **c** 2005. *Black contours* indicate potential density (σ_θ), and their intervals are 0.2 σ_θ. *Green contours* in **b** and **c** indicate mean potential density averaged from 2001 to 2008 (color figure online)

of the outcrop line are not discussed here. In 2003, PV anomalies were negative to the north of the HLCC at 20°N (Fig. 6a, c). To determine how the anomalies extended north of the HLCC, we plotted PV anomalies along a long-term mean acceleration potential isoline (Fig. 7). The horizontal gradient of the acceleration potential corresponds approximately to the geostrophic current on an isopycnal surface (Montgomery and Stroup 1962). Although the acceleration potential isolines varied interannually in the zonal direction approximately 5° east of Hawaii at approximately 25°N (not shown), the PV variability can be roughly traced along the mean isoline. At the

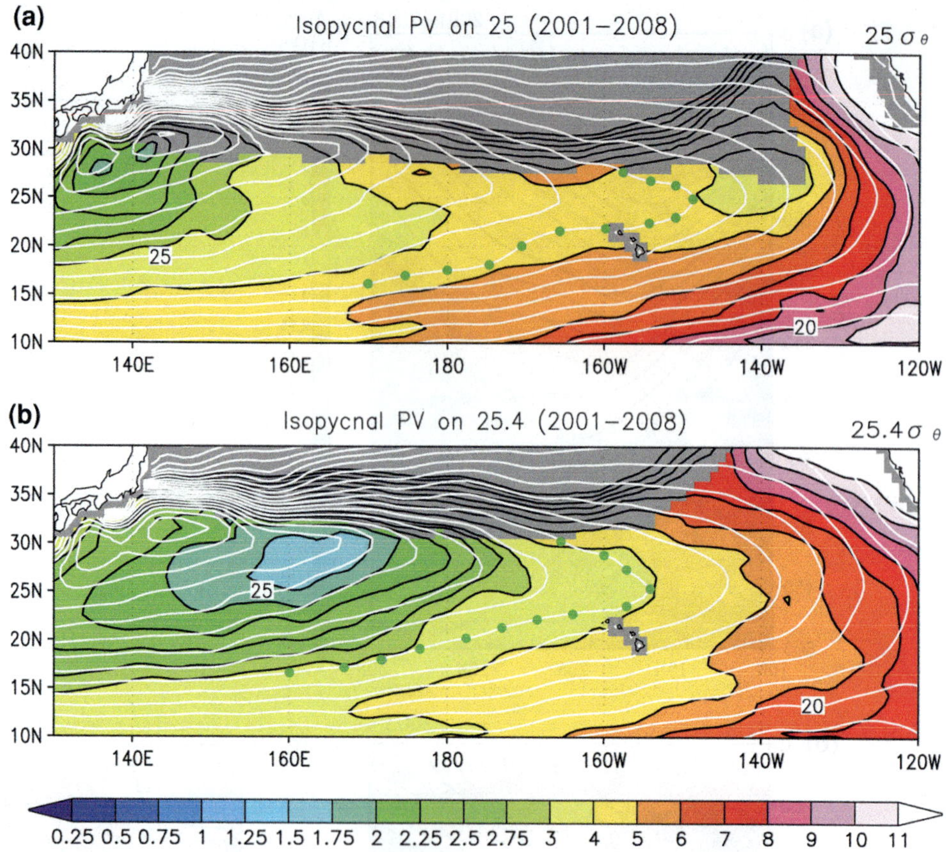

Fig. 5 Mean PV (10^{-10} m^{-1} s^{-1}) averaged from 2001 to 2008 at the **a** 25.0 σ_θ and **b** 25.4 σ_θ surfaces based on the MOAA GPV. *White contours* indicate acceleration potential (m^2 s^{-2}) at the surfaces relative to 2,000 dbar. Contour interval is 0.5 m^2 s^{-2}. *Gray shade* indicates the area with a winter surface density greater than each isopycnal. Along the isolines of acceleration potential with *green dots*, time–distance plots of isopycnal PV anomaly are shown in Fig. 7 (color figure online)

25.0 σ_θ surface, the negative anomalies between 170°W and 160°W from 2003 to early 2004 can be traced back to approximately 150°W from late 2002 to early 2003 (Fig. 7a), suggesting a southwestward intrusion of low PV water from the western edge of the ESTMW. The PV anomalies enhanced the meridional negative PV gradients to the south of the anomalies. At the 25.4 σ_θ surface, the negative PV anomalies were distributed between 175°W and 155°W west of Hawaii in 2003 but do not extend further east (Fig. 6c). These negative PV anomalies cannot be traced back clearly along the acceleration potential isoline (Fig. 7b).

In 2005, negative PV anomalies were distributed west of Hawaii and north of 18°N at the 25.0 σ_θ and 25.4 σ_θ surfaces (Fig. 6b, d). The negative anomalies at the 25.0 σ_θ surface between 180°W and 170°W in late 2005 appear to have migrated westward from the east (Fig. 7a). At the 25.4 σ_θ surface (Fig. 7b), the negative anomalies at approximately 170°W in 2005 appear to have been induced by an eastward expansion of the STMW to approximately 160°W in the same period (Fig. 6d). These negative PV anomalies at the 25.0 σ_θ and 25.4 σ_θ surfaces both enhanced meridional PV gradients to the south of the PV anomalies (Fig. 7).

From late 2004 to early 2005, the negative PV anomalies at both the 25.0 σ_θ and the 25.4 σ_θ surfaces were relatively weak at approximately 170°W–160°W near Hawaii compared to before and after this period (Fig. 7). The anomalously weak trade winds and meridional gradient of the wind curl near Hawaii in this period (Fig. 3) likely reduced the pycnocline slanting along the HLCC by the wind-driven mechanism via Rossby waves and then weakened the negative PV anomalies and their meridional gradients near Hawaii. After passing the Rossby waves, the pycnocline slanting along the HLCC could get back to the former state. We also found positive PV anomalies to the south of the HLCC at the 25.4 σ_θ surface in 2003 and at the 25.0 σ_θ and 25.4 σ_θ surfaces in 2005 (Fig. 6). This result suggests that not only PV anomalies north of the HLCC but also PV anomalies to the south and a wind forcing around Hawaii influence interannual variations of the HLCC.

3.3 Contributions of subsurface PV variability to the HLCC

The eastward countercurrent is associated with a slanting pycnocline based on the thermal wind balance theory. This section examines the contributions of PV anomalies to the pycnocline slope following the method of Kobashi et al. (2006), who showed that the STFs and STCCs are anchored by large negative PV gradients. Under the

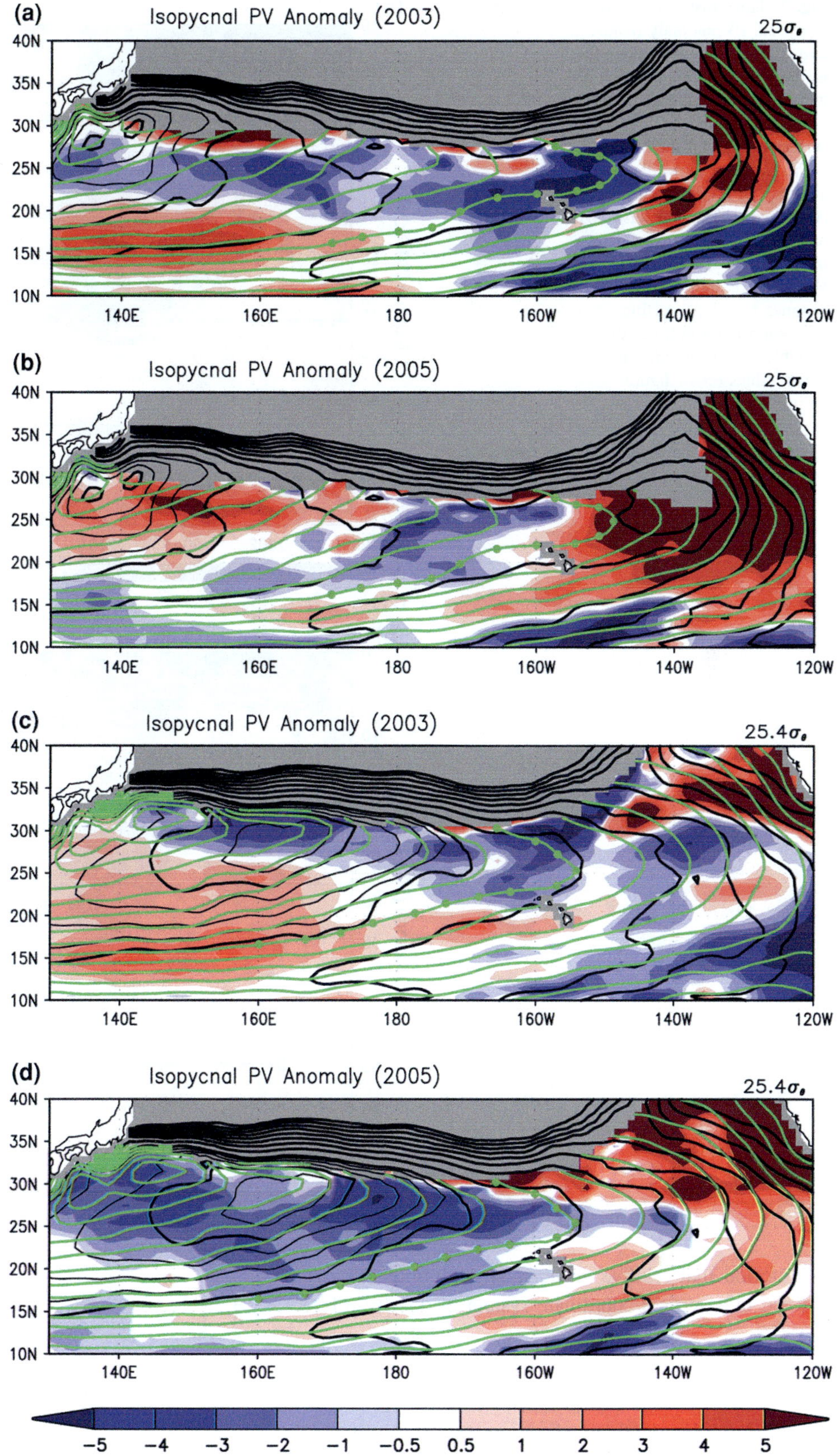

Fig. 6 Yearly PV anomalies (*color*, unit is 10^{-10} m^{-1} s^{-1}) in **a** 2003 and **b** 2005 from the mean averaged from 2001 to 2008 at the 25.0 σ_θ surface based on the MOAA GPV; **c**, **d** are identical to **a**, **b** but at the 25.4 σ_θ surface. *Black* and *green contours* show the mean PV and acceleration potential (m^2 s^{-2}), respectively, averaged from 2001 to 2008. *Gray shade* indicates the area with a winter surface density greater than each isopycnal (color figure online)

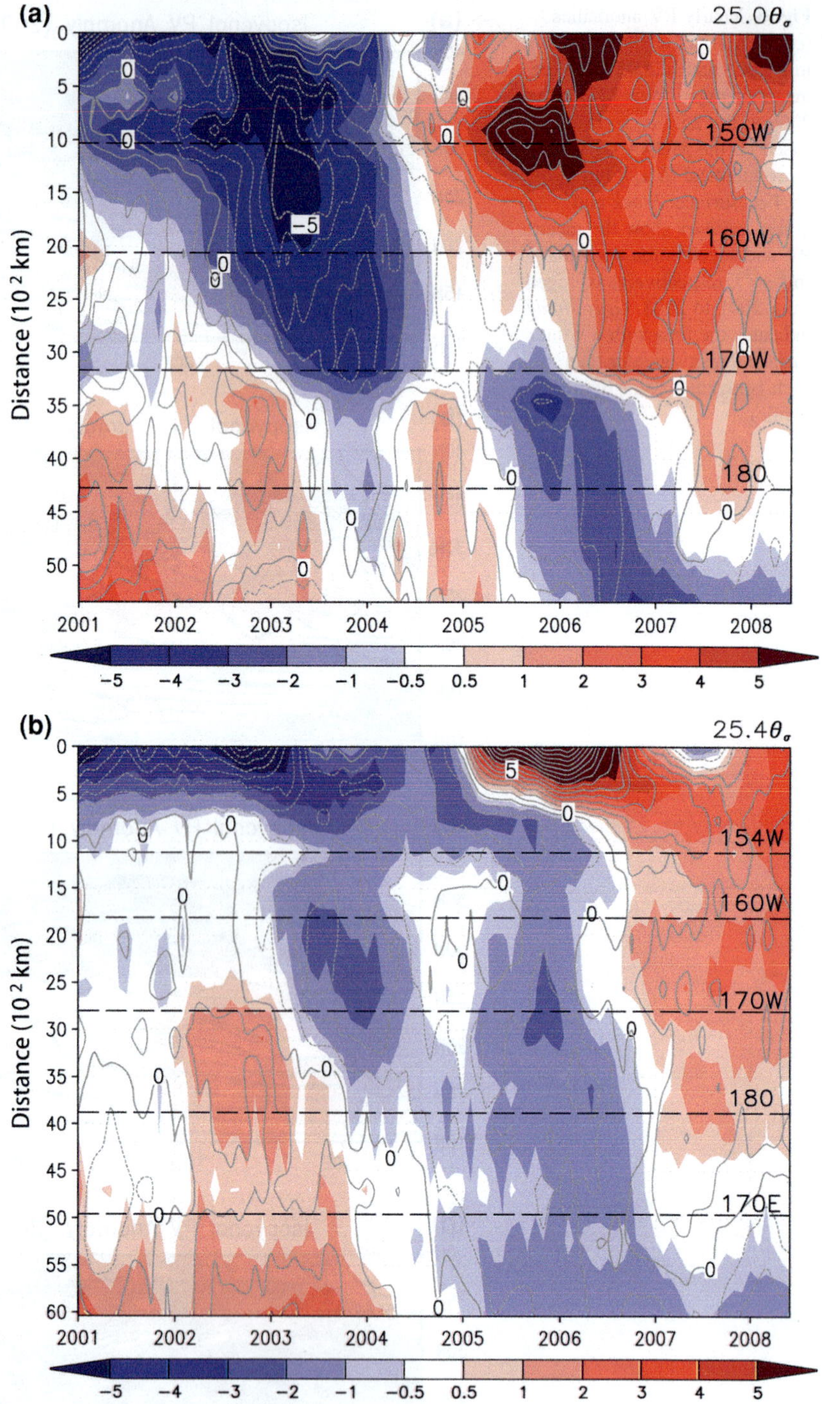

Fig. 7 Time–distance plot of isopycnal PV anomaly (*color*, unit is 10^{-10} m^{-1} s^{-1}) along the isoline of acceleration potential with *green dots* in Figs. 5, 6 at the **a** 25.0 σ_θ and **b** 25.4 σ_θ surfaces based on the MOAA GPV. *Contour* shows the anomaly of PV difference compared to 5° south (*contour*, unit is 10^{-10} m^{-1} s^{-1}). Contour interval is 10^{-10} m^{-1} s^{-1}. *Vertical axis* indicates the distance (10^2 km) starting from the winter outcrop line. All variables are 13-month running mean values (color figure online)

assumption that relative vorticity is negligible, PV (q) is defined in a vertical coordinate system of density ρ by

$$q(\rho) = -\frac{f}{\rho_0 \, \partial Z(\rho)/\partial\rho}, \tag{1}$$

where Z is the depth of the isopycnal surface and is negative below sea level, f is the planetary vorticity, and ρ_0 is the reference density. Solving Z and taking the meridional derivative, the isopycnal slope is estimated by

$$\left(\frac{\partial Z(\rho)}{\partial y}\right)_\rho = -\frac{1}{\rho_0}\int_{\rho_b}^{\rho} \frac{1}{q(\rho')}\left(\beta - \frac{f}{q(\rho')}\left(\frac{\partial q(\rho')}{\partial y}\right)_\rho\right)d\rho' + \left(\frac{\partial Z_0(\rho_b)}{\partial y}\right)_\rho, \tag{2}$$

which is identical to Eq. 3 in Kobashi et al. (2006). Here, Z_0 is the depth of the reference isopycnal surface ρ_b ($\geq\rho$), and β is the meridional gradient of f. The subscript ρ denotes a partial derivative on a constant density surface. For the isopycnal surface of density ρ to slope up toward the north, the PV gradient deviation $(\beta - (f/q)(\partial q/\partial y)_\rho)$ should be a large positive value above the reference isopycnal surface.

Figure 8 shows the meridional section of PV, meridional PV gradients, and meridional density gradients averaged between 180°W and 165°W. The mean plot (Fig. 8a) averaged from 2001 to 2008 is similar to the hydrographic observations (Fig. 7c in Kobashi et al. 2006). Large negative PV gradients ($<-2.0 \times 10^{-11}$ m^{-1} s^{-1}) in the density range of 24.6–25.7 σ_θ are distributed below the strong density front ($>15 \times 10^{-7}$ kg m^{-4}) at approximately 20°N, corresponding to the HLCC. The negative PV gradients in the density range lighter than 25.0 σ_θ are distributed south of the latitude of HLCC. No local minimum PV corresponding to the mode water is found north of the front, consistent with results reported by Kobashi et al. (2006). The large negative PV gradients are probably induced by a pycnocline slanting along the HLCC through Rossby waves under the wind-driven mechanism.

Large negative PV anomalies north of the HLCC seem to have accelerated the current in 2003 and 2005, as suggested in Sect. 3.2. To test this hypothesis, we examined the contributions of PV anomalies to HLCC acceleration based on Eq. 2. The negative PV gradients were larger ($<-2.0 \times 10^{-11}$ m^{-1} s^{-1}) in the density range around 25.2 σ_θ (25.0 σ_θ) at the latitude of the HLCC in 2003 (2005) than those averaged from 2001 to 2008 (Fig. 8). Above the negative PV gradients, large meridional density gradients ($>2.0 \times 10^{-6}$ kg m^{-4}) were found at densities less than 23.8 σ_θ (24.2 σ_θ) in 2003 (2005) at approximately 19°N, leading to a strong eastward HLCC. Figure 9 shows a comparison of profiles of minimum PV gradients between 18°N and 21°N around the latitude of the HLCC. The negative PV gradients in both 2003 and 2005 were larger in the density range of 24.3–25.5 σ_θ compared to the mean values from 2001 to 2008. These results confirm that large negative PV anomalies north of the HLCC accelerate the HLCC in 2003 and 2005.

We compared the time series of the vertical integration of the meridional PV gradient from the 24.5 σ_θ to 25.5 σ_θ surfaces and the meridional slope of the 24.2 σ_θ isopycnal (Fig. 10) in order to observe contributions of interannual variations of PV anomalies north of the HLCC on the HLCC variations in detail. The slope is linked to the HLCC speed by the thermal wind relationship and was steeper in 2003, 2005, and 2007 than in other years. Although the standard deviation of the 13-month running mean of the pycnocline slope (1.54×10^{-5} m/m) is about double that of the PV gradient integration (0.86×10^{-5} m/m), the high correlation between the time series suggests that the PV gradient contributed to the isopycnal slope variations. However, the contribution seems to have been small in 2007. The strong trade winds in that year (Fig. 3) may have caused the isopycnal to slope upward locally along the HLCC through the wind-driven mechanism. The results suggest that the HLCC is mainly enhanced by low PV water intrusions to its north, but the wind-driven

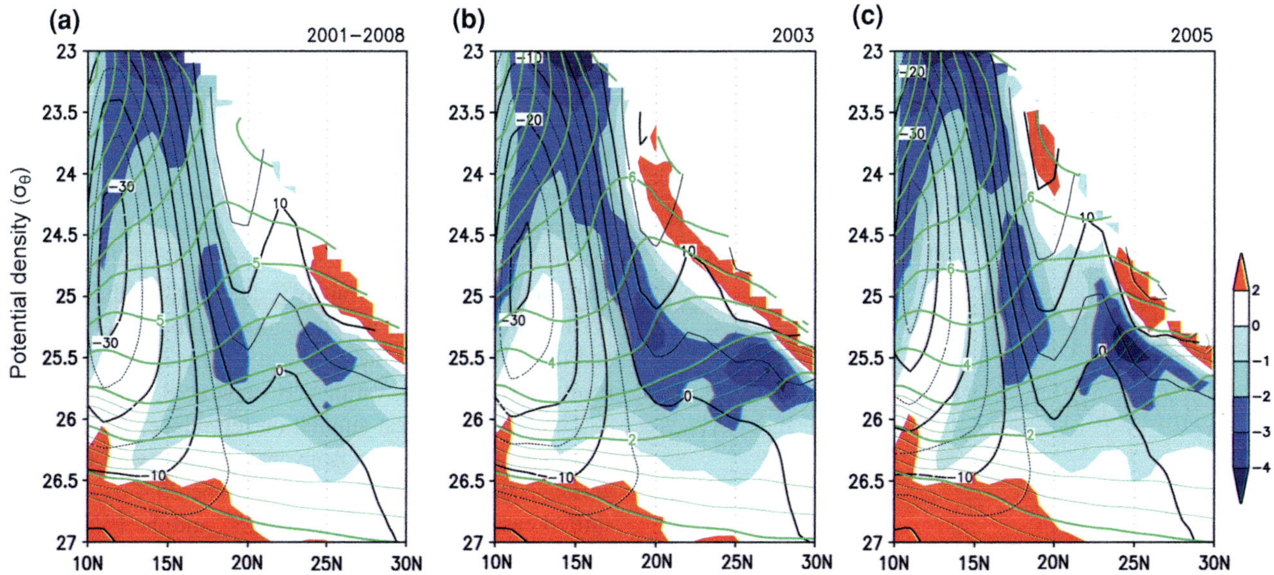

Fig. 8 Meridional gradients of PV $((f/q)(dq/dy)_\rho)$ (*color* unit is 10^{-11} m^{-1} s^{-1}) and of density $(d\rho/dy)$ (*black contour*, unit is 10^{-7} kg m^{-4}). *Green contours* show PV (10^{-10} s^{-1} m^{-1}) averaged between 180°W and 165°W based on the MOAA GPV. **a** Mean averaged from 2001 to 2008 and annual mean in **b** 2003 and **c** 2005. Intervals of *black* and *green contours* are 5×10^{-7} kg m^{-4} and 10^{-10} s^{-1} m^{-1}, respectively (color figure online)

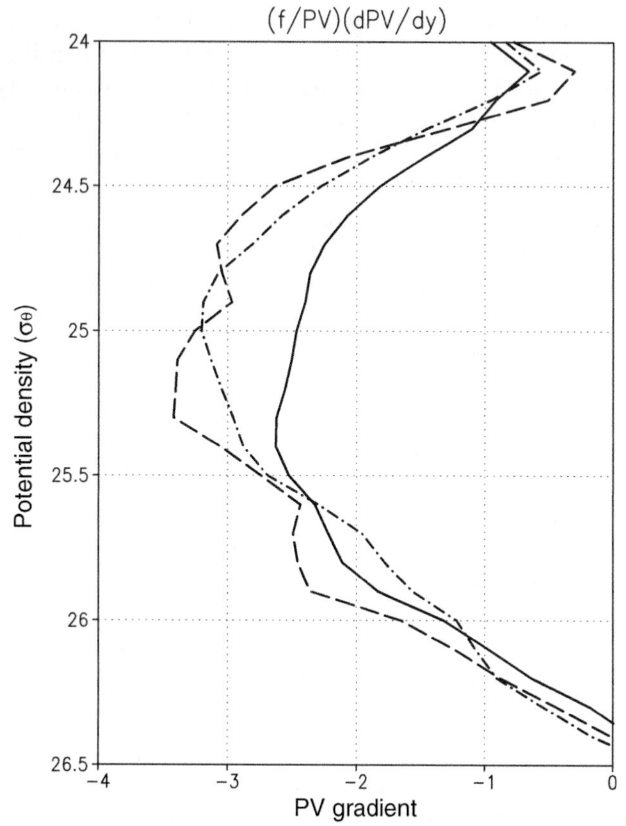

Fig. 9 Minimum meridional PV gradients $((f/q)(dq/dy)_\rho)$ (unit is 10^{-11} m^{-1} s^{-1}) between 18°N and 21°N averaged between 180°W and 165°W based on the MOAA GPV. Mean averaged from 2001 to 2008 (*solid line*) and annual means in 2003 (*dashed line*) and 2005 (*dashed–dotted line*)

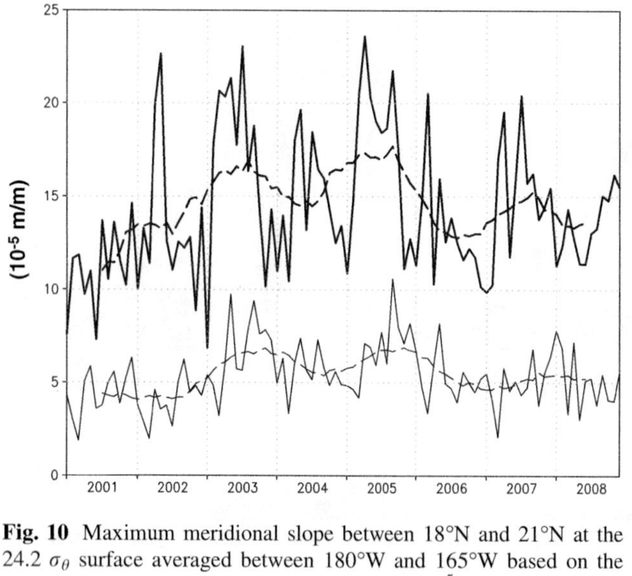

Fig. 10 Maximum meridional slope between 18°N and 21°N at the 24.2 σ_θ surface averaged between 180°W and 165°W based on the monthly MOAA GPV (*thick curve*, unit is 10^{-5} m/m). *Thin curve* shows the integral of meridional PV gradients $(f/q^2)(dq/dy)_\rho)/\rho_0$ from the 24.5 σ_θ to 25.5 σ_θ surface (10^{-5} m/m). *Dash curve* indicates 13-month running mean

mechanism associated with trade wind variations may also influence interannual variations of the HLCC.

4 Mechanism of interannual HLCC variations in the OGCM

4.1 Interannual variations in the hindcast simulation

Sasaki et al. (2010) described interannual variations of HLCC speed based on the OFES hindcast simulation driven by daily mean QuikSCAT wind stress. Figure 2e shows that the HLCC away from Hawaii was relatively strong from mid 2001 to late 2006, with two peaks in 2003 and 2005 in the hindcast simulation as seen in Sasaki et al. (2010). Near Hawaii, the simulated HLCC was strong not only in 2003 and 2005 but also in 2001 and 2007 (Fig. 2f). The accelerations in 2001 and 2007 and the deceleration in 2004 seem to be due to the wind-driven mechanism, as seen in Sect. 3.

There are some discrepancies between the interannual variations of the HLCC seen in the hindcast simulation and those in satellite observations. The durations of the HLCC accelerations in the simulation (Fig. 2e) are different by several months from those seen in the observational data (Fig. 2a), although the periods with the peaks of HLCC speed in 2003 and 2005 almost overlap with those in the observations. The HLCC acceleration in 2005 was relatively weak in the simulation (Fig. 2e), whereas near Hawaii, the HLCC accelerations in 2003 and 2007 were stronger (Fig. 2f) than those in the observations (Fig. 2b). Several factors contribute to these disagreements. The wind forcing was switched from NCEP/NCAR reanalysis to QuikSCAT satellite observations in 1999, and it takes several years to adjust to this wind forcing change. Local eddy–mean flow interactions may also play a role (e.g., Kobashi and Kawamura 2001). Regardless of these deficiencies, it is worth observing how low PV water induced interannual variations of the HLCC from 2001 to 2008 in the simulation that provides an oceanic field dynamically consistent with the primitive equations of the OGCM, because the Argo grid data are smoothed by the OI method, and the error ratio of the grid data is large before 2002 as a result of sparse observations (Hosoda et al. 2008).

The OFES hindcast simulation reproduces the observed mean PV distribution on isopycnal surfaces well (Fig. 11), although the center of the STMW at the 25.4 σ_θ surface is shifted eastward by approximately 5°, and the PVs become too large eastward in the eastern Pacific at both the 25.0 σ_θ and the 25.4 σ_θ surfaces compared with the Argo observations (Fig. 5). Figure 12 shows that negative PV anomalies north of the HLCC at the 25.0 σ_θ (25.4 σ_θ) surface in 2003 (2005) spread eastward across Hawaii (between

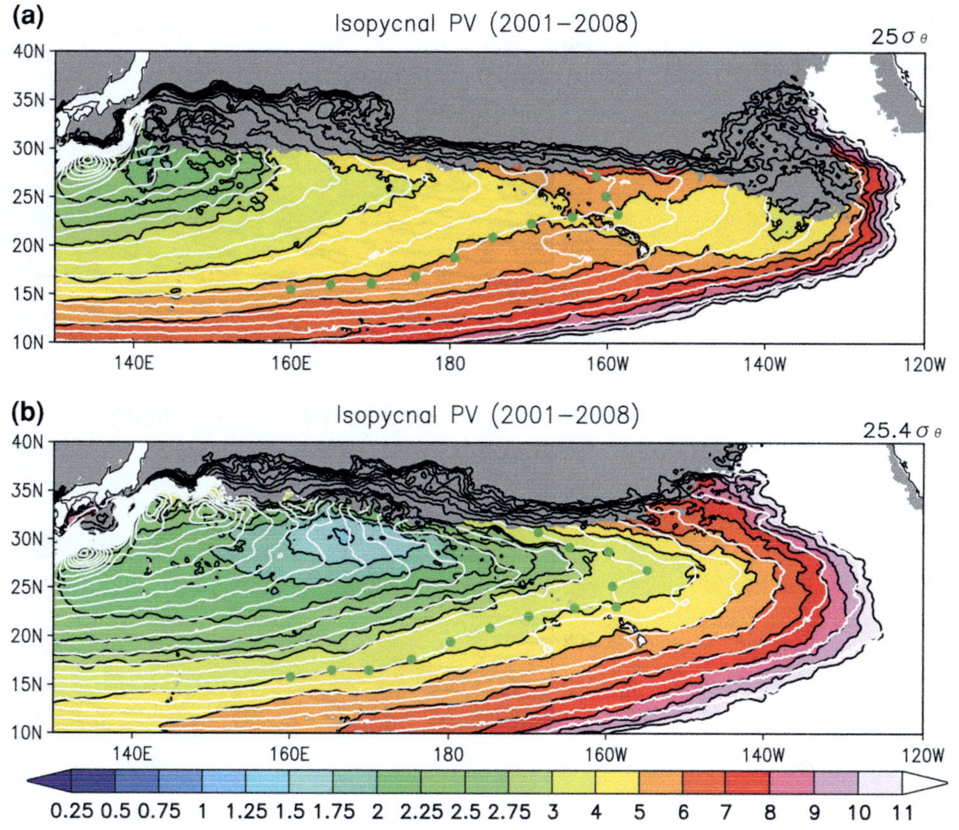

Fig. 11 Same as Fig. 5, but based on the OFES hindcast simulation

170°E and Hawaii). Both negative PV anomalies can be traced back to just south of the winter outcrop line along the mean acceleration potential isolines (Fig. 13a, b). The paths of the PV anomalies seem clearer than those in Argo observations (Fig. 7), probably because of the scarce observations and the smoothing of the dataset by the OI method. These results suggest influences of ESTMW (STMW) variations in 2003 (2005) on the HLCC accelerations, similar to what is seen in Argo observations.

In 2004, the negative PV anomalies between 180°W and 170°W were small, and the meridional gradient anomalies south of the acceleration potential isoline were positive at the 25.0 σ_θ surface (Fig. 13a). The weak negative PV anomalies seem to have been caused by weakening of the pycnocline slope localized along the HLCC that is probably due to the wind-driven mechanism through Rossby waves associated with anomalously weak trade winds in 2004 (Fig. 3). In 2006, the negative PV anomalies were large between 180°W and 160°W at the 25.4 σ_θ surface (Fig. 13b), but the meridional PV gradient anomalies were mostly positive because of negative PV anomalies spreading to the south of the HLCC (not shown). In addition to low PV intrusions north of the HLCC, the wind-driven mechanism and the low PV south of the HLCC may influence interannual variations of the HLCC in the simulation, as evaluated using the observations presented in Sect. 3.

4.2 Sensitivity simulation without interannual wind variations around Hawaii

A sensitivity experiment that does not include interannual wind variations around Hawaii was conducted to determine whether the wind-driven mechanism or other mechanisms, including low PV water intrusions, cause the interannual variations of the HLCC. If the HLCC variations away from Hawaii were similar between the sensitivity and hindcast simulations, the wind-driven mechanism could be excluded as a mechanism causing HLCC variations. The sensitivity simulation (Fig. 2g) captures the HLCC acceleration away from Hawaii in 2003 similarly to the hindcast simulation (Fig. 2e), suggesting that the HLCC acceleration was not caused by the wind-driven mechanism. At the 25.0 σ_θ surface, negative PV anomalies between 175°W and 160°W, which migrated from east-northeast in 2003 (Fig. 13c), seem to have caused negative meridional PV gradients and pushed up the pycnocline above the PV anomalies, accelerating the HLCC.

However, the acceleration of the HLCC in 2005 was less clear in the sensitivity simulation (Fig. 2g). The negative PV anomalies migrated along the acceleration potential isoline at the 25.4 σ_θ surface (Fig. 13d), but they were weaker between 180°W and 160°W in 2005 than those in the hindcast simulation (Fig. 13b). The weak negative PV anomalies did not result in large meridional PV gradient

Fig. 12 Same as Fig. 6, but based on the OFES hindcast simulation. Note that the color scale is different from that in Fig. 6

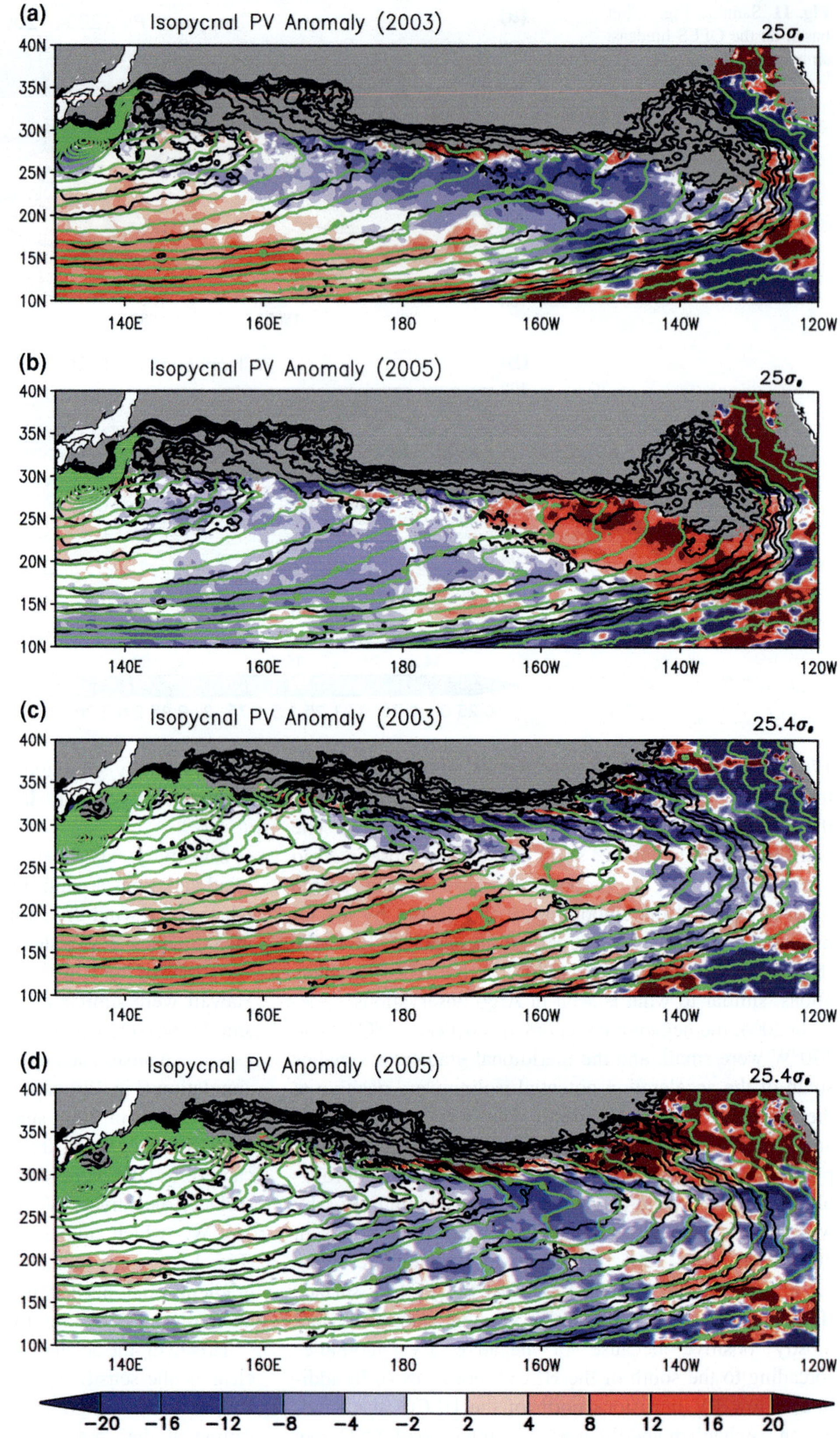

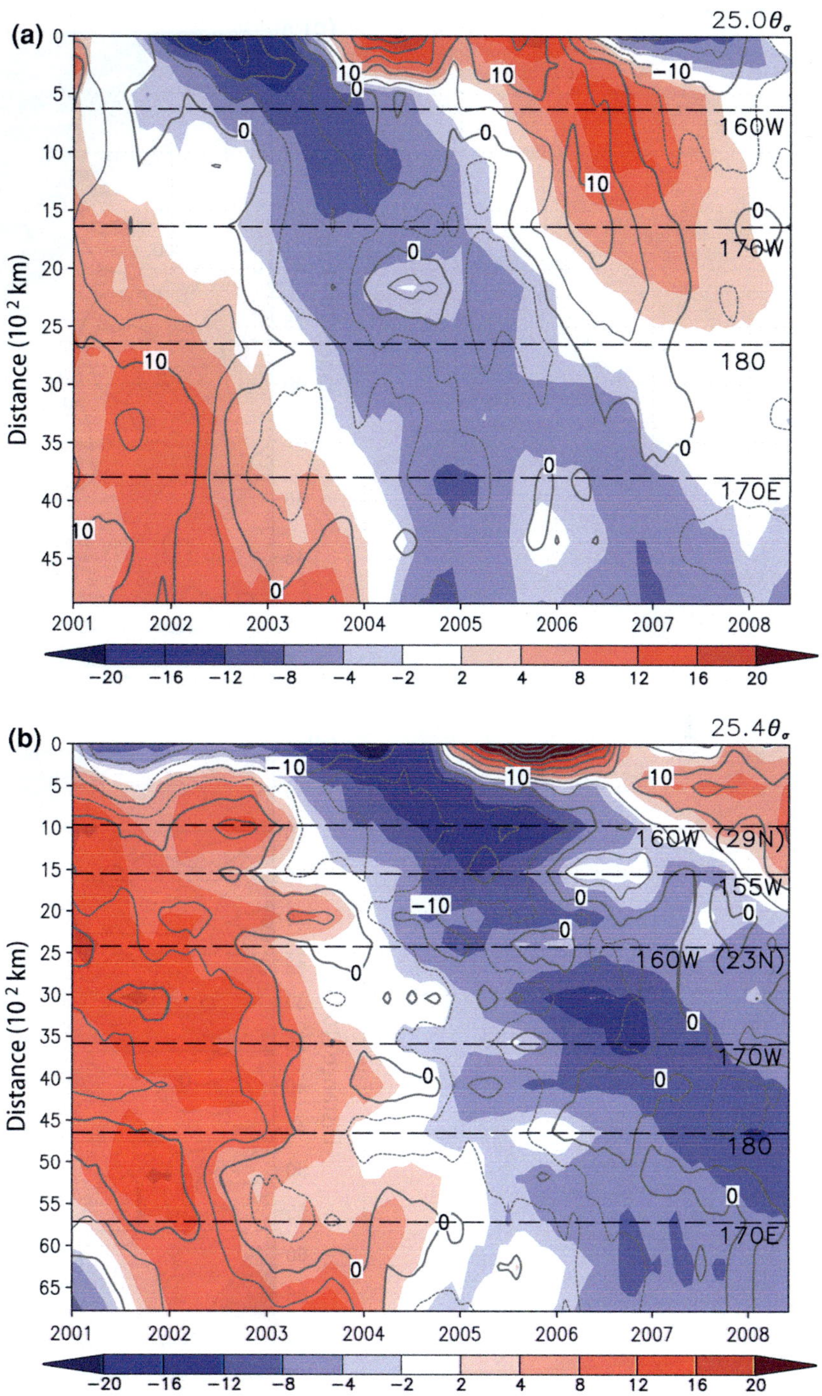

Fig. 13 **a**, **b** Same as Fig. 7, but based on the OFES hindcast simulation. *Vertical axis* indicates distance along the acceleration potential with *green dots* starting from the outcrop line (Fig. 11). **c**, **d** Based on OFES sensitivity simulation. Note that the color scale is different from that in Fig. 7 and the contour interval is 5×10^{-10} m^{-1} s^{-1} (color figure online)

along the HLCC (Fig. 13d). The spreading of low PV water to the south of the HLCC in that year also seems to have weakened the negative PV gradient (not shown).

The interannual variations of the HLCC near Hawaii are weakened in the sensitivity simulation but do not disappear completely (Fig. 2h). The remaining HLCC variations could arise from mechanisms other than wind-driven mechanism, including low PV water intrusions and wind variability east of Hawaii via Rossby wave propagations. In contrast, the acceleration (deceleration) in 2007 (2004) does not occur in the sensitivity simulation, supporting that these variations are induced by a wind curl forcing around Hawaii. The acceleration in 2001, which is also suggested to result from a wind-driven mechanism as well as in 2007, still remains, but weakens compared to before and after the year. These results suggest that interannual variations of the HLCC near Hawaii are partially induced by the wind-driven mechanism.

Fig. 13 continued

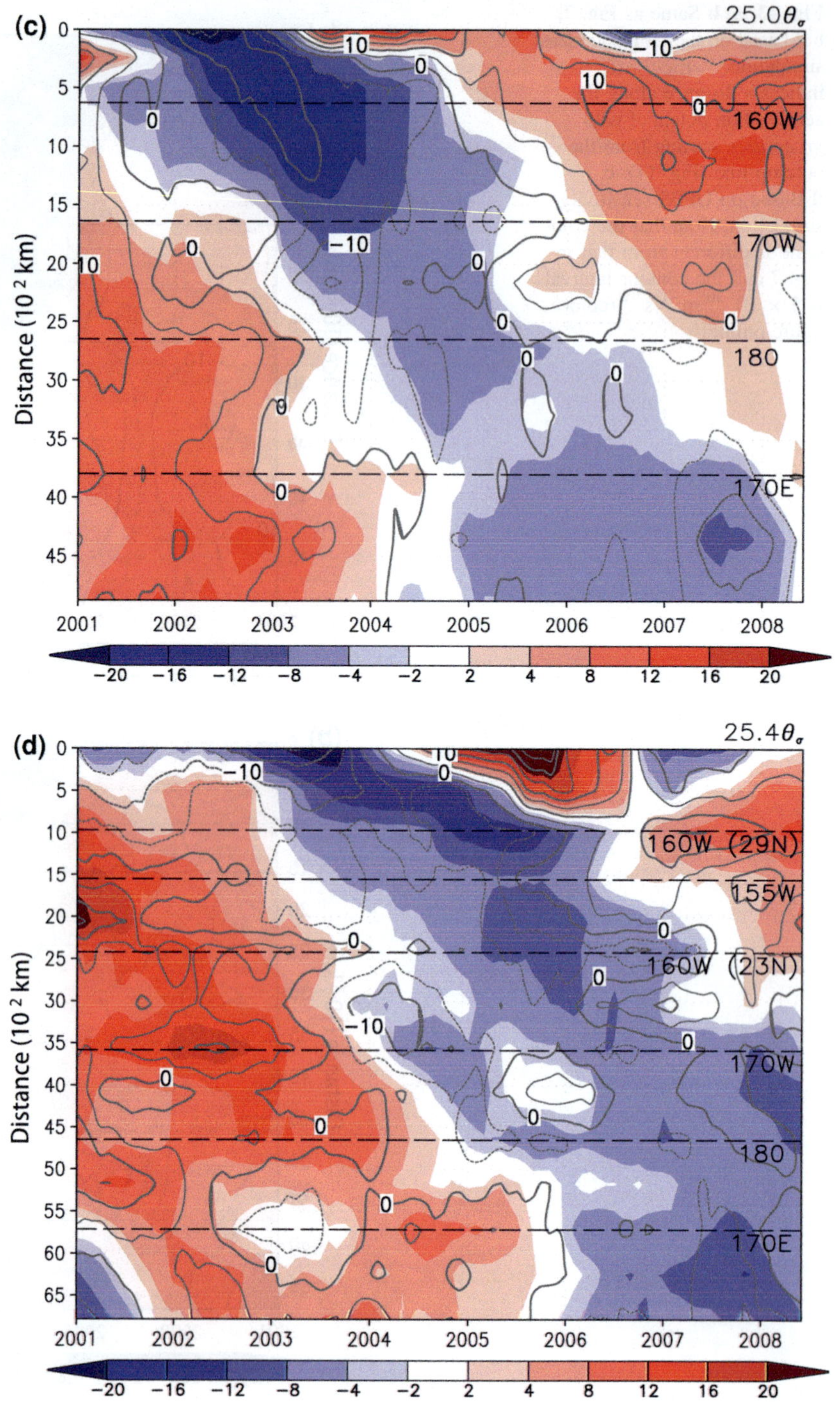

5 Discussion

This section discusses the formation mechanism of the low PV anomalies that influence the interannual variations of the HLCC. Our analyses suggest that variations of the ESTMW (STMW) could contribute to low PV water intrusions north of the HLCC in 2003 (2005). In the Argo observations (Fig. 7a) and the OFES hindcast simulation (Fig. 13a), negative PV anomalies at the 25.0 σ_θ surface north of the HLCC accelerating the current in 2003 can be traced back to the western edge of the ESTMW in 2002. The anomalously deep mixed layer in the western portion of the ESTMW in 2002 (not shown) could enhance ESTMW formation. Toyoda et al. (2011) studied the interannual variability of ESTMW formation in the 1990s and suggested that winter atmospheric cooling, Ekman convergence of salty water, and insolation change due to low-level cloud cover are important in this process. Detailed analyses of ESTMW interannual variations in the 2000s are beyond the scope of our study.

In the hindcast simulation, the negative PV anomalies at the 25.4 σ_θ surface north of the HLCC in 2005 can be traced back to the northeastern edge of the STMW in 2003 (Fig. 13b). The HLCC acceleration in 2005 was relatively weak in the simulation compared with the Argo observations (Fig. 2). However, it is worth discussing the interannual variations of STMW formation in the simulated oceanic field, since a dynamical interpretation is available using outputs from the OGCM. The simulated mixed layer in the STMW formation region was anomalously deep in 2003 in the winter (Fig. 14), consistent with historical observations (Qiu and Chen 2006; Sugimoto and Hanawa 2010). Previous studies have suggested that wind variations (Suga and Hanawa 1995), stability of the Kuroshio Extension path (Qiu and Chen 2006), and the meridional migration of the Aleutian Low (Sugimoto and Hanawa 2010) lead to interannual variations in mixed layer depth and STMW formation. We found anomalously strong westerly wind stress in the region of the anomalously deep mixed layer in 2003 in the winter (Fig. 14). The northwest monsoon was anomalously strong during this winter, probably because of a strong Aleutian Low, based on the North Pacific index (NPI; Trenberth and Hurrell 1994). It appears that strong westerly winds cause a deeper mixed layer via strong surface cooling that intensifies STMW formation. The subsequent southward intrusion of low PV water from the northeastern edge of the STMW enhanced the HLCC in 2005 in the hindcast simulation. This mechanism is consistent with findings from previous studies on decadal changes in the STCC by Yamanaka et al. (2008) using an OGCM and Xie et al. (2011) using a coupled GCM.

In the ocean hindcast simulation forced by observed wind variability, mode waters display considerable decadal variability, especially in OGCMs of moderate resolution ($\sim 1°$) (Xie et al. 2000; Hosoda et al. 2004). In an eddy-resolving simulation, such mode water variability has been found to induce significant changes in STCC (Yamanaka et al. 2008). Nonaka et al. (2011, this issue) examined this mode water effect on the interannual variability of STCC in a hindcast simulation using the OFES forced by NCEP/NCAR reanalysis. In coupled ocean–atmosphere simulations, mode water changes are the dominant mechanism for STCC variability, with obvious effects on the atmosphere owing to thermal advection by anomalous surface currents (Xie et al. 2011). Accumulation of Argo observations will enable critical examination of this mode water-induced mechanism for surface current variability, including its relationship to STCCs and the HLCC.

It appears that low PV water is important for interannual variations of the HLCC in the 2000s. Other mechanisms, including the wind-driven mechanism, might cause HLCC acceleration on interannual to decadal timescales. The Pacific decadal oscillation (PDO; Mantua et al. 1997) could induce HLCC variation. Variations of mixed layer depth (Carton et al. 2008) and low PV water subduction (Qu and Chen 2009; Chen et al. 2010) associated with the PDO may influence the HLCC via variations of pycnocline slope. Mean-flow acceleration through eddies (Kobashi and Kawamura 2001; Liu and Li 2007) may also play a role. As the present study covers a short period of time, it does not rule out a role for the wind-driven mechanism in HLCC variability, which is considerable near Hawaii. In the OFES hindcast simulation forced by the NCEP/NCAR reanalysis for 1950–2008, the wind-driven mechanism is dominant in the interannual and decadal variability in the HLCC (not shown). However, the simulation does not capture the HLCC accelerations in 2003 and 2005 (not shown). The orographic wind stress

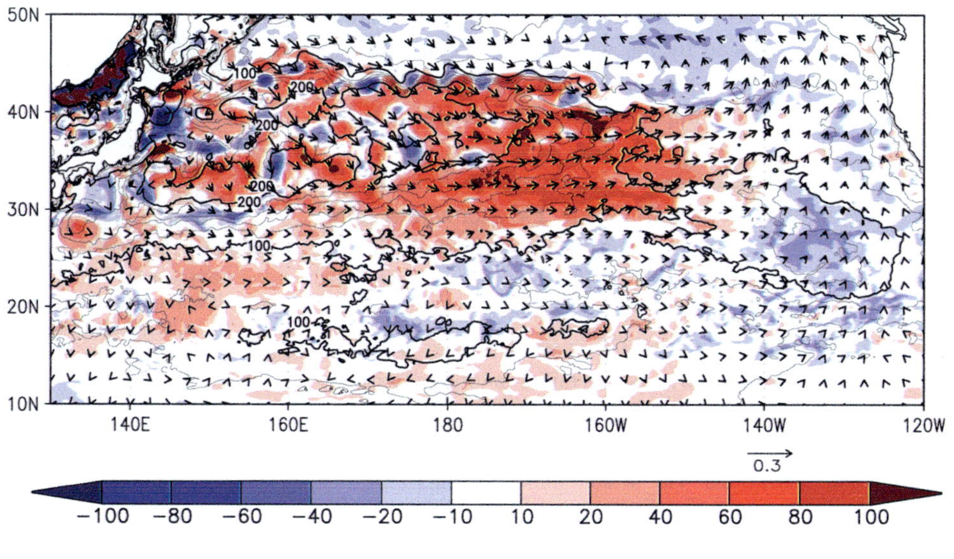

Fig. 14 Winter (from January to March) mixed layer depth anomaly (*color*, unit is m) in 2003 based on the OFES hindcast simulation. *Contours* indicate the mean winter mixed layer depth averaged from 2001 to 2008. The *thick* (*thin*) contour intervals are 100 m (20 m). *Vectors* represent anomalies of winter wind stress (N m^{-2}). The mixed layer depth is defined as the depth at which the potential density differs from the sea surface density by 0.125 σ_θ (color figure online)

curl dipole west of Hawaii in the reanalysis is too strong, and its distributions are too broad.

It is possible that air–sea interactions accompanied by the HLCC with high SST (Xie et al. 2001) play a role in interannual variations of the HLCC. In 2003 and 2005, an elongated positive wind stress curl extended westward from Hawaii at approximately 19°N, tilting slightly southward in a more distinct manner than in other years based on QuikSCAT observations (not shown). In both years, the HLCC based on the AVISO observations was strong, and its position almost overlapped that of the positive wind stress curl, suggesting that the distinct curl away from Hawaii could further enhance the strong HLCC. These results suggest that acceleration of the HLCC due to local air–sea interactions (Sasaki and Nonaka 2006) could further amplify the interannual variations of the HLCC.

6 Conclusions

We investigated the interannual variations of the HLCC in the 2000s using satellite and Argo observations and high-resolution OGCM simulations. The current speed was high in 2003 and 2005, when the trade winds and the orographic wind stress curl dipole in the lee of Hawaii were not particularly strong. Furthermore, the HLCC acceleration took place more strongly away from Hawaii. The results suggest that the wind-driven mechanism is not the leading factor causing the HLCC accelerations in these 2 years. Subsurface PV variations appear to be important in these accelerations. We found that negative PV anomalies in subsurface layers, which were advected to the north of the HLCC, pushed the upper pycnocline upward and enhanced the eastward HLCC through the thermal wind relation, following the mechanism proposed by Kubokawa (1999). The OFES sensitivity simulation forced by wind stress that does not include the interannual variability near Hawaii confirms that the HLCC acceleration away from Hawaii in 2003 was not induced by the wind-driven mechanism. In addition to low PV anomalies, the wind-driven mechanism is also suggested to be important in variations of the HLCC near Hawaii.

We examined the causes of interannual variations in low PV water intrusions. Subsurface low PV anomalies can be traced back to the winter outcrop line, especially in the simulations. ESTMW and STMW intrusions appear to have driven the HLCC accelerations in 2003 and 2005, respectively, although the HLCC is located away from the cores of these mode waters. Detailed analyses will be needed as Argo observations accumulate in the future. The atmospheric variations responsible for HLCC interannual to decadal variations via water mass subduction also remain to be investigated.

Acknowledgments The OFES simulations were conducted on the Earth Simulator under support from JAMSTEC. We thank Drs. B. Qiu and F. Kobashi for valuable discussions and the reviewers for valuable comments. We thank T. Ohira for his help in constructing the MOAA GPV product. QuikSCAT wind stress data in the J-OFURO dataset were provided by Dr. K. Kutsuwada. This study was supported in part through Grant-In-Aid for Scientific Research 22106006 and that on Innovative Areas #2205 by the Japanese Ministry of Education, Culture, Sports, Science and Technology, and also through 23340139 by the Japan Society for the Promotion of Science. IPRC publication # 809.

References

Aoki Y, Suga T, Hanawa K (2002) Subsurface subtropical fronts of the North Pacific as inherent boundaries in the ventilated thermocline. J Phys Oceanogr 32:2299–2311

Bingham F (1992) The formation and spreading of subtropical mode water in the North Pacific. J Geophys Res 97:11177–11189

Boyer TP, Stephens C, Antonov JI, Conkright ME, Locarnini RA, O'Brien TD, Garcia HE (2002) World ocean atlas 2001, vol 2: salinity, NOAA atlas NESDIS 50. US Government Printing Office, Washington, DC

Carton L, Grodsky SA, Liu H (2008) Variability of the oceanic mixed layer, 1960–2004. J Clim 21:1029–1047

Chen J, Qu T, Sasaki YN, Schneider N (2010) Anti-correlated variability in subduction rate of the western and eastern North Pacific Oceans identified by an eddy-resolving ocean GCM. Geophys Res Lett 37:L23608. doi:10.1029/2010GL045239

Flament P, Kennan S, Lumpkin R, Sawyer M, Stroup E (1998) The ocean. In: Juvik SP, Juvik JO (eds) Atlas of Hawaii. University of Hawaii Press, Honolulu, pp 82–86

Hafner J, Xie SP (2003) Far-field simulation of the Hawaiian wake: sea surface temperature and orographic effects. J Atmos Sci 60:3021–3032

Hanawa K, Talley LD (2001) Mode waters. In: Church J (ed) Ocean circulation and climate. Academic, London, pp 373–386

Hautala S, Roemmich D (1998) Subtropical mode water in the Northeast Pacific Basin. J Geophys Res 103(C6):13055–13066

Hosoda S, Xie S-P, Takeuchi K, Nonaka M (2004) Interdecadal temperature variations in the North Pacific central mode water simulated by an OGCM. J Oceanogr 60:865–877

Hosoda S, Ohira T, Nakamura T (2008) A monthly mean dataset of global oceanic temperature and salinity derived from Argo float observations. JAMSTEC Rep Res Dev 8:47–59

Kalnay E, Kanamitsu M, Kistler R, Collins W, Deaven D, Gandin L, Iredell M, Saha S, White G, Woollen J, Zhu Y, Chelliah M, Ebisuzaki W, Higgins W, Janowiak J, Mo KC, Ropelewski C, Leetmaa A, Reynolds R, Jenne R (1996) The NCEP/NCAR 40-year reanalysis project. Bull Am Meteorol Soc 77:437–471

Kobashi F, Kawamura H (2001) Variation of sea surface height at periods of 65–220 days in the subtropical gyre of the North Pacific. J Geophys Res 106:26817–26831

Kobashi F, Kawamura H (2002) Seasonal variation and instability nature of the North Pacific Subtropical Countercurrent and the Hawaiian Lee Countercurrent. J Geophys Res 107:3185. doi: 10.1029/2001JC001225

Kobashi F, Mitsudera H, Xie S-P (2006) Three subtropical fronts in the North Pacific: observational evidence for mode water-induced subsurface frontogenesis. J Geophys Res 111:C09033. doi:10.1029/2006JC003479

Kubokawa A (1999) Ventilated thermocline strongly affected by a deep mixed layer: a theory for subtropical countercurrent. J Phys Oceanogr 29:1314–1333

Kubokawa A, Inui T (1999) Subtropical countercurrent in an idealized ocean GCM. J Phys Oceanogr 29:1303–1313

Kubota M, Iwasaka N, Kizu S, Konda M, Kutsuwada K (2002) Japanese ocean flux data sets with use of remote sensing observations (J-OFURO). J Oceanogr 58:213–225

Kutsuwada K (1998) Impact of wind/wind-stress field in the North Pacific constructed by ADEOS/NSCAT data. J Oceanogr 54:443–456

Liu Q, Li L (2007) Baroclinic stability of oceanic Rossby wave in the North Pacific Subtropical Countercurrent. Chin J Geophys 50(1):84–93

Mantua NJ, Hare SR, Zhang Y, Wallace JM, Francis RC (1997) A Pacific interdecadal climate oscillation with impacts on salmon production. Bull Am Meteorol Soc 78:1069–1079

Masumoto Y, Sasaki H, Kagimoto T, Komori N, Ishida A, Sasai Y, Miyama T, Motoi T, Mitsudera H, Takahashi K, Sakuma H, Yamagata T (2004) A fifty-year eddy-resolving simulation of the world ocean: preliminary outcomes of OFES (OGCM for the Earth Simulator). J Earth Simulator 1:35–56

Montgomery RB, Stroup ED (1962) Equatorial waters and currents at 150°W in July–August 1952. Johns Hopkins Oceanographic Studies, no 1. Johns Hopkins University Press, Baltimore

Nonaka M, Xie S-P, Sasaki H (2011) Interannual variations in low potential vorticity water and the subtropical countercurrent in an eddy-resolving OGCM. J Oceanogr. doi:10.1007/s10872-011-0042-3

Oka E (2009) Seasonal and interannual variation of North Pacific Subtropical Mode Water in 2003–2006. J Oceanogr 65:151–164. doi:10.1007/s10872-009-0015-y

Pacanowski RC, Griffies SM (1999) The MOM 3 manual, GFDL Ocean Group technical report no 4. NOAA/Geophysical Fluid Dynamics Laboratory, Princeton

Qiu B, Chen S (2006) Decadal variability in the formation of the North Pacific Subtropical Mode Water: oceanic versus atmospheric control. J Phys Oceanogr 36:1365–1380. doi:10.1175/JPO2918.1

Qiu B, Koh DA, Lumpkin C, Flament P (1997) Existence and formation mechanism of the North Hawaiian Ridge Current. J Phys Oceanogr 27:431–444

Qu T, Chen J (2009) A North Pacific decadal variability in subduction rate. Geophys Res Lett 36:L22602. doi:10.1029/2009GL040914

Sakamoto TT, Sumi A, Emori S, Nishimura T, Hasumi H, Suzuki T, Kimoto M (2004) Far-reaching effects of the Hawaiian Islands in the CCSR/NIES/FRCGC high-resolution climate model. Geophys Res Lett 31:L17212. doi:10.1029/2004GL020907

Sasaki H, Nonaka M (2006) Far-reaching Hawaiian Lee Countercurrent driven by wind-stress curl induced by warm SST band along the current. Geophys Res Lett 33:L13602. doi:10.1029/2006GL026540

Sasaki H, Sasai Y, Nonaka M, Masumoto Y, Kawahara S (2006) An eddy-resolving simulation of the quasi-global ocean driven by satellite-observed wind field: preliminary outcomes from physical and biological fields. J Earth Simulator 6:35–49

Sasaki H, Nonaka M, Masumoto Y, Sasai Y, Uehara H, Sakuma H (2008) An eddy-resolving hindcast simulation of the quasi-global ocean from 1950 to 2003 on the Earth Simulator. In: Ohfuchi W, Hamilton K (eds) High resolution numerical modelling of the atmosphere and ocean. Springer, New York, pp 157–186

Sasaki H, Xie S-P, Taguchi B, Nonaka M, Masumoto Y (2010) Seasonal variations of the Hawaiian Lee Countercurrent induced by the meridional migration of the trade winds. Ocean Dyn 60(3):705–715

Smith RB, Grubišic V (1993) Aerial observation of Hawaii's wake. J Atmos Sci 50:3728–3750

Stephens C, Antonov JI, Boyer TP, Conkright ME, Locarnini RA, O'Brien TD, Garcia HE (2002) World ocean atlas 2001, vol 1: temperature, NOAA atlas NESDIS 49. US Government Printing Office, Washington, DC

Suga T, Hanawa K (1995) Interannual variations of North Pacific Subtropical Mode Water in the 137E section. J Phys Oceanogr 25:1012–1017

Sugimoto S, Hanawa K (2010) Impact of Aleutian Low activity on the STMW formation in the Kuroshio recirculation gyre region. Geophys Res Lett 37:L03606. doi:10.1029/2009GL041795

Toyoda T, Awaji T, Masuda S, Sugiura N, Igarashi H, Mochizuki T, Ishikawa Y (2011) Interannual variability of North Pacific eastern subtropical mode water formation in the 1990s derived from a 4-dimensional variational ocean data assimilation experiment. Dyn Atmos Oceans 51(1–2):1–25

Trenberth KE, Hurrell JW (1994) Decadal atmosphere–ocean variations in the Pacific. Clim Dyn 9:303–319

Uda M, Hasunuma K (1969) The eastward subtropical countercurrent in the western North Pacific Ocean. J Oceanogr Soc Jpn 25:201–210

Xie S-P, Kunitani T, Kubokawa A, Nonaka M, Hosoda S (2000) Interdecadal thermocline variability in the North Pacific for 1958–1997: a GCM simulation. J Phys Oceanogr 30:2798–2813

Xie SP, Liu WT, Liu Q, Nonaka M (2001) Far-reaching effects of the Hawaiian Islands on the Pacific Ocean–atmosphere system. Science 292:2057–2060

Xie S-P, Xu L-X, Liu Q, Kobashi F (2011) Dynamical role of mode-water ventilation in decadal variability in the central subtropical gyre of the North Pacific. J Clim 24:1212–1225

Yamanaka G, Ishizaki H, Hirabara M, Ishikawa I (2008) Decadal variability of the subtropical front of the western North Pacific in an eddy-resolving ocean general circulation model. J Geophys Res 113:C12027. doi:10.1029/2008JC005002

Yang Y, Xie S-P, Hafner J (2008a) Cloud patterns lee of Hawaii Island: a synthesis of satellite observations and numerical simulation. J Geophys Res 113:D15126. doi:10.1029/2008JD009889

Yang Y, Ma J, Xie S-P (2008b) Observations of the trade wind wakes of Kauai and Oahu. Geophys Res Lett 35:L04807. doi:10.1029/2007GL031742

Yoshida S, Qiu B, Hacker P (2010) Wind-generated eddy characteristics in the lee of the island of Hawaii. J Geophys Res 115:C03019. doi:10.1029/2009JC005417

Yu Z, Maximenko N, Xie S-P, Nonaka M (2003) On the termination of the Hawaiian Lee Countercurrent. Geophys Res Lett 30:1215. doi:10.1029/2002GL016710

J Oceanogr (2012) 68:139–150
DOI 10.1007/s10872-011-0042-3

SPECIAL SECTION: SHORT CONTRIBUTION

New developments in mode-water research: Dynamic and climatic effects

Interannual variations in low potential vorticity water and the subtropical countercurrent in an eddy-resolving OGCM

Masami Nonaka · Shang-Ping Xie · Hideharu Sasaki

Received: 9 February 2011 / Revised: 27 April 2011 / Accepted: 6 May 2011 / Published online: 9 June 2011

Abstract Interannual-to-decadal variations in the subtropical countercurrent (STCC) and low potential vorticity (PV) water and their relations in the North Pacific Ocean are investigated on the basis of a 60-year-long hindcast integration of an eddy-resolving ocean general circulation model. Although vertically coherent variations are dominant for STCC interannual variability, a correlation analysis shows that an intensified STCC vertical shear accompanies lower PV than usual to the north on 25.5- to 26.1-σ_θ isopycnal surfaces, and intensified meridional density gradient in subsurface layers, consistent with Kubokawa's theory (J Phys Oceanogr 29:1314–1333, 1999). The low-PV signals appear at least 2 years before peaks of STCC, propagating southwestward from the subduction region.

Keywords Subtropical countercurrent · Low potential vorticity water · North Pacific · Interannual variability · Eddy-resolving ocean general circulation model

M. Nonaka (✉)
Research Institute for Global Change, Japan Agency for Marine-Earth Science and Technology (JAMSTEC), 3173-25 Showa-machi, Kanazawa-ku, Yokohama 236-0001, Japan
e-mail: nona@jamstec.go.jp

S.-P. Xie
International Pacific Research Center (IPRC) and Department of Meteorology, School of Ocean and Earth Science and Technology (SOEST), University of Hawaii at Manoa, Honolulu, HI, USA

H. Sasaki
Earth Simulator Center, JAMSTEC, Yokohama, Japan

1 Introduction

In the late 1960s, a weak eastward current embedded in the broad westward current in the southern part of the North Pacific subtropical gyre was discovered (Uda and Hasunuma 1969). As its direction is opposite to that expected from the gyre circulation, it is named the subtropical countercurrent (STCC). Although its whole distribution had not been clarified for a long time because of high eddy activity in the region (Qiu 1999), the accumulation of observational data recently allowed its detailed distributions to be described (Kobashi and Kawamura 2001, 2002; Kobashi et al. 2006). Although STCC is not a strong current, the associated temperature front, the subtropical front (STF), can influence atmospheric fields aloft (Kobashi et al. 2008; Kobashi and Xie 2011). Coupled model results suggest positive coupled feedback associated with the atmospheric response (Xie et al. 2011). In coupled model projections, changes in STCC under global warming can significantly affect distributions of sea surface temperature (SST) changes in the North Pacific (Xie et al. 2010).

Formation mechanisms for STCC have been debated since its discovery. Yoshida and Kidokoro (1967) and Roden (1975) discuss the possible importance of detailed distributions of zonal wind stress, and meridional Ekman convergence, respectively. On the basis of ocean general circulation model (OGCM) experiments, Takeuchi (1984), however, shows that STCC can be formed without narrow scale structures in zonal wind and meridional Ekman convergence. In a theoretical study, Kubokawa (1999) proposes that STCC is associated with meridional density gradient caused by the accumulation of low potential vorticity (PV), thick water layers subducted to the north. A numerical simulation supports this hypothesis (Kubokawa and Inui 1999). Specifically, the mode water pushes the upper pycnocline upward. On the southern flank of the mode water, this creates a northward shoaling of the upper pycnocline, which sustains an eastward current shear by thermal wind. Observational data also suggest that relative distributions between STCC and low-PV water are consistent with Kubokawa's (1999) hypothesis (Aoki et al. 2002). Furthermore, from detailed investigation of historical observed data, Kobashi et al. (2006) show that the northern and eastern branches of STCC locate at the southern edge of the low-PV regions corresponding to the North Pacific Subtropical Mode Water (NPSTMW, Masuzawa 1969; Hanawa and Talley 2001), and to the North Pacific Central Mode Water (NPCMW, Nakamura 1996; Suga et al. 1997; Oka and Suga 2005), respectively.

Although these observed climatological fields strongly support Kubokawa's (1999) hypothesis, it is still not clear if interannual-to-decadal variations in STCC are caused by variations of low-PV waters. On decadal timescales, in a 300-year simulation of a coupled model, Xie et al. (2011) show that STCC variations are caused by changes in low-PV water subduction. Although the resolution of the coupled model is limited and low-PV water tends to be exaggerated in the model, on the basis on an eddy-resolving OGCM simulation Yamanaka et al. (2008) show that decadal differences in STCC are associated with those in low-PV waters, consistent with Kubokawa's hypothesis. On interannual timescales, Qiu and Chen (2010) and Kobashi and Xie (2011) indicate the importance of Ekman convergence for STF and STCC, but the possibility of contribution of low-PV water variations has not been explored. The present study investigates if variations in low-PV water subduction can affect STCC on interannual timescales, based on an eddy-resolving OGCM simulation that is different from Yamanaka et al.'s.

In Sect. 2, we introduce the model and describe simulated low-PV waters. Simulated STCC is described in Sect. 3, and the relationship between interannual-to-decadal variations in STCC and low-PV waters is investigated in Sect. 4. Section 5 gives a summary and discussion.

2 Model

2.1 OFES

We used the Modular Ocean Model 3 (MOM3) OGCM (Pacanowski and Griffies, 2000) with substantial modification for the vector-parallel hardware system of Japan's Earth Simulator. Our ocean model for the Earth Simulator (OFES; Masumoto et al. 2004) covers a near-global domain of 75°N–75°S, with a horizontal resolution of 0.1°. The model has 54 vertical levels with resolutions of 5 m at the surface, and the maximum depth is 6065 m. For horizontal mixing of momentum and tracers, we adopted scale-selective damping with a bi-harmonic operator (Smith et al. 2000). The nonlocal K-profile parameterization (KPP) boundary layer mixing scheme (Large et al. 1994) was adopted for the vertical mixing.

The surface heat flux and evaporation were calculated by the bulk formula with atmospheric variables based on National Centers for Environmental Prediction/National Center for Atmospheric Research (NCEP/NCAR) reanalysis data (Kalnay et al. 1996) and the simulated SST field. The freshwater flux was evaluated from the evaporation field and daily precipitation rate data, under the constraint that sea surface salinity (SSS) is restored to the observed monthly climatology at a timescale of 6 days. Wind stress is also obtained from the NCEP/NCAR reanalysis. For details, see Masumoto et al. (2004) and Sasaki et al. (2008).

Following a 50-year integration with climatological monthly-mean forcing from the annual-mean temperature and salinity climatological fields without motion, we conducted a 60-year hindcast integration with daily mean atmospheric fields of the NCEP/NCAR reanalysis data from 1950 to 2009. This hindcast simulation successfully captures variability with intraseasonal-to-decadal timescales (Sasaki et al. 2008) and has been used to investigate interannual-to-decadal variability in the western North Pacific region (Nonaka et al. 2006, 2008; Taguchi et al. 2007, 2010). In the following analyses, the resolution of our model output is reduced to 0.5° for analytical convenience by selecting the data at every five grid points both in the zonal and meridional directions.

2.2 Simulated low-PV water distributions

In the present study, we investigate the relationship between interannual variations of low-PV waters and STCC on the basis of the OFES hindcast integration. Then, first we describe simulated summertime (July–September) low-PV water distributions based on meridional and zonal sections of PV (Fig. 1). In the western part of the North Pacific basin, at 150°E, low-PV water corresponding to the NPSTMW is found around 25.2–25.6 σ_θ (Fig. 1d), and its distribution is well represented in the model (Fig. 1a). In the central part of the basin in the 170°E meridional transect, low-PV water with density around 25.5 σ_θ develops more strongly in the model (Fig. 1b) than in observations (Fig. 1e), although there is a PV minimum in the observation around 30°N. In the zonal section of PV at 28°N, it is confirmed that the simulated low-PV water on 26.0 σ_θ is displaced eastward compared to observations (Fig. 1c, f).

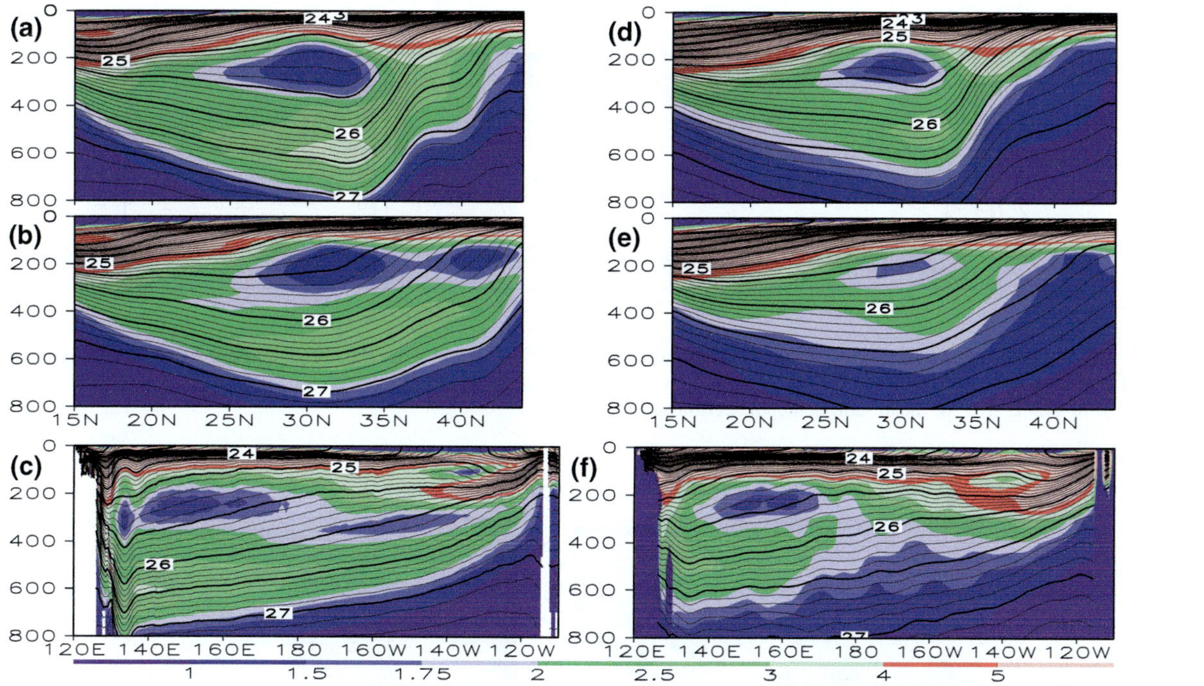

Fig. 1 Long-term mean summertime (July–September) potential vorticity (*shades* as indicated at the *bottom* of *panels* in 10^{-12} cm^{-1} s^{-1}) and density (*contours*, with intervals of 0.1 σ_θ) distributions in OFES (**a–c**) and observational data [**d–f**; *World Ocean Atlas 2001* (Stephens et al. 2002; Boyer et al. 2002)]. **a**, **d** Latitude-depth section of 145°–155°E mean. **b**, **e** The same as **a**, **d**, but for 165°–175°E mean. **c**, **f** Longitude-depth section of 27°–29°N mean

Also, in the central to the eastern part of the basin, the simulated low-PV water is isolated from the lower low-PV layer by relatively high-PV water of around 26.5 σ_θ. Although the simulated low-PV water develops more strongly than the counterpart in the observation in the central to eastern part of the basin, PV distributions are well represented to the west of the international dateline. On isopycnal surfaces (figures not shown but suggested in Fig. 5), it is confirmed that the simulated PV minimum exists about 10° longitude eastward of the observed one in layers of 25.6 σ_θ or higher density (e.g., Fig. 9 of Kobashi et al. 2006), although it is well represented in the shallower layers.

Distributions of the simulated late winter (February–March) mixed layer (ML) depth and sea surface density (Fig. 2) indicate that the deep ML develops in the western and central part of the basin to form a sharp frontal structure of ML depth (MLD). The ML depth front is somewhat sharper than in the observations (Suga et al. 2004). The sharp ML front intersects outcrop lines of around 25.5-σ_θ layers, and induces subduction of the aforementioned lower-PV water in the model, as the intersection of the ML front and outcrop line is a source of low-PV water on the isopycnal layer, and a deeper ML at the intersection can induce lower PV (Inui et al. 1999; Kubokawa and Inui 1999; Xie et al. 2000).

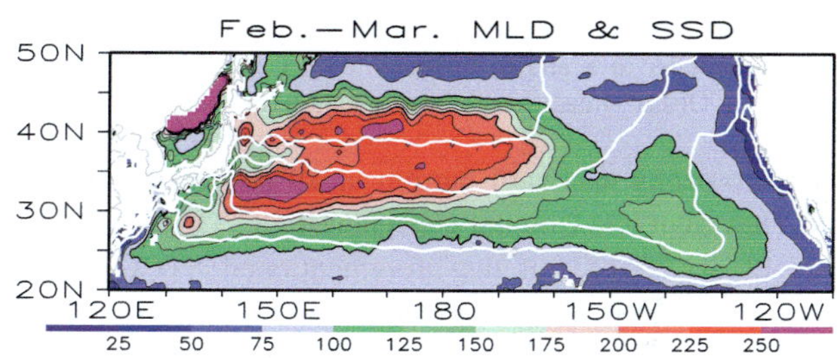

Fig. 2 Long-term mean late winter (February–March) mixed layer depth (MLD; *shades* in m) and sea surface density (*white contours* for 24.5, 25.0, 25.5, and 26.0 σ_θ) fields simulated in OFES. The color scale is given at the *bottom* of the *panel*. MLD is defined as the depth where density difference from the sea surface exceeds 0.125 σ_θ

3 Simulated STCC

3.1 Mean STCC

Figure 3 shows the long-term mean summertime subsurface zonal current velocity field as it has a maximum in summer (not shown), consistent with observations in the eastern part of STCC (White et al. 1978; Kobashi and Kawamura 2002). STCC appears as a weak eastward flow extending in a southwest–northeast direction from 25°N, 150°E to 30°N, 170°W. At the northeastern end, STCC merges with a broad eastward flow corresponding to the northern part of the anticyclonic subtropical gyre. There is

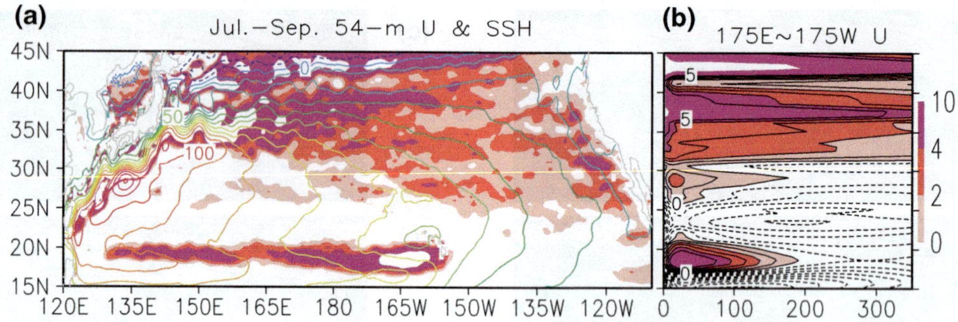

Fig. 3 Long-term mean summertime (July–September) zonal current velocity (*shades* as indicated to the *right* of **b** in cm s^{-1}) **a** at 54-m depth and **b** in 175°E–175°W mean latitude-depth section in OFES. *Contours* in **a** are the corresponding sea surface height field (contour intervals are 10 cm)

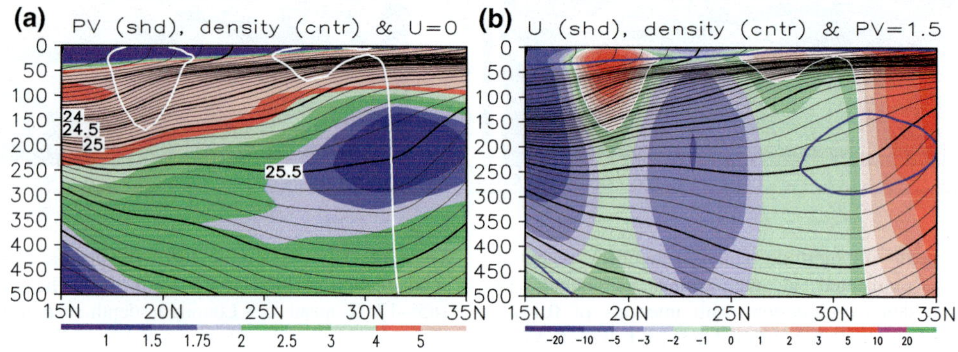

Fig. 4 Latitude-depth sections of summertime (July–September) **a** potential vorticity (*shades*), **b** zonal velocity (*shades*), and density (*black contours* in **a** and **b**; intervals are 0.1 σ_θ). All variables are based on OFES simulation, and are long-term mean and zonally averaged in 160°–180°E. *White* (*blue*) *contours* in **a** (**b**) are for zonal velocity of 0 cm s^{-1} (potential vorticity of 1.5 × 10^{-12} cm^{-1} s^{-1}). *Shades* are indicated as shown at the *bottom* of each *panel*

another strong eastward current at 19°N west of Hawaii, the Hawaiian Lee Counter Current (HLCC, Qiu et al. 1997; Xie et al. 2001), but the aforementioned STCC clearly separates from it. The simulated STCC is limited to the east of 150°E, and the northern STCC that extends in the western basin in observations (Kobashi et al. 2006) is not represented in the model. Also, the simulated STCC is weaker than in observations (Kobashi et al. 2006). These discrepancies may be due to strong dissipation in the model. The latitude-depth section of zonal velocity around the international dateline (Fig. 3b) indicates that STCC extends from the surface to less than 200-m depth as in observations (Fig. 3 of Kobashi and Kawamura 2002). Whereas the observed geostrophic STCC has the maximum eastward velocity at the sea surface, the simulated STCC has the vertical maximum just below the surface mixed layer.

3.2 Relation between low-PV water and STCC in long-term mean and interdecadal variations

We next investigate summertime STCC and low-PV water around 170°E (Fig. 4), focusing on the eastern STCC as the western northern STCC is not represented in the model. In the latitude-depth section of long-term mean summertime PV field, there is low-PV water in the northern part of the subtropical gyre, extending southward along isopycnal surfaces around 25.3–25.7 σ_θ. Above the southern edge of the low-PV water (Fig. 4a), the upper pycnocline shoals northward, and STCC forms just below the sea surface (Fig. 4b). The relative position of STCC to the low-PV water is consistent with observational data (e.g., Kobashi et al. 2006), and Kubokawa's (1999) theory. To the south of STCC, there is the broad and deep westward North Equatorial Current (NEC), the southern branch of the subtropical gyre. Surrounded by the westward current of NEC, HLCC exists just to the south of 20°N. While it is also trapped in the upper layer as STCC, no low-PV water is found in the lower layer to the north of HLCC.

Horizontal distributions of layer thickness (Fig. 5) show that the maximum thickness corresponding to low-PV water appears in the western part of the basin in upper layers (Fig. 5a, b) and in the eastern/northeastern region in deeper layers (Fig. 5c–e). The thick, low-PV water extends from its maximum southwestward or southeastward and then southwestward, but directions of low-PV water

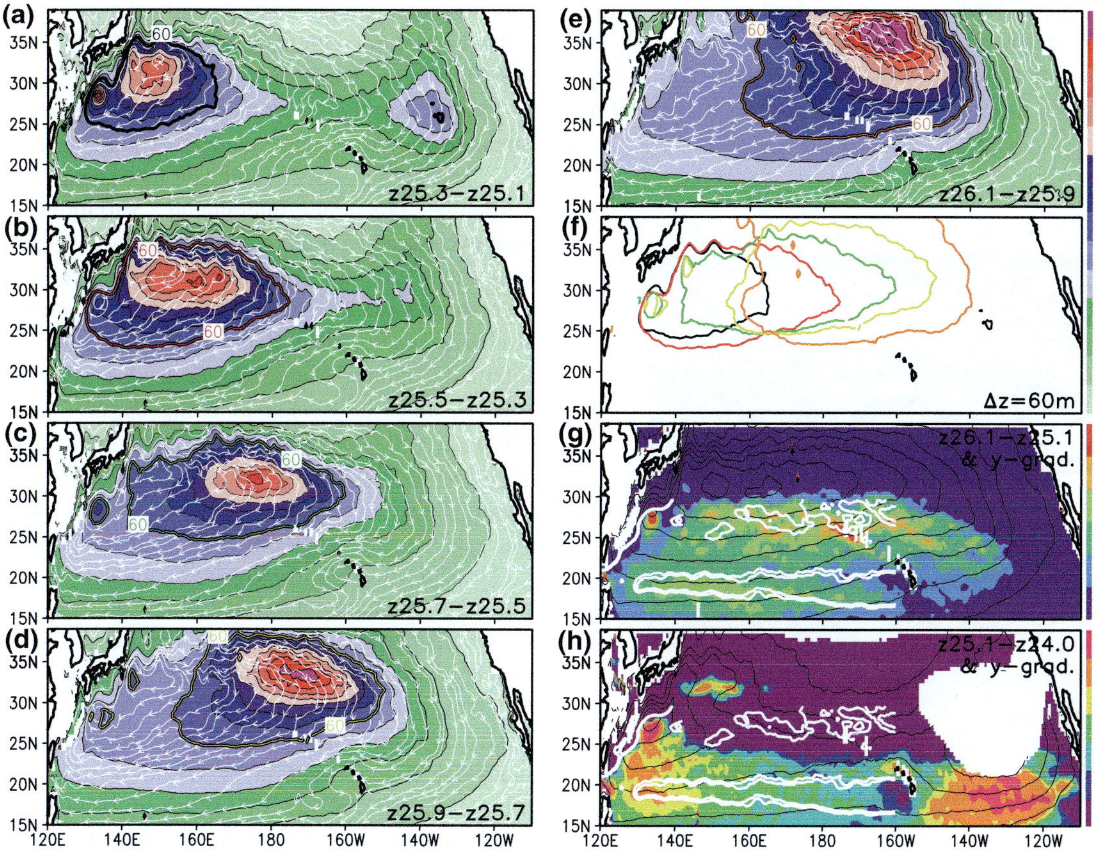

Fig. 5 Horizontal distributions of summertime (July–September) layer thickness (*shades* as indicated to the *right* of **e–f**) for **a** 25.1–25.3 σ_θ, **b** 25.3–25.5 σ_θ, **c** 25.5–25.7 σ_θ, **d** 25.7–25.9 σ_θ, **e** 25.9–26.1 σ_θ, **g** 25.1–26.1 σ_θ, and **h** 24.0–25.1 σ_θ. **f** The same as **a** but for thickness = 60 m for each layer in **a–e**. *Shades* in **g–h** show the meridional gradient of the corresponding thickness (m latitude^{-1} as indicated to the *right* of each *panel*), and *white contours* show zonal velocity = 0, and 2 cm s^{-1} at 54-m depth. In each *panel* of **a–e**, streamlines on the isopycnal surface at the center of each layer are superimposed. All are the long-term mean simulated fields

ventilation are different among the layers as indicated by the streamlines. As a result, thick waters accumulate to the north of STCC (Fig. 5f), as suggested theoretically by Kubokawa (1999), and the meridional gradient of the layer thickness from 25.1 to 26.1 σ_θ shows a local maximum almost along STCC around 27°N (Fig. 5g). In contrast, in the shallower layers (from 24.0 to 25.1 σ_θ), the meridional thickness gradient is small (Fig. 5h) below STCC, indicating that the northward shoaling isopycnal surfaces in the upper layer and the corresponding eastward flow (STCC) is associated with the meridional thickness gradient of the layer from 25.1 to 26.1 σ_θ. It should be also noted that there is a local maximum of the meridional thickness gradient of the 25.1- to 26.1-σ_θ layer below the eastward current to the west of 160°E around 20°N, which may correspond to the southern STCC (Kobashi et al. 2006).

To study interdecadal variations, we plot the same zonal mean fields for the periods of 1975–1979 and 1990–1994 (Fig. 6). STCC is stronger and more organized in structure at 25°–27°N in the 1970s than in the 1990s, consistent with Yamanaka et al. (2008). Low-PV water is confined mostly to the STMW density range of 25.2–25.6 σ_θ in the 1990s but its production increases markedly in the CMW range of 25.5–26.0 σ_θ in the 1970s. The low-PV water intrudes much more southward in the 1970s than 1990s. These changes in low-PV water ventilation are consistent with the STCC change from the thermal wind relationship. Our results show that the association between the STCC and PV distribution holds in both the climatology and interdecadal variability in OFES, consistent with observations and the previous study. This encouraged us to investigate interannual variations in STCC and its relationship to low-PV water as discussed in the next section.

4 Interannual variations in STCC and low-PV water

4.1 Interannual variations in simulated STCC

For investigation of interannual variations in STCC, we plot area-averaged summertime mean zonal velocity at several depths in Fig. 7a. As represented by the speed at

54-m depth, STCC has strong interannual variations with a weak mean eastward current. Also, it is clear that the interannual variations are very similar to those at 404- and 604-m depths. Indeed, the vertical profile of correlation with the area-averaged 54-m zonal velocity (Fig. 7c) shows a deep structure, with correlation coefficient, *r*, higher than 0.9 (0.7) in the upper 600 (1000) m.

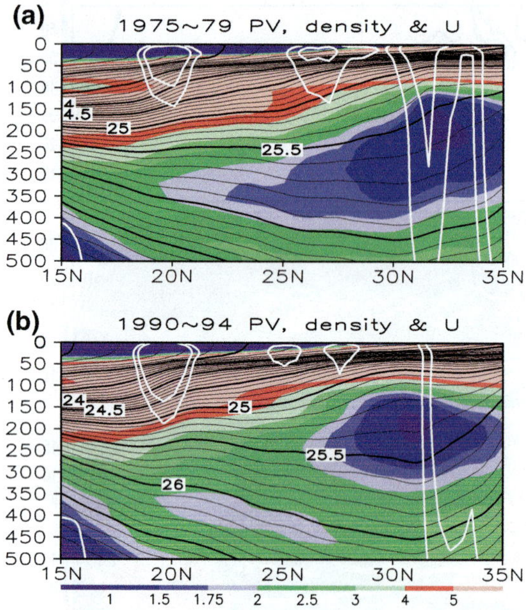

Fig. 6 The same as Fig. 4a, but for the fields averaged in summer time in **a** 1975–1979 and **b** 1990–1994. *White contours* for the zonal current field show 0 and 2 cm s^{-1}

Lagged correlation maps (Fig. 8) between 404-m-depth summertime zonal current velocity and its area-mean at (26°–28°N, 160°–180°E) indicate that narrow bands of the vertically coherent interannual variations propagate southwestward with a few degrees in meridional width and more than 20° in longitudinal length. These properties resemble those of zonal jets or striations examined by Richards et al. (2006), likely not wind-driven but oceanic internally induced variability. Indeed, similar lagged correlation maps of Ekman pumping do not show coherent forcing for the zonal current signals indicated in Fig. 8 (figures not shown). Mechanisms for the formation of the zonal jets/striations are still under debate (Melnichenko et al. 2010), and beyond the scope of the present study.

These vertically coherent interannual variations are not related to variations in low-PV water. As the purpose of the present paper is to investigate the possible influence of low-PV variations on STCC, in the following analyses, we focus on interannual variations that deviate from the vertically uniform structure. To extract upper layer variations independent of deeper layer variations, we compute the linear regression of the time series of 54-m zonal velocity upon velocity at 404-m depth, and then subtract the regression from the original 54-m zonal velocity time series (Fig. 7b). In the following analyses, we use this new

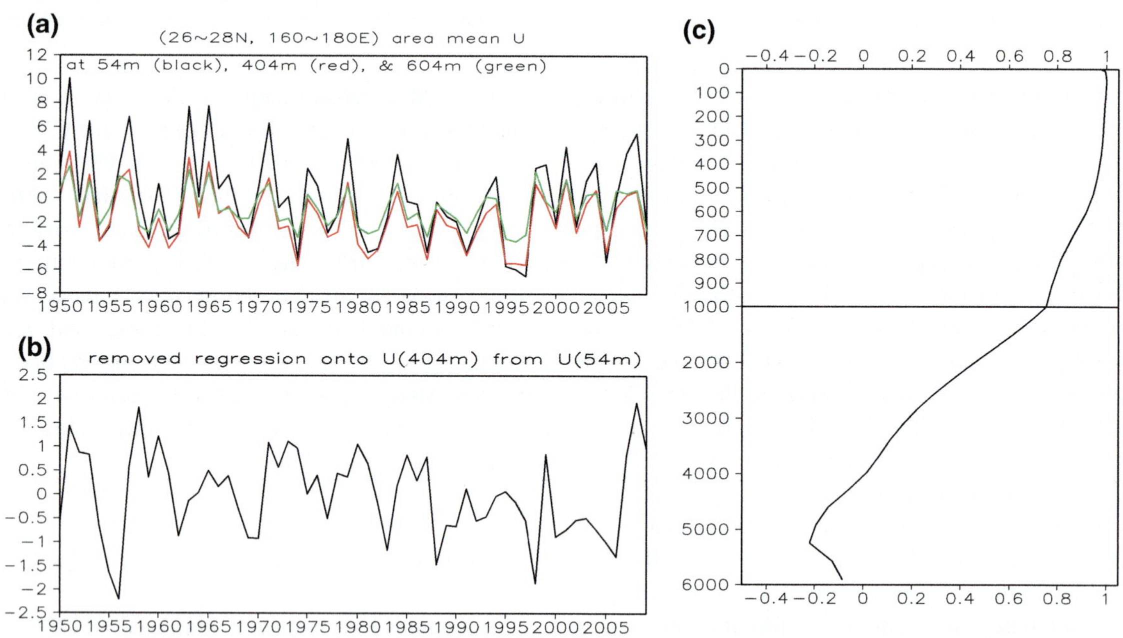

Fig. 7 **a** Time series of area mean simulated summertime (July–September) zonal current velocity at (26°–28°N, 160°–180°E) at 54-m (*black*), 404-m (*red*), and 604-m (*green*) depths. **b** Time series of the index of STCC vertical shear that is defined in the text. **c** Vertical profile of correlation coefficients between the area mean simulated zonal current velocity at (26°–28°N, 160°–180°E) at 54-m depth and those at each vertical level

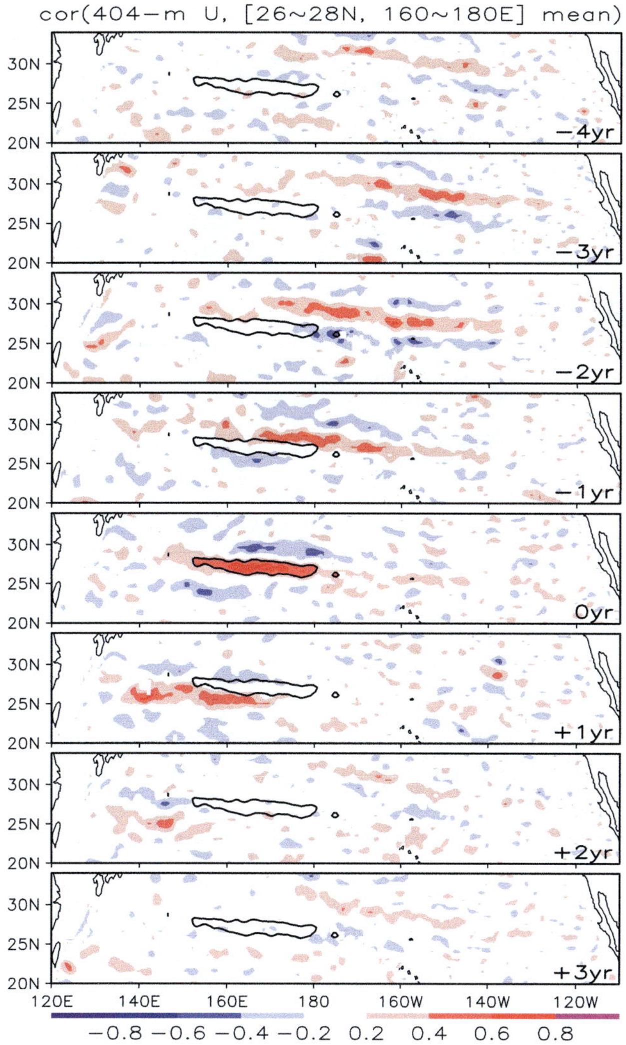

Fig. 8 Lagged correlation (*shades*) maps of summertime (July–September) simulated 404-m-depth zonal current onto the same field averaged at (26°–28°N, 160°–180°E) with lags of −4 to +3 years from the *top* to *bottom*. *Black contours* show the simultaneous correlation = 0.4

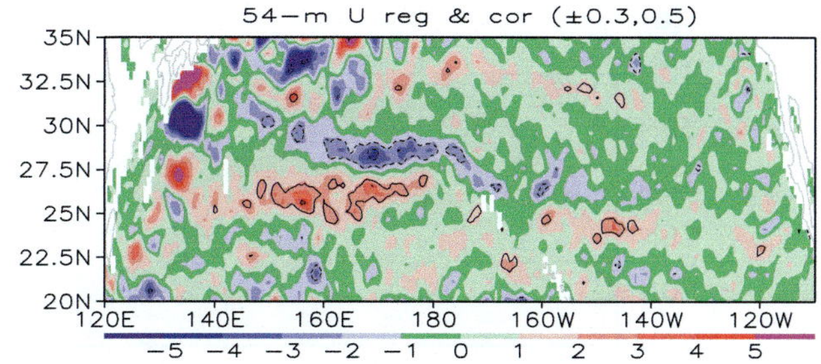

Fig. 9 Simultaneous correlation (contours for ± 0.3, 0.5) and regression (*shades* as indicated at the *bottom* of the *panel*) coefficients between summertime (July–September) 54-m-depth zonal current velocity and the STCC shear index (Fig. 7b)

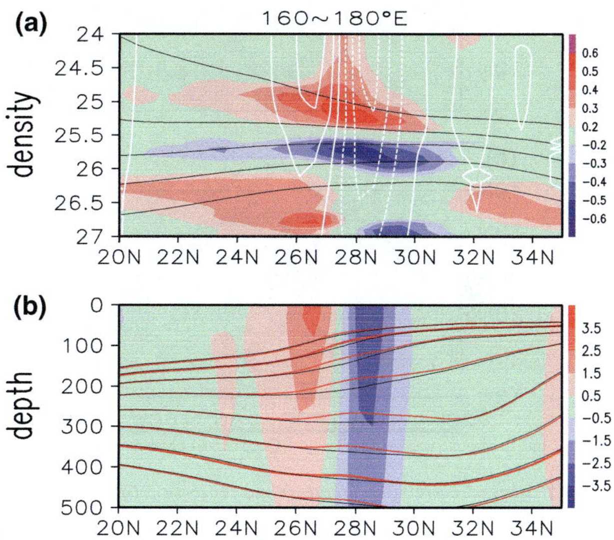

Fig. 10 **a** Correlation of 160°–180°E mean PV (*shades*) and regression map of 160°–180°E mean zonal velocity (*white contours* with 1 cm s^{-1} intervals) onto the STCC shear index as a function of latitude and density. **b** The same as the *top panel*, but for the meridional-depth regression maps for zonal velocity (*shades*) and density (*red contours* with climatology added). The regression maps correspond to 1 cm s^{-1} anomaly of the STCC index. *Black contours* in **a** and **b** are long-term mean isopycnal depth (100–500 m, with 100-m intervals) and density (24.8–26.2 σ_θ, with 0.2-σ_θ intervals). All variables are summertime (July–September) simulated fields

time series averaged at (26°–28°N, 160°–180°E) (Fig. 7b) to represent interannual variations in STCC that is trapped to the near-surface layer and independent of the coherent variations in the upper 1000 m. Selection of the depth for the regression, 404 m, is arbitrary, but the resultant time series does not strongly depend on this choice of depth; almost the same results are obtained if regression upon time series at a different depth, say 604 m, is removed. This STCC shear index is a measure of current shear and a good representation of the surface-trapped current in climatology.

Simultaneous correlation map between the STCC shear index and summertime 54-m-depth zonal velocity (Fig. 9) indicates that the peak of the correlation is about 0.5, and the index explains about 20% of interannual variance in zonal velocity. In other words, major parts of interannual variations in the zonal velocity near the surface are vertically coherent as suggested in Fig. 7. The correlation map shows a dipole pattern with positive and negative correlation zones to the south and north of 27.5°N, respectively, around 160°–180°E, suggesting meridional shift of STCC on interannual timescales.

4.2 Correlation between STCC and low-PV water

To investigate if the STCC interannual variations are related to variations in low-PV water below, we conduct a correlation/regression analysis between the STCC shear

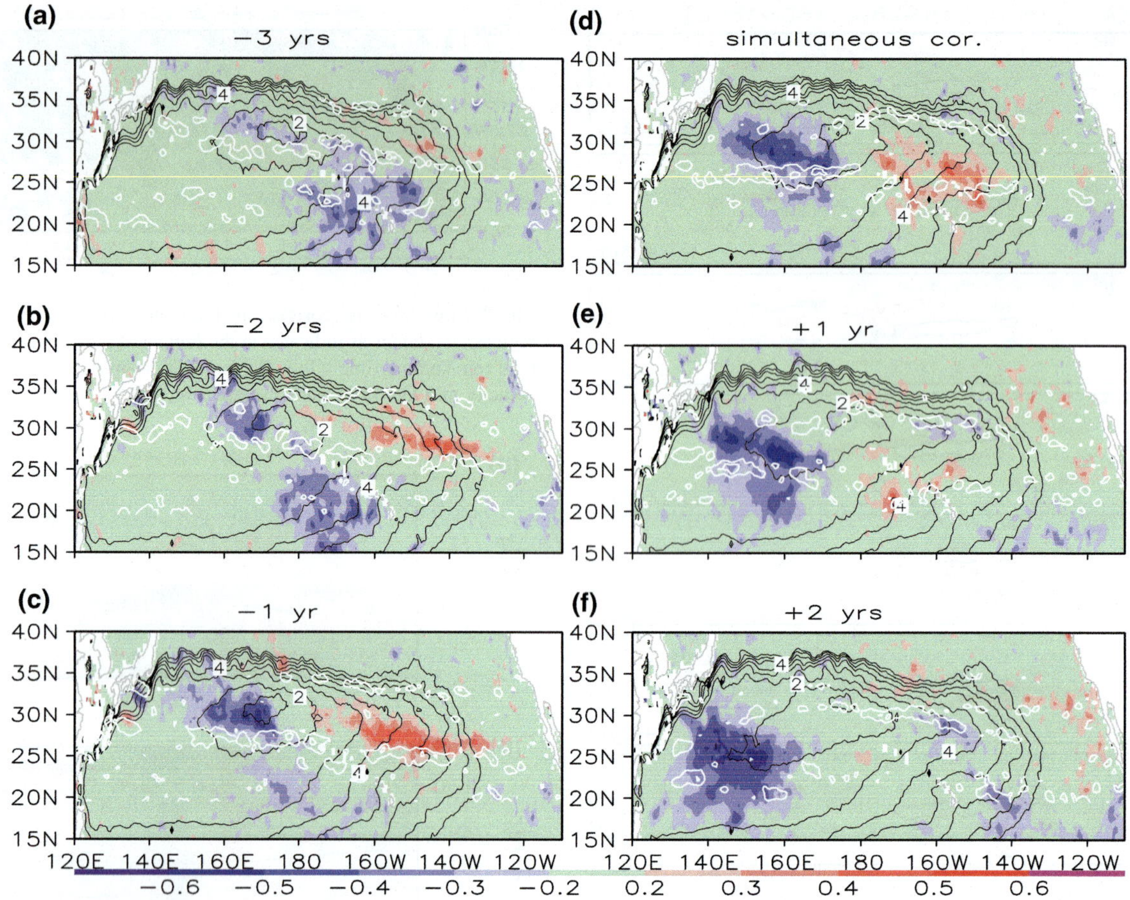

Fig. 11 Lagged correlation (*shades*) and regression (*black contours*) maps of PV on the 25.6-σ_θ surface onto the STCC shear index with lags of **a** −3, **b** −2, **c** −1, **d** 0, **e** 1, and **f** 2 years. *White contours* show the corresponding correlation (0.2, and 0.4) map of 54-m-depth zonal velocity onto the STCC shear index. The regression maps correspond to 1 cm s^{-1} anomaly of the index with climatology added. All variables are summertime (July–September) simulated fields

index defined above and subsurface fields. Simultaneous correlation maps for PV on a meridional-density section (Fig. 10a) indicate that negative correlations appear to the north of STCC around 25.5–26.1 σ_θ, whereas positive correlations are found in shallower layers (shallower than about 25.3 σ_θ). Those negative correlations, representing the intrusion of lower-PV water to the north of STCC, are accompanied by a steeper northward shoaling of isopycnals (Fig. 10b, contours). The stronger northward gradient of density corresponds to stronger vertical shear of eastward geostrophic current, consistent with the stronger STCC (shades in Fig. 10b).

Figure 11 shows lagged correlation maps between the STCC shear index and PV on the 25.6-σ_θ isopycnal surface, on which strongest correlations are found in Fig. 10. Negative PV correlations appear at least 2 years before the peak of STCC, develop and shift south and southwestward. In association with this southwestward shift of negative correlations [which indicate the southwestward development of lower-PV water (black contours)], positive correlations of zonal velocity also shift southward and are found to the south of the negative PV correlation region (white contours in Fig. 11), consistent with the meridional shift suggested in Fig. 8. These results strongly suggest that the STCC variations are induced through a subduction process of low-PV water that develops more than usual.[1]

The association between STCC and PV anomalies is further examined in a latitude-time section of zonally averaged (in 160°–180°E) layer thickness in 25.5- to 26.1-σ_θ layers (Fig. 12). The layer thickness (contours in Fig. 12a) shows significant interannual variations especially in its subduction region (around 32°–36°N), from which the thick layer and its interannual variations extend southward. At the southern edge of a thicker layer, stronger

[1] Around 160°–140°W, positive PV correlations are found to the north of positive zonal velocity anomalies. In contrast to that found in Fig. 10a, these are not accompanied by negative PV anomalies below. The positive PV anomalies are probably due to advection by westward current anomalies found there (not shown) on a background of strong zonal PV gradients.

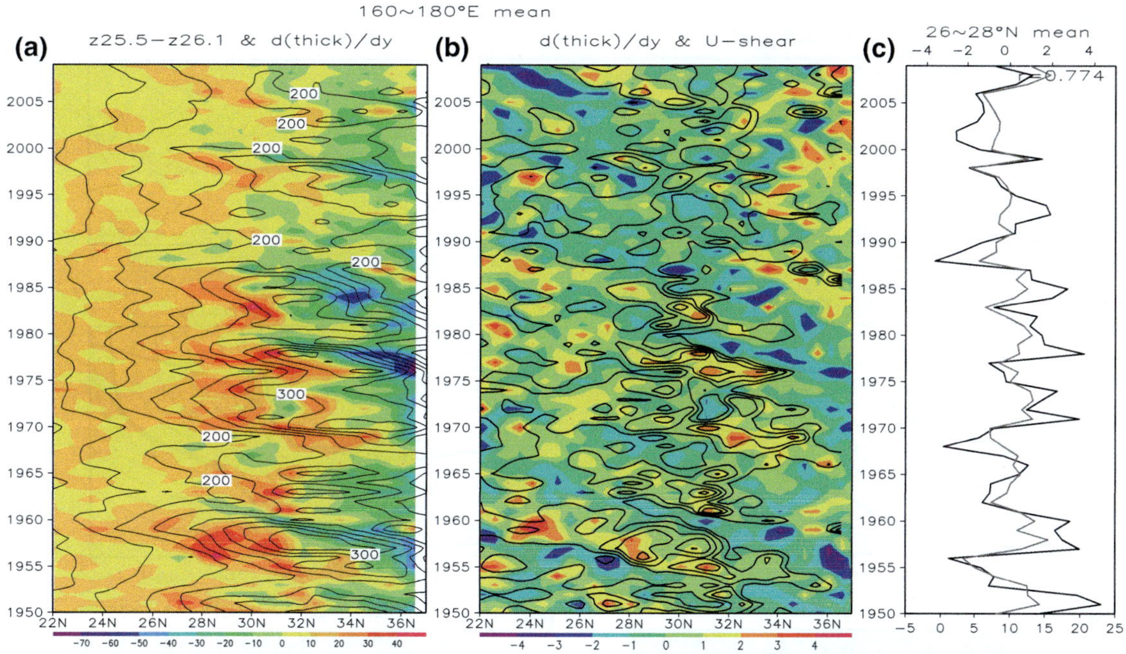

Fig. 12 **a** Latitude-time section of 25.5- to 26.1-σ_θ layer thickness (*contours*; intervals are 25 m) and its meridional gradient (shades; in m latitude^{-1}). **b** The same as **a**, but for vertical shear of zonal velocity (*shades*). *Contours* are the same field as that shown in shades in **a**. **c** 26°–28°N mean vertical shear of zonal velocity (*black*, *top axis*), which is identical to the STCC shear index (Fig. 7b), and meridional gradient of thickness shown in **a**, **b** (*grey*, *bottom axis*). All variables are 160°–180°E mean of summertime (July–September) simulated fields. The vertical shear in **b** is defined in the same way as the STCC shear index (Fig. 7b)

meridional gradients of the thickness (shades in Fig. 12a) induce stronger meridional density gradient and vertical zonal current shear in the upper layer. Indeed, zonal current vertical shear (shades in Fig. 12b) is intensified above the strong meridional gradient of the thickness in 25.5- to 26.1-σ_θ layers (contours in Fig. 12b), and together they propagate southward coherently. For example, at 26°–28°N, the meridional gradient of the thickness and vertical shear of zonal current, the STCC shear index, has high correlation $r = 0.77$ (Fig. 12c). These results confirm the lagged correlation analysis (Fig. 11).

4.3 Causes of thickness variations

As interannual variations in the thickness of the 25.5- to 26.1-σ_θ layer can induce interannual variations in STCC, we further investigate what causes the thickness anomalies in the subduction region, which corresponds to 160°–180°E, 32°–34°N in the climatology. In this subduction region, the interannual variations in the thickness highly correlate with the local SST ($r = -0.80$) (Fig. 13a), and MLD to the north (34°–36°N) ($r = 0.65$) in the late winter. MLD in the subduction region of 32°–34°N, however, has lower correlation (Fig. 13e), and the correlation between SST and thickness there is not caused by the deepening of the winter mixed layer. To see the relation between SST and thickness variations, we compare latitude-depth sections of wintertime and summertime density between a cold (1995–1997) and a warm (2000–2002) SST period (Fig. 13b, c). When SST is cold (Fig. 13b), the outcrop line of 25.5-σ_θ layer shifts southward, and the 25.5- to 26.1-σ_θ layer is exposed to the sea surface at 32°–34°N. Although the 25.5-σ_θ isopycnal surface submerges in summer (blue curve), the 25.5- to 26.1-σ_θ layer remains thick. In the warm years (Fig. 13c), the outcrop of the 25.5-σ_θ surface shifts northward and the 25.5- to 26.1-σ_θ layer does not outcrop in the region of 32°–34°N, is kept thin in winter, and remains thin in summer. SST anomalies in the source region (32°–34°N, 160°–180°E) significantly correlate with eastward wind stress ($r = -0.52$) and are likely influenced by mixing, Ekman cooling, and wind-induced evaporation (but do not significantly correlate with the net surface heat flux probably due to cancellation among influences of atmospheric and SST anomalies on the net flux). Additionally, correlation maps of SST and MLD with the thickness in the source region (Fig. 13d, e) indicate that thicker layer tends to be associated with cooler SST anomalies to the south of the Kuroshio Extension Current (KEC) and deeper MLD near the KEC axis, around which meridional local minimum of MLD forms (Fig. 13e). These anomaly patterns in SST and MLD resemble those with weakened KEC (Nonaka et al. 2011, manuscript in preparation). Indeed, zonally averaged KEC speed (green curve in Fig. 13a) negatively correlates with the thickness

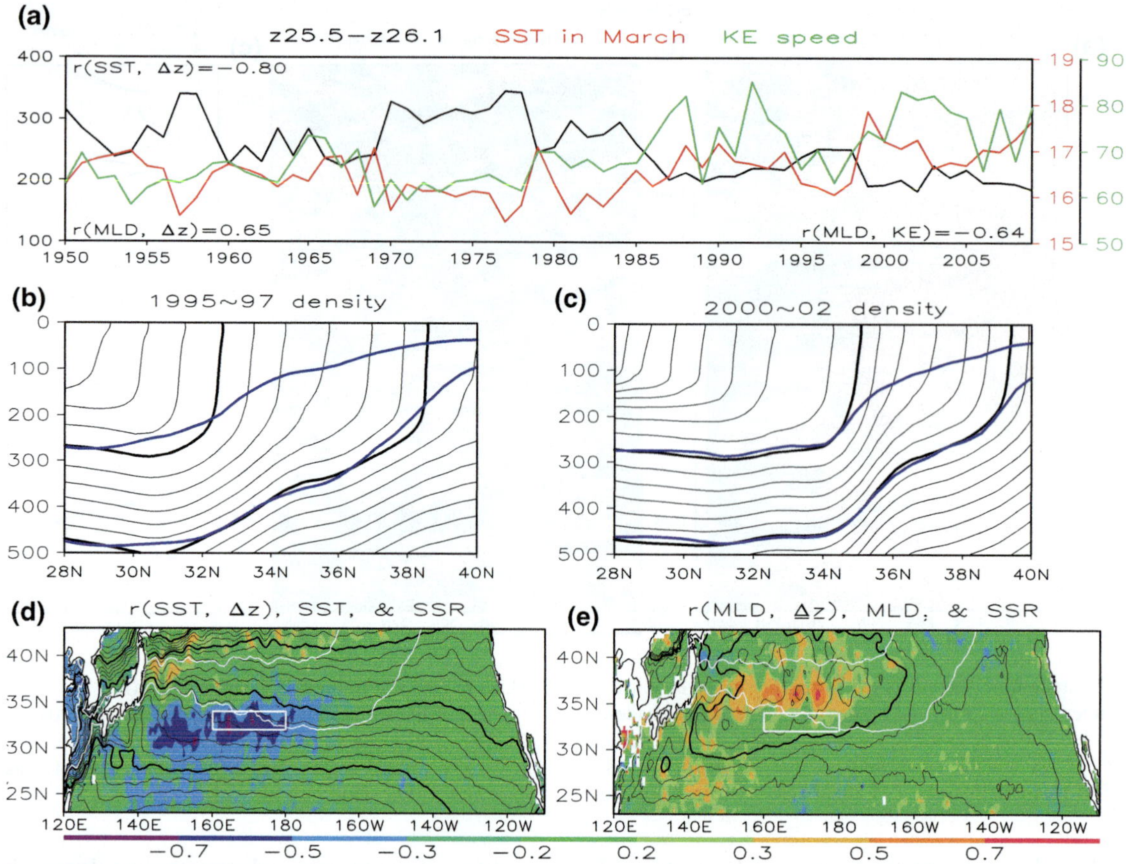

Fig. 13 **a** Time series of 25.5- to 26.1-σ_θ layer thickness (*black*; *left axis*) in summertime (July–September), SST (*red*; *right axis*) in March both averaged at (32°–34°N, 160°–180°E), and the surface Kuroshio Extension Current (KEC) speed (*green*; *right-most axis* in cm s^{-1}) in March averaged at 140°–180°E. The KEC speed is detected at each zonal grid point as a maximum zonal current speed at 30°–40°N. **b**, **c** Meridional-depth sections of 160°–180°E mean density in March (*black contours* with 0.1-σ_θ intervals and *thickened* for 25.5 and 26.1 σ_θ), and in summertime (*blue* for 25.5 and 26.1 σ_θ). **b** Average fields in 1995–97 and **c** 2000–2002. **d**, **e** Correlation map of SST (**d**) and MLD (**e**) in March onto the time series of layer thickness shown in **a**. *Black contours* indicate the long-term mean of the corresponding field, and *white contours* show sea surface density (25.5 and 26.1 σ_θ). All variables are simulated fields

($r = -0.64$). The relation among variations in KEC, low-PV water subduction, and STCC is suggested by Xie et al. (2011) in their analysis of climate models, and this is supported by the results from this eddy-resolving OGCM hindcast that can represent interannual-to-decadal variations of KEC realistically (Nonaka et al. 2006; Taguchi et al. 2007).

5 Summary and discussion

On the basis of a 60-year-long hindcast integration of an eddy-resolving OGCM, OFES, we have investigated interannual-to-decadal variations in STCC and low-PV water and their relations in the North Pacific Ocean. In the model climatology, STCC is found on the south edge of low-PV water as predicted by Kubokawa's (1999) theory and consistent with observations. This close relationship holds for interdecadal variations in STCC and low-PV water in the model. The specific purpose of the present study is to investigate if the relationship between STCC and low-PV water also holds in interannual variations.

In the OFES hindcast, STCC appears from 150°E to around the international dateline, and is trapped to near the surface. As it peaks in summer near the dateline, we examine interannual variations in summertime STCC and low-PV water. The STCC variability on interannual timescales is dominated by a vertically deep structure (>1000 m) that seems to relate to narrow bands of zonal currents or striations. We then define an STCC shear by removing vertically coherent variations from variations in the near-surface layer zonal velocity, to represent the surface-trapped structure of STCC.

A correlation analysis shows that intensified STCC is accompanied by lower PV than usual to the north on 25.5- to 26.1-σ_θ isopycnal surfaces, and by intensified meridional density gradient in subsurface layers (Fig. 10). The low-PV signals appear at least 2 years before peaks of STCC, develop and shift southwestward (Fig. 11). These low-PV signals can be traced back to subduction processes. The results of the correlation analysis are further confirmed by latitude-time sections (Fig. 12).

These results strongly suggest that interannual-to-decadal variations in STCC are associated with variations of low-PV water ventilation. In other words, the relation between STCC and low-PV water suggested by Kubokawa (1999) and found in the observed climatology (Kobashi et al. 2006) holds in interannual-to-decadal variations at least in this particular model. This mechanism is at work for decadal changes in STCC (Yamanaka et al. 2008), and our study demonstrates that it works also on interannual timescales in addition to the influence of Ekman flow variations shown by Qiu and Chen (2010) and Kobashi and Xie (2011). Sasaki et al. (2011) suggest a similar influence of low-PV water ventilation on interannual variations of HLCC. The influence of Ekman flow on STCC is, however, not strong in this model. This may be due to the difference in the region of analyses: the studies of Qiu and Chen (2010) and Kobashi and Xie (2011) focus upon the western northern STCC, whereas the eastern STCC is investigated in this study. Also, the weaker meridional temperature gradient in this model than in the observations can weaken the impact of the Ekman flow.[2]

Low-PV water distributions in OFES are slightly different from observations as discussed in Sect. 2.2. Indeed, the low-PV waters in the layers around 25.6 σ_θ that correlate with STCC tend to develop and extend more eastward than in observations. This suggests that the relationships between variations in the low-PV water and STCC found in the model may be stronger than in the real ocean. Recently, subsurface observations by Argo profilers (Argo Science Team 2001; Hosoda et al. 2010) have accumulated rapidly, and it becomes possible to investigate subsurface interannual variations. Indeed, on the basis of Argo observations, Oka (2009) shows that interannual temperature anomalies in NPSTMW can be advected southward by at least 2° latitude with a half-year lag. The low-PV water related to STCC variability in the present study has higher density and subducts in higher latitudes than NPSTMW. Also, as shown in Fig. 12, thickness variability extends southward from the source region (32°–34°N) to STCC (26°–28°N) with a few years lag. Further accumulation of Argo observations will make it possible in the near future to investigate the propagation of PV anomalies in relation to STCC variability in the real ocean.

[2] Indeed, simulated STF in early spring also show weak correlation ($r \sim 0.4$ at the maximum) between the Ekman convergence and meridional temperature gradient to about 100-m depth. This is consistent with the result of Qiu and Chen (2010) but the correlation is weaker and spatially incoherent (not shown).

Acknowledgments The OFES simulations were conducted on the Earth Simulator under the support of JAMSTEC. We thank the members of the OFES group, including Drs. H. Sakuma, Y. Masumoto, and T. Yamagata, for their efforts and support in the model development. Our thanks are extended to Drs. A. Kubokawa, F. Kobashi, and E. Oka for useful discussions. This study is partially supported by Grand-In-Aid for Scientific Research defrayed by the Ministry of Education, Culture, Sports, Science and Technology of Japan (22106006, 23340139). IPRC/SOEST publication #790/8196.

References

Aoki Y, Suga T, Hanawa K (2002) Subsurface subtropical fronts of the North Pacific as inherent boundaries in the ventilated thermocline. J Phys Oceanogr 32:2299–2311

Argo Science Team (2001) Argo: the global array of profiling floats. In: Koblinsky CJ, Smith NR (eds) Observing the oceans in the 21st century. GODAE Project Office, Bureau of Meteorology, Melbourne, pp 248–258

Boyer TP, Stephens C, Antonov JI, Conkright ME, Locarnini RA, O'Brien TD, Garcia HE (2002) World ocean atlas 2001. Salinity, vol 1. In: Levitus S (ed) NOAA atlas NESDIS 50. US Government Printing Office, Washington, DC, p 176

Hanawa K, Talley L (2001) Mode waters, in ocean circulation and climate. In: Siedler G, Church J, Gould J (eds) Int Geophys Ser. Elsevier, New York, pp 373–386

Hosoda S, Ohira T, Sato K, Suga T (2010) Improved description of global mixed-layer depth using Argo profiling floats. J Oceanogr 66:773–787

Inui T, Takeuchi K, Hanawa K (1999) A numerical investigation of the subduction process in response to an abrupt intensification of the westerlies. J Phys Oceanogr 29:1993–2015

Kalnay E et al (1996) The NCEP/NCAR 40-year reanalysis project. Bull Am Meteorol Soc 77:437–471

Kobashi F, Kawamura H (2001) Variation of sea surface height at periods of 65–220 days in the subtropical gyre of the North Pacific. J Geophys Res 106:26817–26831

Kobashi F, Kawamura H (2002) Seasonal variation and instability nature of the North Pacific Subtropical Countercurrent and the Hawaiian Lee Countercurrent. J Geophys Res 107:3185. doi: 10.1029/2001JC001225

Kobashi F, Xie SP (2011) Interannual variability of the North Pacific Subtropical Countercurrent: role of local ocean-atmosphere interaction. J Oceanogr (submitted)

Kobashi F, Mitsudera H, Xie SP (2006) Three subtropical fronts in the North Pacific: observational evidence for mode water-induced subsurface frontogenesis. J Geophys Res 111:C09033. doi: 10.1029/2006JC003479

Kobashi F, Xie SP, Iwasaka N, Sakamoto T (2008) Deep atmospheric response to the North Pacific oceanic subtropical front in spring. J Clim 21:5960–5975

Kubokawa A (1999) Ventilated thermocline strongly affected by a deep mixed layer: a theory for subtropical countercurrent. J Phys Oceanogr 29:1314–1333

Kubokawa A, Inui T (1999) Subtropical countercurrent in an idealized ocean GCM. J Phys Oceanogr 29:1303–1313

Large WG, McWilliams JC, Doney SC (1994) Oceanic vertical mixing: a review and a model with a nonlocal boundary layer parameterization. Rev Geophys 32:363–403

Masumoto Y, Sasaki H, Kagimoto T, Komori N, Ishida A, Sasai Y, Miyama T, Motoi T, Mitsudera H, Takahashi K, Sakuma H, Yamagata T (2004) A fifty-year eddy-resolving simulation of the world ocean—preliminary outcomes of OFES (OGCM for the Earth Simulator). J Earth Simulator 1:31–52

Masuzawa J (1969) Subtropical mode water. Deep Sea Res 16:436–472
Melnichenko OV, Maximenko NA, Schneider N, Sasaki H (2010) Quasi-stationary striations in basin-scale oceanic circulation: vorticity balance from observations and eddy-resolving model. Ocean Dyn 60:653–666
Nakamura H (1996) A pycnostad on the bottom of the ventilated portion in the central subtropical North Pacific: its distribution and formation. J Oceanogr 52:171–188
Nonaka M, Nakamura H, Tanimoto T, Kagimoto T, Sasaki H (2006) Decadal variability in the Kuroshio–Oyashio Extension simulated in an eddy-resolving OGCM. J Clim 19:1970–1989
Nonaka M, Nakamura H, Tanimoto Y, Kagimoto T, Sasaki H (2008) Interannual-to-decadal variability in the Oyashio Current and its influence on temperature in the subarctic frontal zone: an eddy-resolving OGCM simulation. J Clim 21:6283–6303
Oka E (2009) Seasonal and interannual variation of North Pacific Subtropical Mode Water in 2003–2006. J Oceanogr 65:151–164
Oka E, Suga T (2005) Differential formation and circulation of North Pacific central mode water. J Phys Oceanogr 35:1997–2011
Pacanowski RC, Griffies SM (2000) MOM 3.0 manual. Geophysical Fluid Dynamics Laboratory/National Oceanic and Atmospheric Administration, Princeton
Qiu B (1999) Seasonal eddy field modulation of the North Pacific subtropical countercurrent: TOPEX/Poseidon observations and theory. J Phys Oceanogr 29:2471–2486
Qiu B, Chen S (2010) Interannual variability of the North Pacific subtropical countercurrent and its associated mesoscale eddy field. J Phys Oceanogr 40:213–225
Qiu B, Koh DA, Lumpkin C, Flament P (1997) Existence and formation mechanism of the North Hawaiian Ridge Current. J Phys Oceanogr 27:431–444
Richards KJ, Maximenko NA, Bryan FO, Sasaki H (2006) Zonal jets in the Pacific Ocean. Geophys Res Lett 33:L03605. doi: 10.1029/2005GL024645
Roden GI (1975) On the North Pacific temperature, salinity, sound velocity fronts, and the relation to the wind and energy flux fields. J Phys Oceanogr 5:557–571
Sasaki H, Nonaka M, Masumoto Y, Sasai Y, Uehara H, Sakuma H (2008) An eddy-resolving hindcast simulation of the quasi-global ocean from 1950 to 2003 on the Earth Simulator. In: Hamilton K, Ohfuchi W (eds) High resolution numerical modelling of the atmosphere and ocean. Springer, New York, pp 157–185
Sasaki H, Xie SP, Taguchi B, Nonaka M, Hosoda S, Masumoto Y (2011) Interannual variations of the Hawaiian Lee Countercurrent induced by low potential vorticity water ventilation in the subsurface. J Oceanogr (submitted)
Smith RD, Maltrud ME, Bryan FO, Hecht MW (2000) Numerical simulation of the North Atlantic Ocean at 1/10°. J Phys Oceanogr 30:1532–1561
Stephens C, Antonov JI, Boyer TP, Conkright ME, Locarnini RA, O'Brien TD, Garcia HE (2002) World ocean atlas 2001. Temperature, vol. 1. In: Levitus S (ed) NOAA atlas NESDIS 49. US Government Printing Office, Washington, DC, p 167
Suga T, Takei Y, Hanawa K (1997) Thermostad distribution in the North Pacific subtropical gyre: the central mode water and the subtropical mode water. J Phys Oceanogr 27:140–152
Suga T, Motoki K, Aoki Y, Macdonald AM (2004) The North Pacific climatology of winter mixed layer and mode waters. J Phys Oceanogr 34:3–22
Taguchi B, Xie SP, Schneider N, Nonaka M, Sasaki H, Sasai Y (2007) Decadal variability of the Kuroshio Extension: observations and an eddy-resolving model hindcast. J Clim 20:2357–2377
Taguchi B, Qiu B, Nonaka M, Sasaki H, Xie SP, Schneider N (2010) Decadal variability in the Kuroshio Extension: mesoscale eddies and recirculations. Ocean Dyn 60:673–691. doi: 10.1007/s10236-010-0295-1
Takeuchi K (1984) Numerical study of the subtropical front and the subtropical countercurrent. J Oceanogr Soc Jpn 40:371–381
Uda M, Hasunuma K (1969) The eastward subtropical countercurrent in the western North Pacific Ocean. J Oceanogr Soc Jpn 25:201–210
White WB, Hasunuma K, Solomon H (1978) Large scale seasonal and secular variability of the Subtropical Front in the Western North Pacific from 1954 to 1974. J Geophys Res 83:4531–4544
Xie SP, Kunitani T, Kubokawa A, Nonaka M, Hosoda S (2000) Interdecadal thermocline variability in the North Pacific for 1958–97: a GCM simulation. J Phys Oceanogr 30:2798–2813
Xie SP, Liu T, Liu Q, Nonaka M (2001) Far-reaching effects of the Hawaiian Islands on the Pacific Ocean-atmosphere. Science 292:2057–2060
Xie SP, Deser C, Vecchi GA, Ma J, Teng H, Wittenberg AT (2010) Global warming pattern formation: sea surface temperature and rainfall. J Clim 23:966–986
Xie SP, Xu L, Liu Q, Kobashi F (2011) Dynamical role of mode water ventilation in decadal variability in the central subtropical gyre of the North Pacific. J Clim 24:1212–1225
Yamanaka G, Ishizaki H, Hirabara M, Ishikawa I (2008) Decadal variability of the subtropical front of the western North Pacific in an eddy-resolving ocean general circulation model. J Geophys Res 113:C12027. doi:10.1029/2008JC005002
Yoshida K, Kidokoro T (1967) A subtropical countercurrent in the North Pacific—an eastward flow near the subtropical convergence. J Oceanogr Soc Jpn 23:88–91

J Oceanogr (2012) 68:113–126
DOI 10.1007/s10872-011-0048-x

SPECIAL SECTION: ORIGINAL ARTICLE

New developments in mode-water research: Dynamic and climatic effects

Interannual variability of the North Pacific Subtropical Countercurrent: role of local ocean–atmosphere interaction

Fumiaki Kobashi · Shang-Ping Xie

Received: 5 January 2011 / Revised: 17 May 2011 / Accepted: 6 June 2011 / Published online: 13 July 2011

Abstract Seasonal and interannual variability of the Subtropical Countercurrent (STCC) in the western North Pacific are investigated using observations by satellites and Argo profiling floats and an atmospheric reanalysis. The STCC displays a clear seasonal cycle. It is strong in late winter to early summer with a peak in June, and weak in fall. Interannual variations of the spring STCC are associated with an enhanced subtropical front (STF) below the surface mixed layer. In climatology, the SST front induces a band of cyclonic wind stress in May north of the STCC on the background of anticyclonic curls that drive the subtropical gyre. The band of cyclonic wind and the SST front show large interannual variability and are positively correlated with each other, suggesting a positive feedback between them. The cyclonic wind anomaly is negatively correlated with the SSH and SST below. The strong (weak) cyclonic wind anomaly elevates (depresses) the thermocline and causes the fall (rise) in the SSH and SST, accelerating (decelerating) STCC to the south. It is suggested that the anomalies in the SST front and STCC in the preceding winter affect the subsequent development of the cyclonic wind anomaly in May. Results from our analysis of interannual variability support the idea that the local wind forcing in May causes the subsequent variations in STCC.

Keywords Subtropical countercurrent · Interannual variability · Ocean–atmosphere interaction · Sea surface temperature front

F. Kobashi (✉)
Faculty of Marine Technology, Tokyo University of Marine Science and Technology, Etchujima 2-1-6, Koto-ku, Tokyo 135-8533, Japan
e-mail: kobashi@kaiyodai.ac.jp

F. Kobashi
Research Institute for Global Change, Japan Agency for Marine-Earth Science and Technology, Yokosuka, Japan

S.-P. Xie
International Pacific Research Center (IPRC) and Department of Meteorology, School of Ocean and Earth Science and Technology (SOEST), University of Hawaii at Manoa, Honolulu, HI, USA

1 Introduction

The Subtropical Countercurrent (STCC) flows eastward in the southern subtropical gyre of the North Pacific where the classical wind-driven circulation theory predicts a broad westward flow. The STCC accompanies a subsurface temperature and density front called the subtropical front (STF; Uda and Hasunuma 1969). Figure 1 shows the temperature section along 155°E from the World Ocean Circulation Experiment (WOCE) Hydrographic Program (WHP), displaying two subsurface STFs which are associated with a northward shoaling of the upper main thermocline. Below the upper thermocline, mode water, a vertically homogeneous water mass characterized by low potential vorticity (PV), exists to the north of the STF.

Recent studies show that mode water plays a central role in the STCC and STF formation. Using an idealized ocean general circulation model (GCM), Kubokawa and Inui (1999) reproduced STCC on the southern flank of low PV water in the thermocline. In their model, the low PV water originates in the deep surface mixed layer in the northwestern subtropical gyre and is subducted at the intersection of the density outcrop line and a mixed layer depth (MLD) front. The MLD front is a narrow transition zone separating shallow and deep mixed layers. The low PV waters subducted on different isopycnals from different locations are advected southward by the subtropical gyre, eventually stacked up in the vertical and forming a thick low PV pool. This thick low PV pool causes the upper pycnocline to rise, forming a density front and eastward countercurrent on the southern edge of the low PV pool. The ventilated thermocline affected by the MLD front and the resultant STCC formation are theoretically discussed by Kubokawa (1999). This mechanism has won support from observations (Aoki et al. 2002; Kobashi et al. 2006). Constructing a long-term mean climatology from enhanced hydrographic data, Kobashi et al. (2006) found double bands of STCC extending almost zonally along 19–21°N and 21–26°N (Fig. 2), and

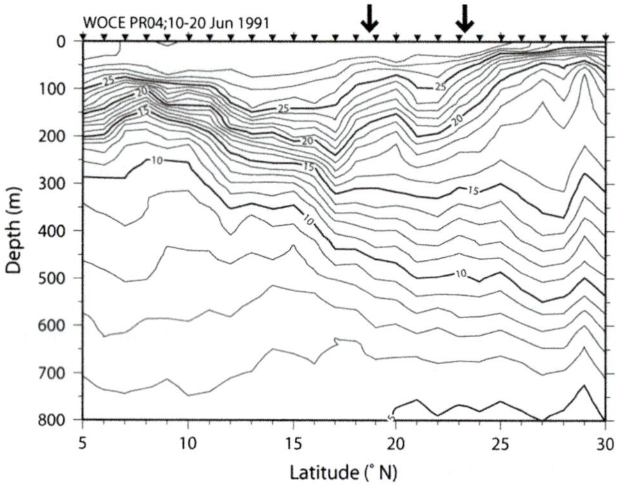

Fig. 1 Temperature section along 155°E, taken by the R/V Ryofu Maru in a WOCE cruise during June 1991. *Arrows* indicate the position of STFs. The location of observation is denoted by *inverted triangles* (from Kobashi et al. 2006)

showed that these two bands are each anchored by mode waters below the STF to the north. They indicated the importance of the Subtropical Mode Water (STMW; Masuzawa 1969) and the Central Mode Water (CMW; Nakamura 1996; Suga et al. 1997) in the formation of STCC.

The STCC is maintained not only by the subsurface front but also partly by a surface front. Along the northern STCC in the western Pacific, the surface front appears with a distinct sea surface temperature (SST) front during winter and early spring (Kobashi and Kawamura 2002). A recent finding is that the SST front of the northern STCC significantly influences the overlying atmosphere. From an analysis of satellite observations and an atmospheric reanalysis, Kobashi et al. (2008) found that, in April to May along the SST front surface, wind stress curls turn weakly cyclonic on the general background of anticyclonic curls that drive the subtropical gyre. The cyclonic wind curls are collocated with a band of high column water vapor, indicative of a deep structure of the moist layer. On the weather timescale, the cyclonic curls are associated with low pressure systems of a subsynoptic scale in space, fueled by atmospheric baroclinicity due to the SST front and latent heat release associated with high SST in April to May. Similar atmospheric effects are observed from coupled GCM simulations, where SST anomalies created by thermal advection of the STCC causes cyclonic surface wind curls and local maximum of precipitation along the STCC (Tokinaga et al. 2009; Xie et al. 2011).

The STCC exhibits variability on various timescales. Analyzing historical temperature profiles in the western North Pacific, White et al. (1978) found seasonal and interannual variations of STF. On the seasonal timescale, the STF is strong in spring and weak in fall. This seasonal cycle has been confirmed by Kobashi and Kawamura (2002) using updated hydrographic observations. They showed that the STCC is clearly identified in winter to

Fig. 2 Annual mean surface zonal geostrophic velocity relative to 400 dbar (*color shade*), and sea surface dynamic height referenced to 1000 dbar (*contours* with an interval of 5 cm). Because 400 dbar corresponds to mostly the deepest level that STF reaches, the zonal velocity represents the vertically integrated strength of STF. *Open circles* denote the position of STF and STCC (from Kobashi et al. 2006)

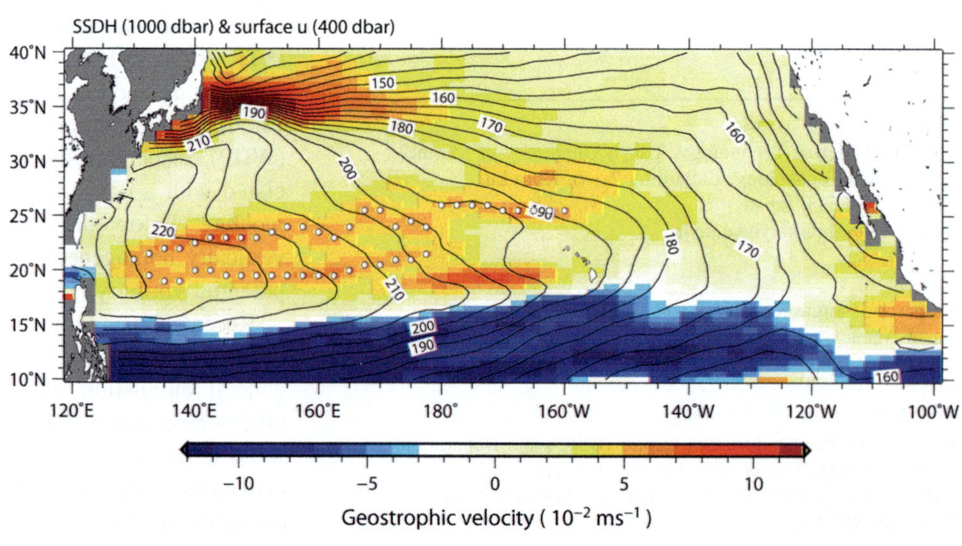

summer with the strongest subsurface STF in spring. The mechanism for the seasonal variations in the subsurface STF has not been clarified yet.

Interannual variations of STCC and STF have not been examined since the study of White et al. (1978). Recently Qiu and Chen (2010) found interannual changes in mesoscale eddy variability in the western STCC region, where enhanced eddy production occurs through baroclinic instability of the vertically sheared current system between the STCC and underlying wind-driven westward flow (Qiu 1999; Kobashi and Kawamura 2002). Analyzing repeat hydrographic surveys across the STCC along the 137°E meridian, which is conducted seasonally by the Japan Meteorological Agency (JMA), they showed that the vertical shear of STCC tends to be larger in eddy-rich years than in eddy-weak years, indicating that the intensity of baroclinic instability controls the interannual changes in the eddy activity. Furthermore, they showed a correlation between the eddy activity and the meridional Ekman convergence associated with basin-scale surface winds, suggesting that the Ekman convergence affects surface temperature frontogenesis along the STCC, causing the changes in the vertical shear of the STCC and thus the eddy activity. Their results indicate the Ekman convergence as a cause of interannual variations in the STCC.

Recent numerical modeling studies revealed STCC variability induced by mode waters (Yamanaka et al. 2008; Nonaka et al. 2011). Analyzing long hindcasts using realistic eddy-resolving ocean GCMs, Yamanaka et al. (2008) showed that the intensity of STF was strong in the late 1970s and weak in the early 1990s. This decadal variability is caused by changes in strength and path of STMW and CMW in the model. The advection of lower PV mode waters strengthens the STF, whereas the weak intensity and reduced southward intrusion of mode water result in the weak STF. A similar effect of mode water on surface circulation is also reported from Argo observations on the Hawaiian Lee Countercurrent by Sasaki et al. (2011). The mode-water-induced STCC variability is also found in coupled GCM simulations, in which decadal variability of mode water ventilation, in particular CMW ventilation, alters the STCC and its thermal advection, inducing SST anomalies and atmospheric response (Xie et al. 2011). Such climate effects of mode water ventilation are found with distinct spatial patterns of SST changes in global warming simulations (Xie et al. 2010; Xu et al. 2011). For details on the changes in the STCC and their relation to mode waters, see the review paper by Kobashi and Kubokawa (2011).

The present study examines seasonal and interannual variations of the northern STCC in the western North Pacific, using observations from satellites and Argo profiling floats and an atmospheric reanalysis. We show that local cyclonic wind curl in May induced by the SST front is a primary cause of interannual variations in the STCC in spring. In a mean state, the northern STCC is maintained by the low PV core of the STMW (Kobashi et al. 2006), which is identified to the north of the STF as a thermostad of about 17°C in Fig. 1. We find no significant correspondence between the variations of the spring STCC and STMW on interannual timescales.

The remainder of the paper is organized as follows. Section 2 describes the data used in this study. Section 3 investigates the seasonal variations of the STCC, the SST front, and the wind stress curl. Section 4 examines the characteristics of interannual variations of the STCC, and then links the STCC variations to local wind curl changes. Section 5 is a summary and discusses the implications of this study.

2 Data

To examine STCC variability, we use a satellite altimetry product, the Maps of Absolute Dynamic Topography from 1993 to 2008, from Archiving, Validation and Interpretation of Satellite Oceanographic data (AVISO), France. The sea surface height (SSH) observed from multiple satellites is merged and mapped onto a 1/3° Mercator grid every 7 days. We use a homogenous dataset based on stable sampling of SSH in time, which is constructed from the observations of two satellites (Topex/Poseidon and European Remote Sensing satellites or Jason-1 and Envisat or Jason-2 and Envisat) with the same ground track.

An optimally interpolated SST dataset developed by the National Climatic Data Center (NCDC) of the National Oceanic and Atmospheric Administration (NOAA) is used to explore the SST front. This dataset is produced from the blend of Advanced Very High Resolution Radiometer (AVHRR) infrared satellite observations and in situ measurements and is processed onto daily SST maps on a 0.25° grid (Reynolds et al. 2007). We analyze the data from 1982 to 2008.

We use the Japanese 25-year Reanalysis (JRA-25), which is produced by JMA and Central Research Institute of Electric Power Industry (CRIEPI). The JRA-25 is computed on a T106 global spectral forecast model with 40 vertical levels using a three-dimensional variational data assimilation method (Onogi et al. 2007). The JRA-25 data cover a period from 1979 to 2004, and the data since 2005 are continually computed using the real-time climatic assimilation system named JMA Climate Data Assimilation System (JCDAS), the same as the JRA-25 data. We use 6-hourly surface wind stress vectors on a T106 Gaussian grid (approximately 1.125° grid) for the period from 1982 to 2008.

Temperature data obtained by Argo profiling floats are used to investigate ocean structures in the STCC region. The dataset is based on Argo's real-time quality controlled data archived by the Argo Global Data Assembly Center, which are further processed with visual inspection for suspect data by Oka et al. (2007). We also use Grid Point Value of the Monthly Objective Analysis using the Argo data (MOAA GPV) produced by Hosoda et al. (2008), where temperature and salinity profiles from real-time quality controlled Argo data and available ship observations are optimally interpolated at standard depths on a 1° grid in space every month since 2001. These data are spatially smoothed, but they can be used to examine changes in mode water broadly distributed to the north of the STCC.

3 Seasonal variability

The monthly climatology maps of SST, meridional SST gradient, wind stress, wind stress curl, and surface geostrophic velocity are calculated from the data from 1993 to 2008. The meridional SST gradient is used to track SST fronts. Because the SST front in the STCC region extends almost zonally, the meridional gradient is a good representation of the strength of the SST front. Surface geostrophic velocity is obtained from absolute SSH maps assuming geostrophy. Figure 3 shows the meridional profiles of these variables zonally averaged between 135° and 150°E. This longitude range corresponds to the area with pronounced cyclonic wind curls induced by the STF (Kobashi et al. 2008).

Figure 3c shows two distinct bands of an eastward flow in the latitude ranges of 18–19°N and 20–26°N. The meridional positions of these two bands are consistent with the hydrographic observations in Fig. 2, and they are referred to as the southern and northern STCCs by Kobashi et al. (2006). The eastward velocity of the northern STCC becomes strong in late winter to early summer with the peak in June, and becomes weak in fall, in agreement with the result of hydrographic observations by Kobashi and Kawamura (2002). The change in the meridional velocity is smaller than in the zonal velocity, but there is an identifiable tendency for the STCC to flow southward in spring to summer (see vectors in Fig. 3c).

Seasonal variation is also evident along the southern STCC, where the STCC appears with eastward velocity in spring to summer. In winter and fall, although no eastward current is present, there is a meridional minimum in westward velocity at almost the same position of the southern STCC. The STF leaves a clear signature in the annual mean map of surface velocity relative to 400 dbar (Fig. 2) that represents the vertically integrated strength of STF. The absolute surface velocity in Fig. 3c reflects density structure integrated over the whole water depth. The lower thermocline below the STF deepens toward the north (Fig. 1), opposing the effect of the STF on the surface velocity field through the thermal wind relation. Thus,

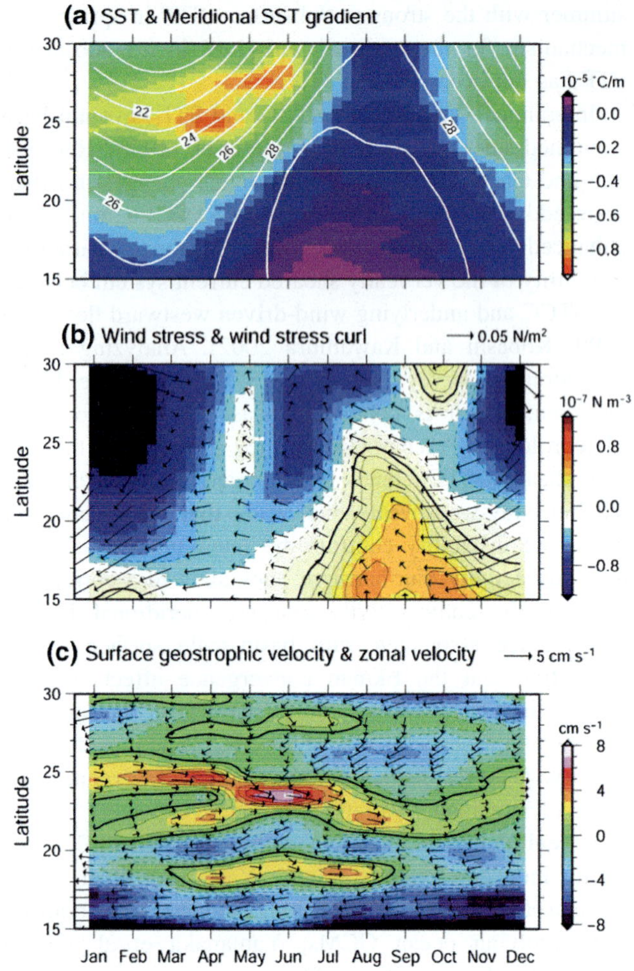

Fig. 3 Mean seasonal cycle as represented in time–latitude sections of **a** SST (*thick contours*) and meridional SST gradient (*color shade*), **b** vector wind stress and wind stress curl (*color shade*), and **c** surface geostrophic velocity vector and zonal velocity (*color shade*), zonally averaged between 135° and 150°E

the weak STF does not result in an eastward current at the surface. This study focuses on the northern STCC.

The SST front appears in late fall to the following spring between 24° and 28°N. In January to April, the SST front is nearly collocated with the STCC along 24.5°N, but in May and June the SST front moves north, while the STCC is slightly shifted to the south. The separation of the SST front from the STCC is also identified from the hydrographic section in Fig. 1, where the SST front is found around 26°N, located further north of the subsurface STF associated with the STCC. The seasonal cycle of the SST front agrees with the results of Kobashi et al. (2008) based on satellite microwave measurements.

Atmospheric effects of the SST front are obvious in the JRA wind stress curl (Fig. 3b). Consistent with the satellite observations of Kobashi et al. (2008), the negative anticyclonic curl over the subtropical gyre decreases locally to the south of the SST front in April to May, indicating cyclonic wind stress curl anomaly induced by the SST front. This cyclonic wind anomaly seems to be followed by the intensification of the STCC in May to June to the south of the cyclonic curl, suggesting that the wind curls in spring have something to do with the subsequent intensification of the STCC. This hypothesis will be tested in Sect. 4 by focusing on interannual variations of the STCC.

The STCC speed is related to both the surface and subsurface STFs. We estimate the contributions of these two fronts, to infer the mechanism for the seasonal variation. The monthly mean STCC velocity is extracted as a meridional maximum between 20° and 26°N in Fig. 3c, plotted for each month in Fig. 4, together with the 90% confidence interval of the zonal mean. The confidence interval is estimated on the basis of a degree of freedom that is calculated from the spatial correlation function used in the mapping procedure of the SSH (Le Trano et al. 1998). From the thermal wind relation, the vertical shear of zonal geostrophic velocity within the surface mixed layer Δu is related to the meridional gradient of temperature $\partial T/\partial y$, as expressed by

$$\Delta u = -\alpha g h(\partial T/\partial y), \qquad (1)$$

where α is the thermal expansion coefficient, g is the gravity constant, and h is the MLD. This study calculated the temperature gradient from the mean meridional profile of the SST in Fig. 3a and used the value at the STCC position. The thermal expansion coefficient was computed using the SST at the STCC position in Fig. 3a and constant salinity (=35). Because this coefficient is almost constant throughout the year, it has little effect on seasonal changes in the vertical shear. The MLD was estimated from the monthly climatology based on Argo observations (Ohno et al. 2009). Averaging the data zonally between 135° and 150°E, we used the value of the MLD at the STCC position. Figure 4 shows the STCC velocity minus the Δu, which represents the effect of the subsurface front below the mixed layer.

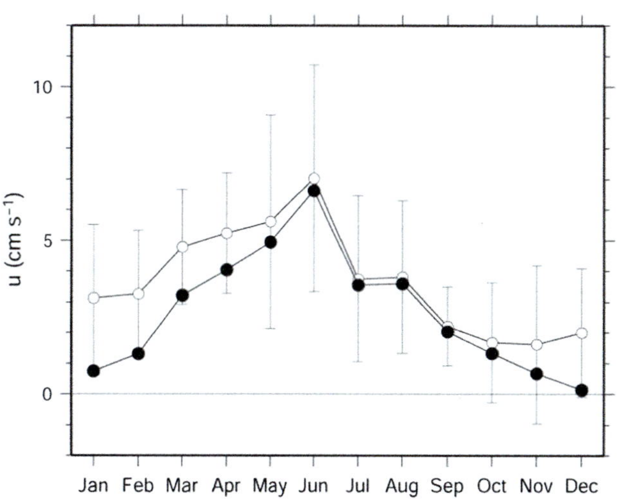

Fig. 4 Monthly mean zonal geostrophic velocity of the STCC extracted as a meridional maximum between 20° and 26°N from the zonally averaged profiles in Fig. 3c (*open circles*). *Error bars* denote 90% confidence interval of the zonal mean. The zonal velocity minus the vertical shear within the surface mixed layer (*filled circles*) represents the subsurface front effect

In November to February, the surface front accounts for the majority of the STCC, explaining about 70% on average. This is due to the formation of a relatively strong surface front (Fig. 3a) within the deep mixed layer. In March and April, although the mixed layer becomes shallower in the STCC region as reported by Takeuchi and Yasuda (2003), the strong surface front (Fig. 3a) makes a significant contribution to the STCC strength (about 30%). From May to October, the STCC is almost maintained by the subsurface front.

4 Interannual variability

The analysis on the seasonal cycle suggests that the wind curls in spring induced by the SST front have some relation to the subsequent variations of the STCC. In this section we test this hypothesis by examining interannual variability.

Before taking a look at the relation between the STCC and the wind stress curl, we describe how the interannual variations of the STCC in each month are related to those in other months, with a focus on late winter to spring when the STCC is strong. For this purpose, the interannual time series of the STCC velocity is calculated for each month, by averaging the zonally averaged zonal geostrophic velocity between 135° and 150°E, within the meridional range of 3° centered at the monthly mean STCC position defined from Fig. 3c. The zonal average suppresses the effect of vigorous mesoscale eddies and isolates the structure of the STCC. Because the meridional position of the STCC is rather stable in time and the STCC is typically situated within 1.5° in latitude north and south of the mean position, the meridional range used here is adequate to capture the STCC variation. Notice that our results are insensitive to the slight change in the meridional range.

Figure 5 shows the correlation coefficients between a month from March to June and all the calendar months. The autocorrelation functions computed from the interannual time series indicate that the dominant temporal scale is almost the same, i.e., about 6 years. On the basis of this result, we estimated the degree of freedom and obtained a confidence limit of the correlation. The STCC variations in March to June are significantly correlated with those in January to July, but the correlation rapidly decreases in August to December, indicating that the variations are

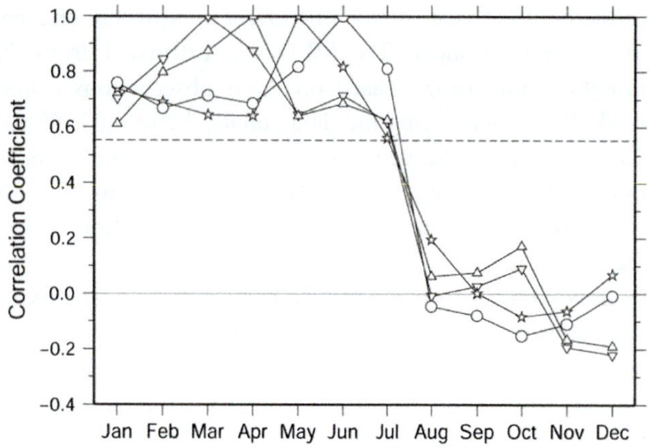

Fig. 5 Correlation coefficients of the interannual variations between the zonal velocity of the STCC in March (*inverted triangles*), April (*triangles*), May (*stars*), and June (*circles*) and in the other months. See the text for details. Correlation exceeding the *dotted line* is significant at the 95% confidence limit (>0.55)

similar during winter to spring, but different from those during summer to fall.

The interannual variations of the STCC, the SST front, and the wind stress curl were examined for each month. We first calculated the zonal averages of the zonal geostrophic velocity, the magnitude of SST gradient, and the wind stress curl between 135° and 150°E from their monthly mean maps. The interannual standard deviations of these zonal mean variables were then computed for each month from the monthly meridional profiles. The result is shown in Fig. 6.

The variance of the wind stress curl shows a local peak in the area of climatological cyclonic wind curls in April and May, which is almost coincident with the high variance of the SST gradient (Fig. 6a, b). This collocation indicates a local interaction between the SST front and the surface wind, though the high variance of the SST gradient occurs to the south of the mean position of the SST front. The zonal velocity shows high variance at 21–22°N on the south flank of the northern STCC from June to September when the STCC is maintained by the subsurface STF.

To explore the local interaction between the SST front and surface wind, we calculated the interannual correlation coefficients between the wind stress curl in May and the SST gradient magnitude, SST anomaly, and SSH anomaly for each month (Fig. 7). The calculation was done at each latitude based on the monthly zonal averages in 135° and 150°E. To highlight signals associated with the narrow band of the STCC, a meridional high-pass filter is applied to SST and SSH anomalies by removing an 8° moving average as the background state. The correlation coefficients were computed for the SSH anomaly from 1993 to 2008 and for the other variables from 1982 to 2008, and their statistical significance levels were calculated on the

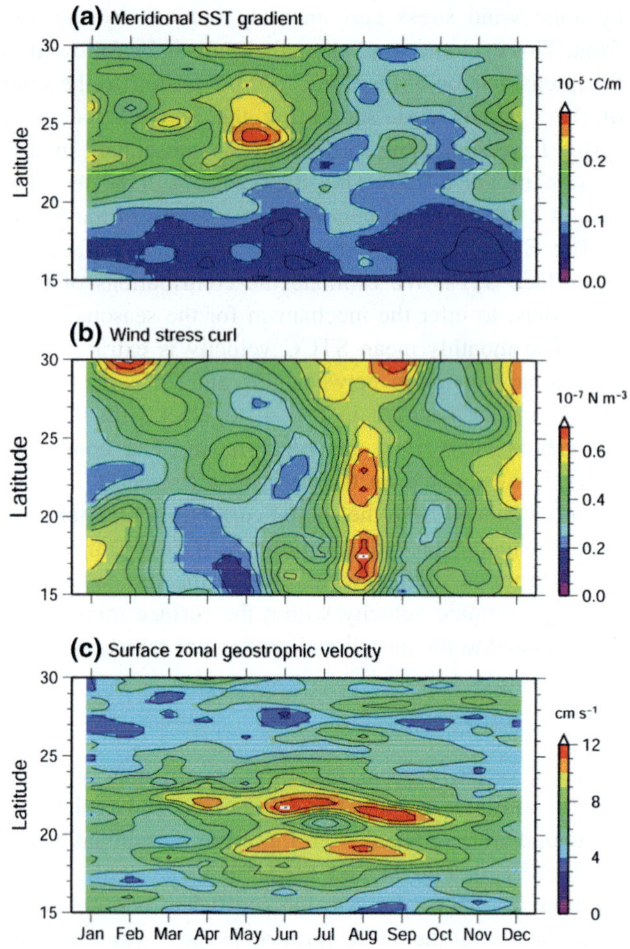

Fig. 6 Monthly distribution of the interannual standard deviations of **a** meridional SST gradient, **b** wind stress curl, and **c** surface zonal geostrophic velocity, calculated from the monthly mean meridional profiles zonally averaged between 135° and 150°E

basis of the degrees of freedom determined from the dominant temporal scale of the STCC variations in spring.

In the band of cyclonic wind curls centered at 25°N in the May climatology, the interannual anomalies of SST and SSH show significant negative correlation, whereas the SST gradient has positive correlation (slightly to the south of the negative SST correlation). This result means that strong (weak) cyclonic wind curl is concurrent with the fall (rise) in SST and SSH below and intensified (weakened) SST front to the south. The negative correlation with the SSH anomalies persists through June along the northern flank of the strong STCC at 23.5°N in May to June (Fig. 3c).

The lag correlation implies causality, as has been discussed in many studies (e.g., Frankignoul 1985). The correlations between the wind curl and the SST gradient and the SSH anomaly reach a peak in May with zero lag (Fig. 7), indicating that their variations are coupled. In addition, the correlations remain high and marginally

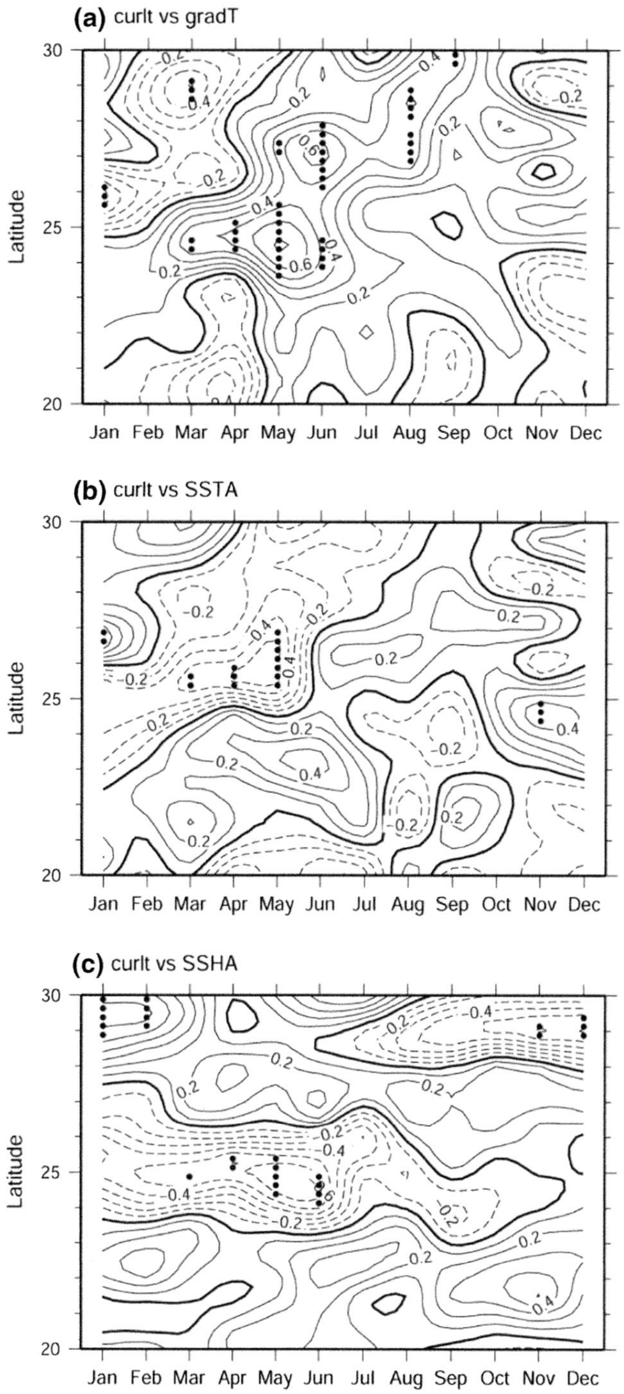

Fig. 7 Correlation coefficients of the interannual variations between the wind stress curl in May and **a** the magnitude of SST gradient, **b** SST anomaly, and **c** SSH anomaly in the other months, calculated at each latitude using the monthly mean meridional profiles zonally averaged between 135° and 150°E. Grid points of significant correlation at the 95% confidence limit are denoted by *black dots*

significant along 24–25°N with the SST gradient and the SSH anomaly leading from late winter. This suggests that the May wind stress curl is induced to some degree by the SST front and STCC in the preceding late winter to early spring, though the mechanism is not clear. We will discuss these results in Sect. 5.

Figure 8 shows time series of the mean wind stress curl and SST gradient in May and the May-to-June mean SSH anomaly averaged in the area of 24–26°N and 135–150°E. To see the relation with the STCC, the May-to-June mean zonal velocity averaged in the south area of the cyclonic wind anomaly, in 22–25°N and 135–150°E, are also calculated. In support of Fig. 7, the wind curl changes almost in synchronization with the SST gradient and the SSH. The SSH is negatively correlated with the zonal velocity of the STCC, indicating that the fall (rise) in the SSH results in the strong (weak) STCC to the south by geostrophic balance.

The interannual variations in Fig. 8 are consistent with satellite observations. Using surface wind velocity measured by the QuikSCAT scatterometer and SST by the Tropical Rainfall Measuring Mission (TRMM) satellite's Microwave Imager (TMI) from 2000 to 2008, we a obtained significant correlation of 0.69 at the 95% confidence limit between the SST gradient and the wind stress curl.

To examine how ocean structure changes in response to the cyclonic wind stress anomaly, we analyzed the Argo profiling float data. The Argo data have increased in number and spatial coverage in the STCC region since 2005. Focusing on the two characteristic years of 2005 with strong cyclonic wind anomaly and 2006 with weak cyclonic anomaly (Fig. 8), we calculated the mean vertical profiles of potential temperature from the observations in May and June in the area of 24–26°N and 135–150°E where the significant correlation between the wind curl and the SSH anomaly is seen in Fig. 7. Figure 9a compares 2005 and 2006 when the cyclonic wind curls are strong and weak, respectively. There were 13 profiles for 2005, and 32 for 2006.

Compared with the composite profile in 2006, the temperature decreases overall and the isotherms are lifted upward in 2005. The difference in depth of isotherms between the 2 years is shown in Fig. 9b. The isotherms are elevated by about 50 m between 7° and 17°C in the lower main thermocline (Fig. 9a). This rise of the thermocline simultaneously with the cyclonic wind stress curl is indicative of enhanced Ekman suction. For comparison, the same analysis was done for the areas south and north of the cyclonic wind anomaly between 21.5–23.5°N and 26.5–28.5°N in the same longitude range (not shown). Unlike the area of the cyclonic anomaly, the temperature profiles do not show large differences between the 2 years, demonstrating that the change in the thermocline is the response to the local wind forcing via Ekman suction.

Argo observations in the STCC region are relatively sparse and the spatial sampling is different between 2005 and 2006. The different sampling could cause apparent

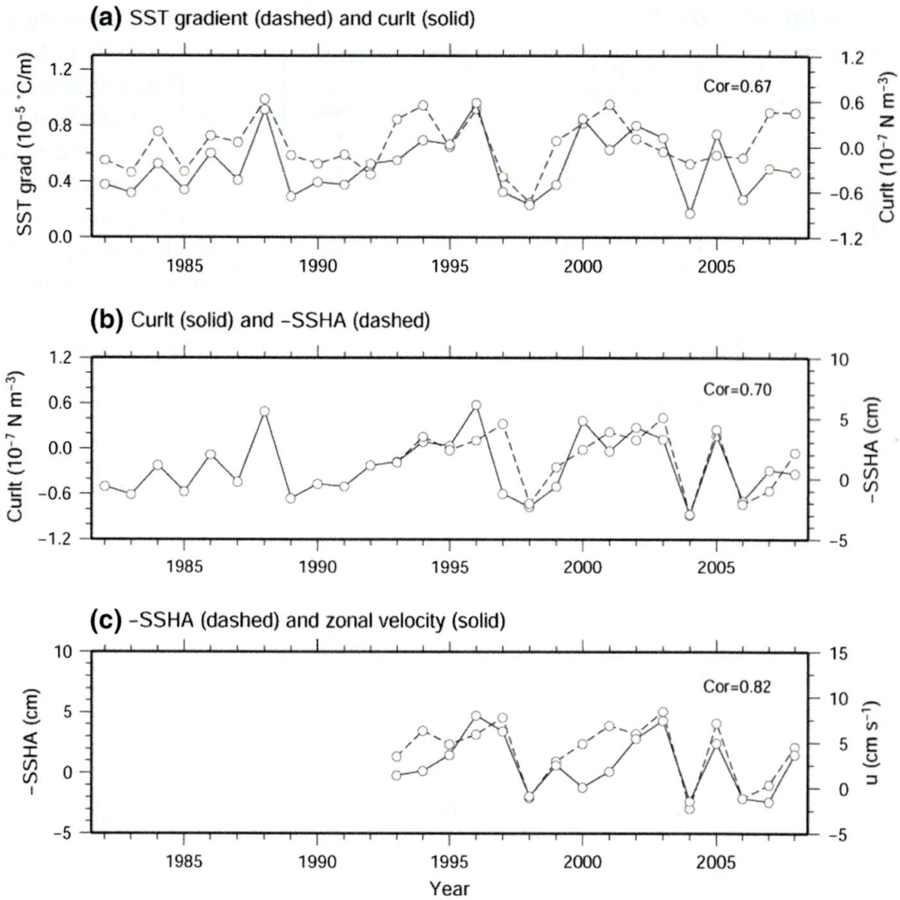

Fig. 8 Time series of May mean wind stress curl and SST gradient and the May-to-June mean SSH anomaly averaged in the area of 24–26°N and 135–150°E, and the May-to-June mean zonal geostrophic velocity of the STCC averaged in 22–25°N and 135–150°E. The SST gradient is shown by a *dashed line* in **a**, the wind stress curl by a *solid line* in **a** and **b**, the SSH anomaly by a *dashed line* in **b** and **c**, the zonal velocity by a *solid line* in **c**. For easy comparison, the SSH anomaly is multiplied by a minus

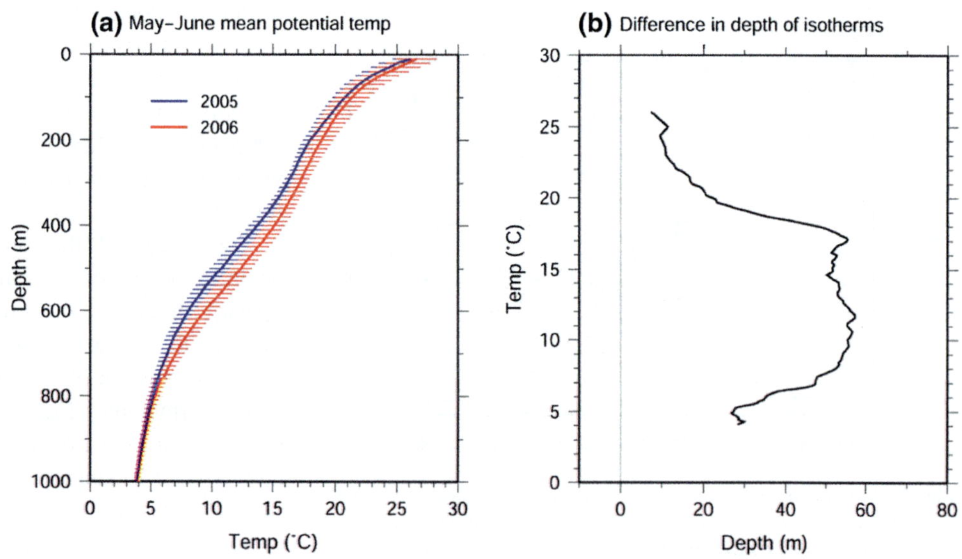

Fig. 9 **a** Mean vertical profiles of potential temperature in 2005 and 2006 together with the standard deviation of the mean, calculated from the Argo profiling floats in May and June in the area of 24–26°N and 135–150°E. **b** Difference in depth of isotherms between 2005 and 2006. The number of profiles used is 13 for 2005 and 32 for 2006

differences in the mean temperature profiles in Fig. 9, when the temperature varies in space. To evaluate this effect, an additional analysis was done using the monthly temperature climatology from the World Ocean Atlas (WOA) 2005 (Locarnini et al. 2006). The climatological temperature profile at the position of the Argo observation is obtained from the monthly WOA and is subtracted from the Argo temperature profile. Then, the anomaly profiles are averaged in the area, added to the area mean profile calculated from the WOA. This procedure reduces the difference in the spatial sampling. The result is quantitatively almost the same as Fig. 9 (not shown), corroborating that the difference in the temperature profile between the 2 years is not due to that in the spatial sampling of Argo observations.

The change in the deep ocean down to the lower thermocline implies that mode water is not involved with the difference in the STCC between 2005 and 2006 springs. As mentioned in Sect. 1, the mean northern STCC is anchored by the low PV core of the STMW between the upper and lower thermoclines that is advected to the north of the STCC by the Kuroshio recirculation gyre (Kobashi et al. 2006). The STMW is found in Fig. 9a with weak vertical gradients of temperature around 17°C. As mentioned above, because the difference between the 2 years appears over a wide range of temperature, it seems unlikely that the change in the STMW causes the STCC difference between the 2 years.

The influence of the STMW on the spring STCC is examined using the Argo gridded dataset from the MOAA GPV during 2001 to 2008. The zonal velocity of the STCC from the MOAA GPV is similar to that from the satellite observations in variability (not shown). The layer thickness in the typical density range of the STMW (1025.0–1025.7 kg m^{-3}) was computed at each grid from the monthly data, and was zonally averaged between 135° and 150°E. Figure 10 depicts the meridional profiles of the May–June mean STMW thickness, together with the area-averaged zonal velocity in Fig. 8c to compare with the STCC changes.

The thick STMW appears around 30.5°N. The change in the STMW is characterized by a gradual thickening from 2001 to 2006, consistent with the result of Qiu and Chen (2006), and likely related to the dynamic state of the Kuroshio Extension. There is no evidence that the southward

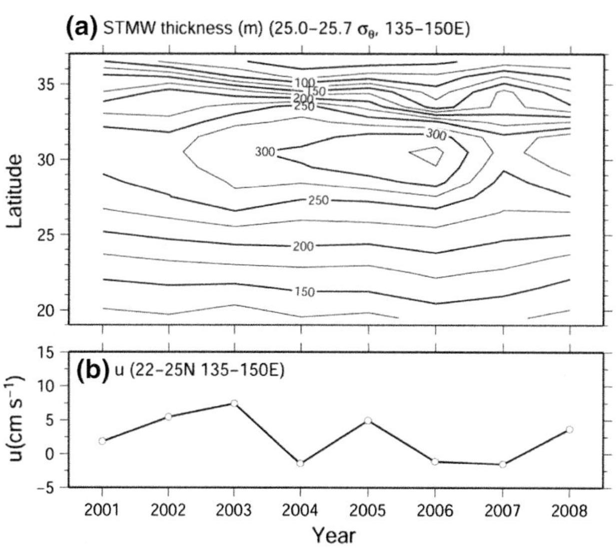

Fig. 10 **a** Interannual changes in May–June mean STMW thickness in the density range 1025.0–1025.7 kg m^{-3}, zonally averaged between 135° and 150°E. **b** Area-averaged zonal velocity of the STCC (the same as Fig. 8c)

intrusion of thick STMW is coincident with the intensification of the STCC velocity, indicating no connection of the STMW with the interannual variations of the spring STCC. This 8-year result, however, cannot rule out a connection between the mode water and the STCC on decadal timescales, as suggested by the model study of Yamanaka et al. (2008). Because the STMW fluctuates dominantly on decadal timescales (Yasuda and Hanawa 1997; Qiu and Chen 2006; Sugimoto and Hanawa 2010; Oka and Qiu 2011), longer observations are required to assess the influence of the STMW on the STCC. Indeed in an analysis of an eddy-resolving ocean GCM hindcast, Nonaka et al. (2011) showed that mode water variability is important for STCC on decadal timescales and has a weaker but still discernible effect on interannual variations in STCC.

5 Summary and discussion

The present study has used observations from satellites and Argo profiling floats and an atmospheric reanalysis to investigate seasonal and interannual variations of the STCC in the western North Pacific. The satellite altimeter data confirm the two bands of the STCC, consistent with hydrographic observations. The present study focuses on the northern STCC in the latitude range of 20–26°N. The STCC has a clear seasonal cycle. It is strong in late winter to early summer (March to June) with a peak in June, and weak in fall. The SST front that forms in the mixed layer appears along the STCC in January to April, contributing to the strong STCC in March to April. In May to June, while the surface front moves northward, the STCC is slightly shifted to the south. For these months, the STCC is maintained by the enhanced subsurface STF.

The SST front induces weak cyclonic wind curls to the north of the STCC in May. The correlation analysis for interannual variations reveals that the strong (weak) cyclonic wind anomaly causes the SST and SSH to fall (rise), accelerating (decelerating) the STCC to the south by geostrophic balance. The changes in the ocean have a deep vertical structure down to the lower thermocline. The thermocline is lifted upward locally in the presence of the cyclonic wind curl, indicative of enhanced Ekman suction. The Ekman suction increases the meridional tilt of the thermocline, accelerating the STCC to the south. Thus, the local wind forcing in May is a major factor for interannual variations in the spring STCC.

These results indicate that in May the ocean affects the atmosphere, which then in turn affects the ocean. Why does such ocean–atmosphere interaction occur in May? First, we address why the cyclonic wind anomaly forms in May. This issue was discussed by Kobashi et al. (2008), who suggested that a strong SST front and high SST in May

(Fig. 3a) both favor the generation of the cyclonic wind anomaly.

Why does the cyclonic wind anomaly in May exert a significant influence on the STCC? Using a 1.5-layer reduced-gravity model, the baroclinic ocean response to surface wind forcing is expressed by the linear vorticity equation

$$\frac{\partial h}{\partial t} = -\frac{g'}{g} w_e, \tag{2}$$

where h is the SSH, and g' the reduced gravity. Ekman pumping velocity w_e is defined by $w_e = \mathrm{curl}_z(\tau/f)/\rho_0$, where τ is the wind stress vector, f the Coriolis parameter, and ρ_0 the reference density. Here the dissipation and SSH propagation due to Rossby waves are omitted for simplicity. Taking the meridional derivative and applying the geostrophic relation yield

$$\frac{\partial u}{\partial t} = \frac{g'}{f} \frac{\partial w_e}{\partial y}, \tag{3}$$

where u is the surface zonal geostrophic velocity. The equation indicates that a positive meridional gradient of Ekman pumping velocity leads to an increase in surface eastward velocity. Using the monthly climatology of wind stress data from QuikSCAT observations, we calculated the mean gradient of Ekman pumping velocity for each month averaged in the STCC region of 20–25°N, 135–150°E. The result is shown in Fig. 11. Except in April and May, the meridional gradient of Ekman pumping is nearly zero or negative, because the STCC is usually situated along the southern flank of the bowl structure of negative Ekman pumping velocity in the subtropical gyre, as inferred from the wind stress curl distribution in Fig. 3b. This indicates that the Ekman pumping forcing acts to weaken the STCC, against which the STCC is maintained by mode water. In April and May, however, the meridional gradient of Ekman pumping changes to be positive as a zonal band of cyclonic wind curls forms to the north of the STCC, causing a subsequent intensification of the STCC in late spring. Thus, the cyclonic wind anomaly in spring has an important effect on the STCC. The same result is also obtained from JRA-25 wind stress, though the magnitude of the meridional gradient is smaller compared to satellite observations because of the coarse resolution of the reanalysis. Note that substituting a typical value of $g' = 0.04$ m s^{-2} in the STCC region in May into Eq. 3 gives rise to a change of eastward velocity of about 0.5 cm s^{-1} for 1 month in May, which is on the same order of magnitude as the actual change in the STCC velocity, about 0.7 cm s^{-1} as estimated from the monthly change in surface velocity in spring averaged in the area of the STCC with eastward velocity in Fig. 3c.

The wind stress curl and SST gradient exhibit large interannual variability in May and are positively correlated

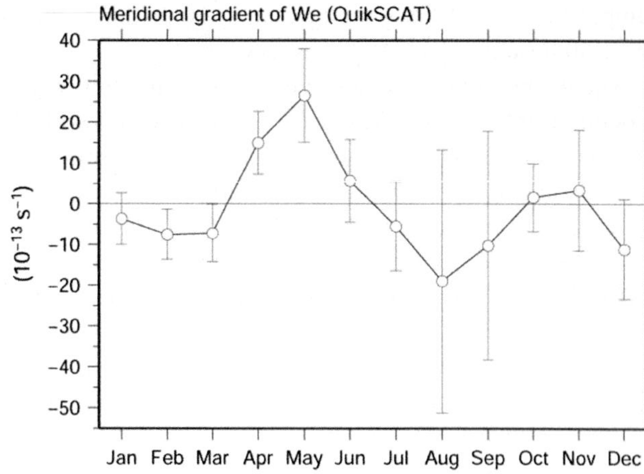

Fig. 11 Monthly mean meridional gradients of Ekman pumping velocity in the STCC area of 20–25°N and 135–150°E, calculated from QuikSCAT observations. *Error bars* denote 95% confidence interval of the area mean

with each other, suggesting a positive feedback of the local interaction between the surface wind and the SST front. The cyclonic wind anomaly lowers the SST below and enhances the SST front to the south. The enhanced SST front intensifies the baroclinicity of the lower atmosphere, energizing subsynoptic lows and resulting in the formation of the band of cyclonic wind anomaly (Kobashi et al. 2008).

The lag correlation analysis suggests that the May wind stress curl is affected by the preceding SST front and STCC in late winter to early spring. In these seasons, the SST front makes a large contribution to the strength of STCC. Interannual variations in meridional SST gradient and surface zonal velocity are negatively correlated along the mean position of the STCC at −0.61 and −0.74 for February and March, respectively, indicating that the STCC velocity in late winter is affected by the SST front. The strong SST front and STCC in late winter possibly gives favorable conditions for the development of a strong SST front in spring, which induces the formation of the cyclonic wind anomaly and results in the subsequent variations of the STCC in spring. Indeed, the STCC variations in spring and winter are highly correlated (Fig. 5). Further analysis will be needed to clarify the mechanism of how the atmosphere in spring is affected by the preceding SST front.

What regulates the interannual variations of the SST front in late winter? In the subtropical gyre, with the westerlies to the north and the northeast trade winds to the south, the Ekman convergence is considered an important process for the changes in the surface front along the STCC (Kazmin and Rienecker 1996; Dinniman and Rienecker, 1999). Indeed, Qiu and Chen (2010) found a correlation between interannual changes in the mesoscale eddy activity in the STCC region and the meridional Ekman convergence. In the STCC region, because the mesoscale eddies are due to baroclinic instability of the STCC in late winter (Qiu 1999; Kobashi and Kawamura 2002), the interannual change in the eddy activity is considered to represent the STCC strength in late winter.

We examined the relation between the meridional Ekman convergence and the late winter SST front. Following Nakamura and Kazmin (2003) and Qiu and Chen (2010), the interannual changes in the meridional SST gradient, $G \equiv -\partial T/\partial y$, are related to the meridional Ekman convergence, as expressed by

$$\frac{\partial G}{\partial t} \approx -\frac{\partial}{\partial y}(V_E \cdot G), \tag{4}$$

where $V_E = -\tau_x/\rho f h$ is the meridional Ekman velocity, τ_x is the zonal wind stress, ρ is the reference density, f is the Coriolis parameter, and h is the depth of the surface mixed layer. Here we have neglected zonal variations and vertical advection/mixing. This is rewritten in the integral form as follows,

$$G \approx \int_{t_1}^{t_2} -\frac{\partial}{\partial y}(V_E \cdot G)\mathrm{d}t. \tag{5}$$

$V_E \cdot G$ was calculated for each month during 1982 to 2008 using JRA-25 wind stress and NOAA SST data and the monthly mean climatology of the MLD (Ohno et al. 2009), and then was averaged in the SST front region of 24–26°N, 135–150°E. The integrated Ekman convergence was computed for February and March by setting t_2 to be February or March and t_1 from the previous October to t_2. The correlation coefficients between the interannual variations of G and the integrated Ekman convergence in February/March are shown in Fig. 12. The statistical significance of the correlation was estimated from the degrees of freedom based on the dominant temporal scale of G. The SST gradient in February shows a significant correlation at the 95% confidence limit with the Ekman convergence integrated from January to February (the correlation coefficient is 0.59). The SST gradient in March shows a weak but positive correlation with the integrated Ekman convergence during winter. These results suggest that the winter Ekman convergence affects the interannual changes in the SST front in February and March. However, the comparison of the interannual variances shows that the winter Ekman convergence explains only at most 10% of the variance of the SST front. We need further investigation to clarify the mechanism of the interannual variations of the SST front in late winter.

Our results suggest the following mechanism for interannual STCC variations (Fig. 13). The SST front and thereby the STCC in late winter affect the subsequent

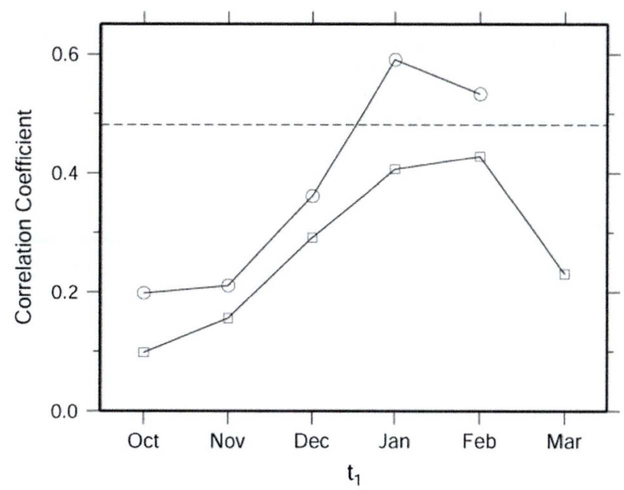

Fig. 12 Correlation coefficients between the interannual variations of the meridional SST gradients (G) in February (*circles*) and March (*squares*) and the meridional Ekman convergence integrated from t_1 to February (*circles*) and March (*squares*). See the text for details. Correlation exceeding the *dotted line* is significant at the 95% confidence limit

development of the cyclonic wind anomaly in spring. The strong (weak) cyclonic wind anomaly enhances (weakens) the Ekman suction and elevates (depresses) the thermocline below, accelerating (decelerating) the STCC to the south. Because the strong Ekman suction acts to enhance the SST front, the positive feedback between the wind curl and the SST front is suggested to occur in spring.

The cyclonic wind curl anomaly in the STCC region was first reported about 40 years ago by Yoshida and Kidokoro (1967a, b) from an analysis of historical ship-based wind charts. They considered that this cyclonic wind drives the mean STCC. However, it proved to be the other way around, with the SST front actually causing the cyclonic wind anomaly. Although the cyclonic wind anomaly is not important for the formation of the mean STCC (Takeuchi 1984), the present study indicates that it is a major factor for interannual variations of the spring STCC.

The interannual variations of the STCC are different between winter to spring and summer to fall (Fig. 5). The present study reveals that the variations in May to June are coupled to the wind stress curl and are affected by the preceding STCC in late winter. By contrast, the STCC in summer to fall shows no correlation with that in winter and spring, suggestive of other mechanisms. The STCC in these seasons is not so strong but exhibits large interannual variability (Fig. 6). We examined but found no relationship to STMW (not shown). Because the eddy activity is high in the STCC region especially in spring, eddy–mean current interaction could be one possible mechanism for the STCC variations in summer to fall.

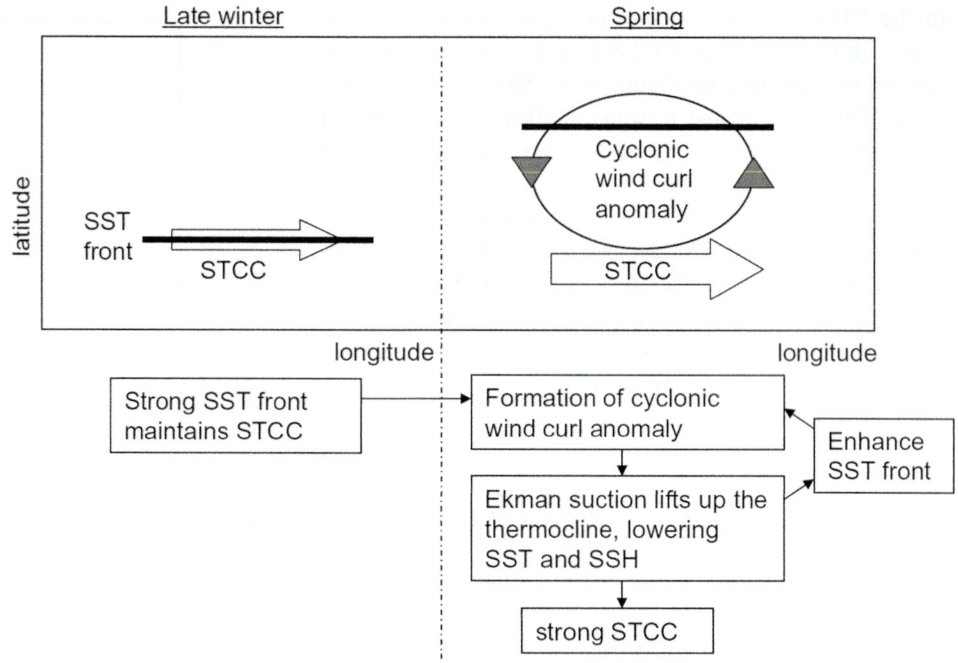

Fig. 13 A schematic summary depicting the interannual variations of the STCC (*open arrows*) and the SST front (*thick horizontal lines*) related to the cyclonic wind curl (*circle*) during late winter (*left panel*) and spring (*right panel*). See the text for details

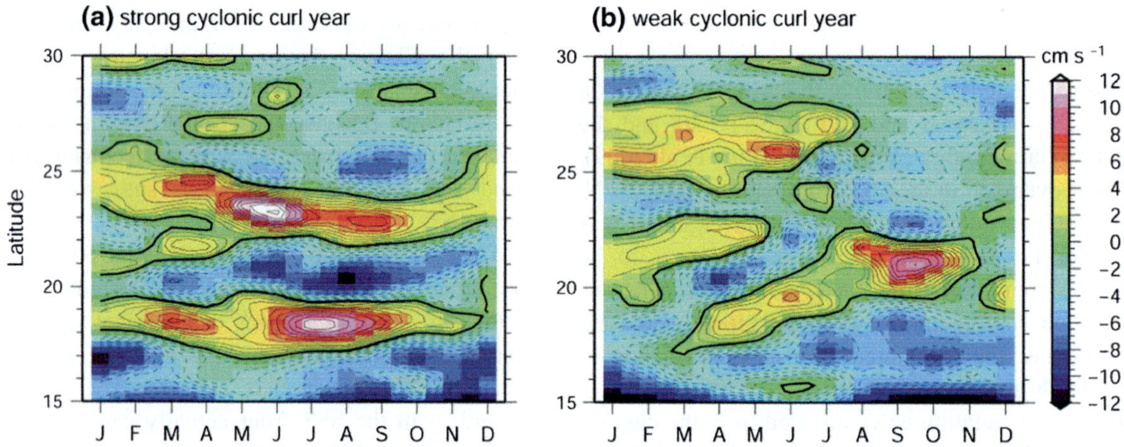

Fig. 14 Composite of seasonal cycle of surface zonal geostrophic velocity in **a** strong cyclonic curl years (1996, 2000, 2002, 2005) and **b** weak cyclonic curl years (1998, 2004, 2006), calculated from the monthly mean meridional profiles zonally averaged between 135° and 150°E

White et al. (1978) suggested from an analysis of temperature data during 1954–1974 that the annual mean STCC tends to be strong in El Niño years. We examined the relation between the STCC in spring and El Niño Southern Oscillation (ENSO), by calculating the lag correlations between the May–June mean zonal velocity of the STCC in Fig. 8c and the monthly climate indices in each month from the previous May to May of the Southern Oscillation index (SOI) and the Niño3 index. These indices are obtained from the JMA website and are smoothed with a 5-month running mean. The maximum correlation coefficients with the SOI and Niño3 are 0.21 and 0.30, respectively, neither being insignificant. Significant correlation was not found between the wind stress curl in the STCC region (Fig. 8a) and the climate indices. These results demonstrate that interannual variations of the spring STCC are induced by the local wind changes rather than by the large-scale circulation changes associated with the ENSO.

Finally, we examine the influence of the cyclonic wind stress anomaly in May on the STCC seasonal cycle. Figure 14 shows the composite profiles of the zonal velocity zonally averaged in the same area as in Fig. 3c, for the years of strong and weak cyclonic wind anomaly. The strong and weak years were defined as those when the wind curl time series exceeds plus and minus one standard deviation, respectively. In the strong cyclonic wind years, the STCC remains as a narrow and sharp band of the eastward flow all year round, and the seasonal cycle is clearly identified with a maximum in spring and a minimum in fall, the same as the mean seasonal cycle in Fig. 3c. In contrast, the STCC is rather blurred and the spring peak disappears in the weak cyclonic curl years. These features strongly suggest that the cyclonic wind anomaly in May is an important factor to form the robust STCC with a clear seasonal cycle.

Acknowledgments Constructive comments from anonymous reviewers greatly helped improve the manuscript. This study is partially supported by Grants-in-Aid for Scientific Research of the Ministry of Education, Culture, Sports, Science and Technology, Japan [20740267, 22106007, 23340139]. The altimeter product was produced by Ssalto/Duacs and distributed by AVISO, with support from CNES. The SST dataset was obtained from the NOAA website, the TMI SST and QuikSCAT wind data from the Remote Sensing Systems website, and the JRA-25 data from the JMA website. The Argo profiling float data is obtained from the website (http://ocg.aori.u-tokyo.ac.jp/member/eoka/data/), and the Grid Point Value of the Monthly Objective Analysis using the Argo data (MOAA GPV) from the Japan Agency for Marine-Earth Science and Technology (JAMSTEC) website. IPRC publication 796.

References

Aoki Y, Suga T, Hanawa K (2002) Subsurface subtropical fronts of the North Pacific as inherent boundaries in the ventilated thermocline. J Phys Oceanogr 32:2299–2311

Dinniman MS, Rienecker MM (1999) Frontogenesis in the North Pacific oceanic frontal zones: a numerical simulation. J Phys Oceanogr 29:537–559

Frankignoul C (1985) Sea surface temperature anomalies, planetary waves and air–sea feedback in the middle latitudes. Rev Geophys 23:357–390

Hosoda S, Ohira T, Nakamura T (2008) A monthly mean dataset of global oceanic temperature and salinity derived from Argo float observations. JAMSTEC Rep Res Dev 8:47–59

Kazmin AS, Rienecker MM (1996) Variability and frontogenesis in the large-scale oceanic frontal zones. J Geophys Res 101:907–921

Kobashi F, Kawamura H (2002) Seasonal variation and instability nature of the North Pacific Subtropical Countercurrent and the Hawaiian Lee Countercurrent. J Geophys Res 107:3185. doi:10.1029/2001JC001225

Kobashi F, Kubokawa A (2011) Review on North Pacific Subtropical Countercurrent and Subtropical Front: role of mode water in ocean circulation and climate. J Oceanogr (this issue)

Kobashi F, Mitsudera H, Xie S-P (2006) Three subtropical fronts in the North Pacific: observational evidence for mode water-induced subsurface frontogenesis. J Geophys Res 111:C09033. doi:10.1029/2006JC003479

Kobashi F, Xie S-P, Iwasaka N, Sakamoto TT (2008) Deep atmospheric response to the North Pacific oceanic subtropical front in spring. J Climate 21:5960–5975

Kubokawa A (1999) Ventilated thermocline strongly affected by a deep mixed layer: a theory for subtropical countercurrent. J Phys Oceanogr 29:1314–1333

Kubokawa A, Inui T (1999) Subtropical countercurrent in an idealized ocean GCM. J Phys Oceanogr 29:1303–1313

Locarnini RA, Mishonov AV, Antonov JI, Boyer TP, Garcia HE (2006) World ocean atlas 2005, vol 1. In: Levitus S (ed) Temperature. NOAA atlas NESDIS 61. US Government Printing Office, Washington, DC

Le Traon PY, Nadal F, Ducet N (1998) An improved mapping method of multi-satellite altimeter data. J Atmos Oceanic Technol 25:522–534

Masuzawa J (1969) Subtropical mode water. Deep Sea Res 16:436–472

Nakamura H (1996) A pycnostad on the bottom of the ventilated portion in the central subtropical North Pacific: its distribution and formation. J Oceanogr 52:171–188

Nakamura H, Kazmin AS (2003) Decadal changes in the North Pacific oceanic frontal zones as revealed in ship and satellite observations. J Geophys Res 108:3078. doi:10.1029/1999JC000085

Nonaka M, Xie S-P, Sasaki H (2011) Interannual variations in low potential vorticity water and the subtropical countercurrent in an eddy-resolving OGCM. J Oceanogr. doi:10.1007/s10872-011-0042-3 (this issue)

Ohno Y, Iwasaka N, Kobashi F, Sato Y (2009) Mixed layer depth climatology of the North Pacific based on Argo observations. J Oceanogr 65:1–16

Oka E, Qiu B (2011) Progress of North Pacific mode water research in the past decade. J Oceanogr. doi:10.1007/s10872-011-0032-5 (this issue)

Oka E, Talley LD, Suga T (2007) Temporal variability of winter mixed layer in the mid- to high-latitude North Pacific. J Oceanogr 63:293–307

Onogi K, Tsutsui J, Koide H, Sakamoto M, Kobayashi S, Hatsushika H, Matsumoto T, Yamazaki N, Kamahori H, Takahashi K, Kadokura S, Wada K, Kato K, Oyama R, Ose T, Mannoji N, Taira R (2007) The JRA-25 reanalysis. J Meteor Soc Japan 85:369–432

Qiu B (1999) Seasonal eddy field modulation of the North Pacific subtropical countercurrent: TOPEX/Poseidon observations and theory. J Phys Oceanogr 29:2471–2486

Qiu B, Chen S (2006) Decadal variability in the formation of the North Pacific subtropical mode water: oceanic versus atmospheric control. J Phys Oceanogr 36:1365–1380

Qiu B, Chen S (2010) Interannual variability of the North Pacific Subtropical Countercurrent and its associated mesoscale eddy field. J Phys Oceanogr 40:213–225

Reynolds RW, Smith TM, Liu C, Chelton DB, Casey KS, Schlax MG (2007) Daily high-resolution blended analyses for sea surface temperature. J Climate 20:5473–5496

Sasaki H, Xie S-P, Taguchi B, Nonaka M, Hosoda S, Masumoto Y (2011) Interannual variations of the Hawaiian Lee Countercurrent induced by low potential vorticity water ventilation in the subsurface. J Oceanogr (this issue)

Suga T, Takei Y, Hanawa K (1997) Thermostad distribution in the North Pacific subtropical gyre: the central mode water and the subtropical mode water. J Phys Oceanogr 27:140–152

Sugimoto S, Hanawa K (2010) Impact of Aleutian low activity on the STMW formation in the Kuroshio recirculation gyre region. Geophys Res Lett 37:L03606. doi:10.1029/2009GL041795

Takeuchi K (1984) Numerical study of the Subtropical Front and the Subtropical Countercurrent. J Oceanogr Soc Japan 40:371–381

Takeuchi E, Yasuda I (2003) Wintertime shoaling of oceanic surface mixed layer. Geophys Res Lett 30(22):2152. doi:10.1029/2003GL018511

Tokinaga H, Xie S-P, Kobashi F, Tanimoto Y (2009) Local and remote influences of the Kuroshio Extension on the atmosphere. US CLIVAR Variations 7:1–4

Uda M, Hasunuma K (1969) The eastward subtropical countercurrent in the western North Pacific Ocean. J Oceanogr Soc Japan 25:201–210

White WB, Hasunuma K, Solomon H (1978) Large-scale seasonal, secular variability of the subtropical front in the western North Pacific from 1954 to 1974. J Geophys Res 83:4531–4544

Xie S-P, Deser C, Vecchi GA, Ma J, Teng H, Wittenberg AT (2010) Global warming pattern formation: sea surface temperature and rainfall. J Climate 23:966–986

Xie S-P, Xu L-X, Liu Q, Kobashi F (2011) Dynamical role of mode-water ventilation in decadal variability in the central subtropical gyre of the North Pacific. J Climate 24:1212–1225

Xu L-X, Xie S-P, Liu Q, Kobashi F (2011) Response of the North Pacific subtropical countercurrent and its variability to global warming. J Oceanogr. doi:10.1007/s10872-011-0031-6 (this issue)

Yamanaka G, Ishizaki H, Hirabara M, Ishikawa I (2008) Decadal variability of the Subtropical Front of the western North Pacific in an eddy-resolving ocean general circulation model. J Geophys Res 113:C12027. doi:10.1029/2008JC005002

Yasuda T, Hanawa K (1997) Decadal changes in the mode waters in the midlatitude North Pacific. J Phys Oceanogr 27:858–870

Yoshida K, Kidokoro T (1967a) A subtropical countercurrent in the North Pacific: an eastward flow near the Subtropical Convergence. J Oceanogr Soc Japan 23:88–91

Yoshida K, Kidokoro T (1967b) A subtropical countercurrent (II): a prediction of eastward flows at lower subtropical latitudes. J Oceanogr Soc Japan 23:231–236

J Oceanogr (2012) 68:127–137
DOI 10.1007/s10872-011-0031-6

SPECIAL SECTION: ORIGINAL ARTICLE

New developments in mode-water research: Dynamic and climatic effects

Response of the North Pacific subtropical countercurrent and its variability to global warming

Lixiao Xu · Shang-Ping Xie · Qinyu Liu · Fumiaki Kobashi

Received: 30 December 2010 / Revised: 11 April 2011 / Accepted: 28 April 2011 / Published online: 21 May 2011

Abstract Response of the North Pacific subtropical countercurrent (STCC) and its variability to global warming is examined in a state-of-the-art coupled model that is forced by increasing greenhouse gas concentrations. Compared with the present climate, the upper ocean is more stratified, and the mixed layer depth (MLD) shoals in warmer climate. The maximum change of winter MLD appears in the Kuroshio–Oyashio extension (KOE) region, where the mean MLD is the deepest in the North Pacific. This weakens the MLD front and reduces lateral induction. As a result of the reduced subduction rate and a decrease in sea surface density in KOE, mode waters form on lighter isopycnals with reduced thickness. Advected southward, the weakened mode waters decelerate the STCC. On decadal timescales, the dominant mode of sea surface height in the central subtropical gyre represents STCC variability. This STCC mode decays as CO_2 concentrations double in the twenty-first century, owing both to weakened mode waters in the mean state and to reduced variability in mode waters. The reduced mode-water variability can be traced upstream to reduced variations in winter MLD front and subduction in the KOE region where mode water forms.

Keywords North Pacific subtropical countercurrent · Global warming · Decadal variability · Mode waters · Subduction rate · Mixed layer depth

L. Xu · S.-P. Xie (✉) · Q. Liu
Physical Oceanography Laboratory, and Ocean–Atmosphere Interaction and Climate Laboratory, Qingdao, China
e-mail: xie@hawaii.edu

S.-P. Xie
Department of Meteorology, International Pacific Research Center, University of Hawaii at Manoa, Honolulu, Hawaii

F. Kobashi
Faculty of Marine Technology, Tokyo University of Marine Science and Technology, Tokyo, Japan

F. Kobashi
Research Institute for Global Change, Japan Agency for Marine-Earth Science and Technology, Yokosuka, Japan

1 Introduction

The subtropical countercurrent (STCC) is a narrow eastward jet in the central North Pacific (20–30°N) where the Sverdrup theory predicts a broad westward flow (Uda and Hasunuma 1969; Yoshida and Kidokoro 1967). According to the hydrographic analysis of Kobashi (2006), the STCC originates in the western North Pacific around 20°N, flows eastward against the northeast trade winds and stretches northeastward to the north of Hawaii. It is accompanied by a subsurface temperature and density front called the subtropical front (STF), in thermal wind relation with the STCC (Uda and Hasunuma 1969; White et al. 1978; Kobashi 2006; Yamanaka et al. 2008). Furthermore, the STCC maintains a sea surface temperature (SST) front during winter and spring. During April and May when the SST front is still strong, the seasonal warming makes the region conductive to atmospheric convection, and surface wind stress curls turn weakly positive along the front on the background of negative curls that drive the subtropical gyre. On the weather timescale, positive wind curls are related to low-pressure systems of a subsynoptic scale in space, energized by surface baroclinicity and latent heat release along the STF front. The SST front also anchors a meridional maximum in column-integrated water vapor, indicating a deep structure of the atmosphere response (Kobashi et al. 2008).

Mode water appears as a minimum in the vertical gradient of temperature and density in the upper thermocline (i.e., a thermostad or pycnostad), and plays a key role in the formation and maintenance of the STCC and STF as illustrated by theoretical (Kubokawa 1997, 1999), model (Takeuchi 1984; Kubokawa and Inui 1999; Yamanaka et al. 2008), and observational (Aoki et al. 2002; Kobashi 2006) studies. Mode waters, including the subtropical mode water (STMW) [Suga et al. 1989] and the central mode water (CMW) (Nakamura 1996; Suga et al. 1997), are subducted from the northwestern subtropical gyre, at the cross point between the outcrop line and the mixed layer front that separates the shallow and deep mixed layers (Kubokawa 1999; Xie et al. 2000). Subducted at different locations along the mixed layer depth (MLD) front, mode waters of low potential vorticity (PV) on different isopycnals are advected southward by the subtropical gyre, and eventually stacked up vertically to form a thick low-PV pool. This thick low-PV pool pushes the isopycnals in the upper thermocline to rise, leading to an eastward countercurrent on the southern flank. Using an ocean general circulation model (GCM), Kubokawa and Inui (1999) first illustrated this mechanism for STCC formation. The hydrographic analysis of Kobashi (2006) shows that the STCC is indeed anchored by mode waters beneath to the north. Thus, mode waters are not only passive water masses but also exhibit an important dynamical effect on ocean circulation.

Using a 300-year control simulation from the Geophysical Fluid Dynamics Laboratory (GFDL) coupled model CM2.1, Xie et al. (2011) showed that on decadal timescales, the dominant mode of sea surface height (SSH) variability in the central subtropical gyre (170°E–130°W, 15–35°N) is characterized by the strengthening and weakening of the STCC as a result of variations in mode-water ventilation. Changes in mode water can be further traced upstream to variability in the MLD and subduction rate in the KOE region. This STCC mode induces significant SST anomalies via thermal advection. Clear atmospheric response is seen in wind stress curls, with patterns suggestive of positive coupled feedback. Moreover, this natural mode of STCC variability is excited by global warming, resulting in banded structures in sea surface warming that slant in a northeast direction.

Climate change in response to increasing green house gas (GHG) concentrations may influence mode-water properties, the STCC, and its variability. Lee (2009) investigated the response of STMW in the North Pacific to climate change, using a coupled climate model (GFDL CM2.1). Those results showed that under CO_2 doubling, surface waters in the formation region and the core layer of STMW become warmer and fresher. Luo et al. (2009) compared North Pacific mode waters between the present climate and a warmer, future climate based on a set of Intergovernmental Panel on Climate Change (IPCC) Forth Assessment Report (AR4) models. In the warmer climate scenario, mode waters are produced on lighter isopycnal surfaces and are significantly weakened in their formation. In an analysis of a CM2.1 future climate projection under the A1B scenario of atmosphere trace composition change, Xie et al. (2010) noted banded structures in SST warming over both the subtropical North Pacific and Atlantic. These bands slant in a northeast direction, which Xie et al. suggest is indicative of the changing mode waters. It remains unclear how the STCC and its variability respond to global warming, and what role mode waters play.

The present study extends the analysis of mode waters and STCC by Xie et al. (2011) by examining the climate projection forced by increasing GHG. We investigate how the STCC and its variability respond to global warming, and the related mechanisms. Specifically, we wish to address the following questions: How will the STCC change in a warmer climate? What role do mode waters play in changes in STCC and its decadal variability? What physical mechanisms are important for mode water and STCC changes? We show that as the climate warms, the mean STCC weakens, and the STCC decadal variability mode decays in amplitude. Both these changes are mainly due to changes in mode water and subduction processes in the KOE region.

The rest of the paper is organized as follows: Section 2 briefly describes the model and simulations; Section 3

mechanisms for the response; Section 4 examines STCC changes in decadal variability in response to global warming; Section 5 is a summary.

2 Model and simulations

This study uses the output from the National Oceanic and Atmospheric Administration (NOAA) GFDL climate model (CM2.1), a global coupled ocean–atmosphere–land–ice model. The GFDL CM2.1 uses the Flexible Modeling System (FMS) to couple the GFDL atmospheric model (AM2.1) with the Modular Ocean Model version 4 (MOM4). The AM2.1 uses a finite volume dynamical core (Lin 2004), with 2.5° × 2.0° horizontal resolution and 24 vertical layers, nine of which are located in the lowest 1.5 km to represent the planetary boundary layer. The MOM4 has a horizontal resolution of 1.0° × 1.0° and 50 vertical layers, 22 of which are in the upper 220 m. In the meridional direction the resolution decreases toward the equator to 1/3° between 30°S and 30°N. The model formulation and simulation are documented in GFDL Global Atmospheric Model Development Team (2004) and Delworth et al. (2006).

present study. It contains two simulations: a 140-year-long climate of the twentieth-century run (20C3M) forced by the historical GHG from 1861–2000, and a 300-year projection under the Special Report on Emissions Scenarios A1B (SRESA1B) with a CO_2 doubling from their present-day level in the year 2100 and then held fixed through year 2300. The initial conditions for the SRESA1B experiment are taken from 1 January 2001 of the 20C3M experiment. For this study, we combine the two experiments to form a 440-year-long dataset to explore the STCC response to global warming.

3 The STCC mean state response to global warming

Previous studies show that the STCC is anchored by low-PV waters to the north (Takeuchi 1984; Kubokawa 1997, 1999; Kobashi 2006; Yamanaka et al. 2008; Xie et al. 2011). This section considers the relationship between mode waters and STCC variability. We first examine changes in MLD and atmosphere forcing in the mode-water formation region, then investigate the relationship between STCC and mode waters, and finally study the STCC-induced SSH and SST response to global warming. The present-day climatology (hereafter 20C3M) is defined as the average from 1861 to 2000. The future, warmer climatology (hereafter SRESA1B) is defined as the average from 2101 to 2300 when the atmospheric CO_2 concentration has doubled and remains constant. The model captures the gross features of the present-day climatology, with notable biases in too strong PV minima of mode waters; see Xie et al. (2011) and Appendix for a more detailed comparison with observation.

3.1 MLD and its related atmosphere-forcing response to global warming

In the present study, the MLD is defined as the depth at which the water density is 0.03 kg m^{-3} denser than at the sea surface. We examined the vertical structure of the upper ocean in the KOE region from 30 randomly selected profiles, and found that this density criterion of 0.03 kg m^{-3} is adequate for MLD definition (not shown here).

For 20C3M (Fig. 1a), it reaches its seasonal maximum in March for the North Pacific, with a deep mixed layer of the order of 300 m in the northwestern subtropical gyre over the KOE. Separating this deep mixed layer region from the rest of the North Pacific is a narrow transition zone called the MLD front that is key to the formation of low-PV waters. The southern MLD front tilts northeastward and intersects the outcrop lines of potential densities between 25.4 and 26.6 σ_θ for 20C3M. The low-PV waters

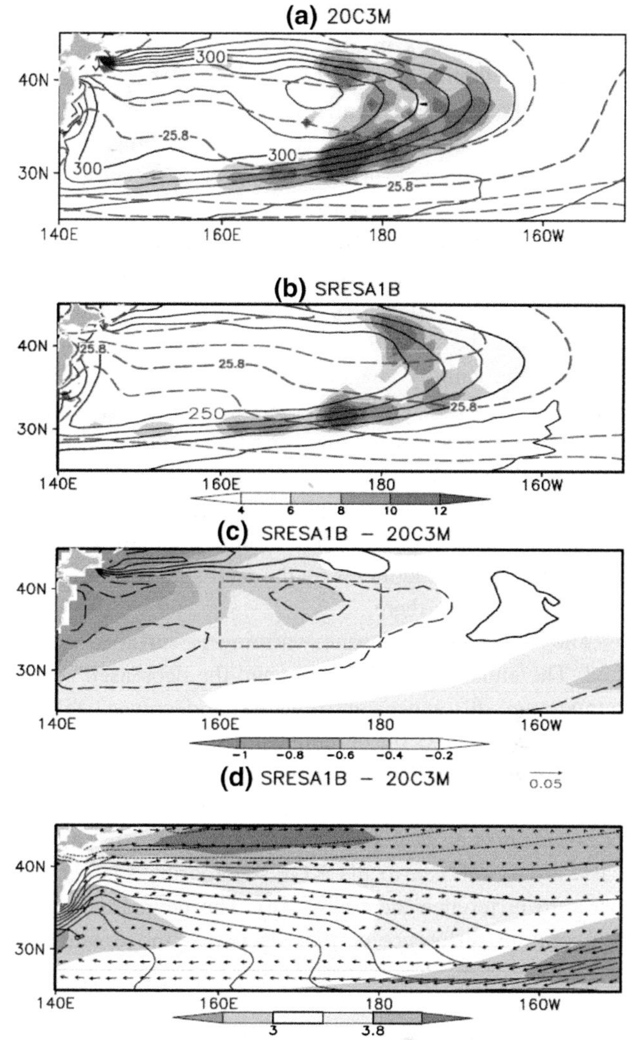

Fig. 1 March mean MLD (*black contours* at 50-m intervals), surface potential density (*blue dashes* at 0.4 kg m^{-3} intervals), and subduction rate (*shading* in 10^{-6} m s^{-1}; only larger than 4.0×10^{-6} m s^{-1} is shown here) in **a** 20C3M from 1861 to 2000, and **b** SRESA1B from 2101 to 2300 when the atmospheric CO_2 concentrations are doubled and held constant thereafter. **c** March mean MLD difference from SRESA1B to 20C3M (*black contours* at 50-m intervals, negative contours *dashed*), superimposed on SRESA1B – 20C3M sea–air temperature difference (*shading* in °C). The sea–air temperature difference is defined as the sea-minus-air surface temperature. **d** March mean sea surface height in 20C3M (*black contours*), the SRESA1B – 20C3M difference in 50-m current and SST (*shading*)

are subducted there at the cross point of the MLD front and the outcrop line (Kubokawa 1999; Xie et al. 2000). Compared to 20C3M, the overall spatial structure of the deep MLD region does not change very much in SRESA1B, where MLD shoals about 100 m. The maximum shoaling happens where the mean MLD is deepest, weakening the MLD front as a result. Surface density is 0.4 kg m^{-3} lighter than in 20C3M, and mode waters are formed at lighter isopycnals (25.4–26.6 σ_θ for 20C3M, and 25.0–26.2 σ_θ for SRESA1B).

Figure 2a shows the sea water temperature and salinity averaged in the box in Fig. 1c; Fig. 2b shows the corresponding potential density. The temperature increases and salinity decreases in a warmer climate, both contributing to a decrease in sea water density. The temperature warming is greater near the surface and decreases with depth, whereas the salinity decrease is almost uniform from surface to ~600 m. Thus, the temperature warming is the major cause of the increased stratification and the decreased MLD (see also, Fig. 2c, d). The temperature change may be determined by local warming processes whereas the salinity change is due to advection, because the local fresh water flux anomalies into the ocean are negative. On the basis of the same GFDL model, Stouffer et al. (2006) and Lee (2009) suggested that this freshening may be induced by increased precipitation in the western equatorial Pacific. The negative salinity anomalies are then advected to higher latitudes by the Kuroshio, leading to the decreased salinity there.

The GHG-induced ocean warming is surfaced-intensified. The enhanced stratification and the decreased sea–air temperature difference contribute to the decreased MLD in the KOE region. As the climate warms, surface warming is much stronger over land because of less-efficient evaporative cooling and a smaller heat capacity than over the ocean (Sutton et al. 2007). The land warming is advected eastward from Asia to the KOE, decreasing sea–air temperature difference there (Fig. 1c). The decreased sea–air temperature difference further decreases the upward net surface heat flux, and gives rise to an ocean warming that is greater near the surface and decreases with depth (Fig. 2a). The MLD shoals as a result. This is supported by changes in March MLD, net heat flux, sensible heat flux, and zonal wind stress (Fig. 3) averaged in the box in Fig. 1c where the maximum MLD change is located. Despite an increase in local wind speed, the net ocean-to-atmosphere heat loss is reduced by up to 40 W m^{-2} in a warmer climate, to which the decreasing sensible heat flux contributes the most (Fig. 3b, c). This illustrates the crucial role of the decreased sea–air temperature difference in the ocean warming and the resultant shoaling of MLD.

The ocean circulation changes, especially those in the Kuroshio and its extension (KE) (Sakamoto et al. 2005; Sato et al. 2006), may also affect SST and MLD distributions. In this model, the KE shifts northward in response to global warming with eastward current anomalies in ~41–45°N, where SST warms and the mixed layer deepens (Fig. 1c, d). Over the climatologically deep MLD region south of 41°N, however, the mixed layer shallows, which seems to be due to increased stratification and reduced surface heat loss.

3.2 Subduction rate and mode waters response to global warming

As the MLD front weakens in a warm climate, the subduction rate is likely to decrease (Fig. 1). Following Williams (1991), the subduction rate may be expressed as

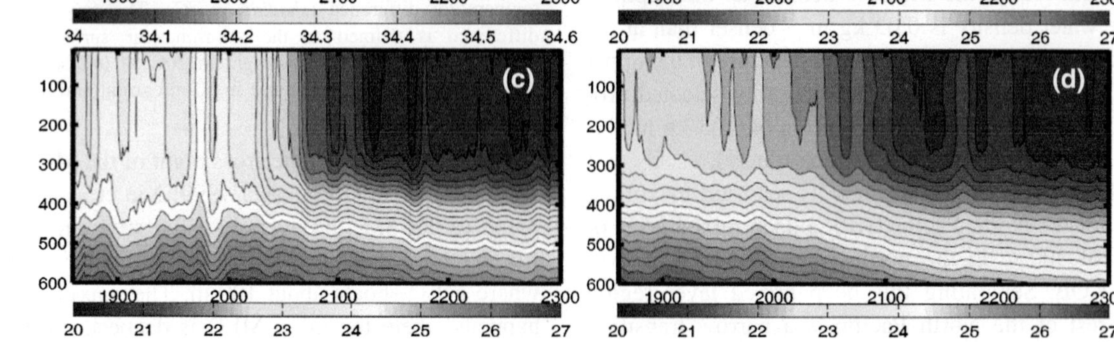

Fig. 2 Area-averaged (*red box* in Fig. 1c: 160–180°E, 33–41°N) **a** March sea water temperature (*contours* in 0.5°C), and salinity (*color shading* in psu) as a function of depth and time; **b** March potential density (kg m^{-3}); **c** the same as **b** except the sea water salinity is fixed at its March climatology; **d** the same as **b** except the sea water temperature is fixed at its March climatology. The statistical period for climatological mean is from 1861 to 2300

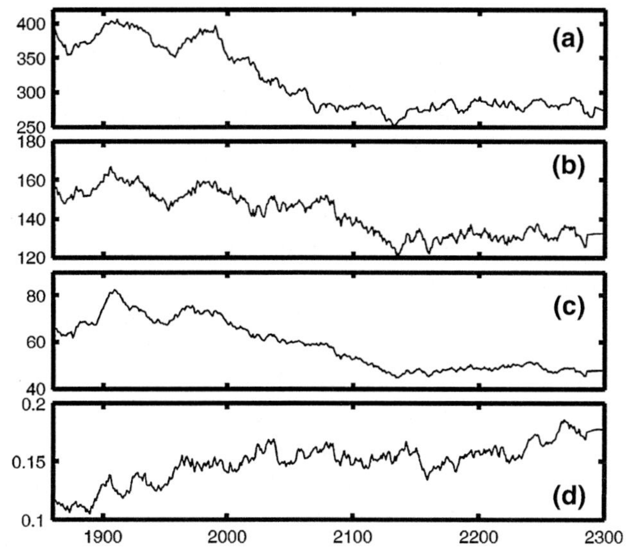

Fig. 3 The area-averaged (*red box* in Fig. 1c: 160–180°E, 33–41°N) **a** MLD (m), **b** net surface heat flux (W m^{-2}), **c** sensible heat flux (W m^{-2}), and **d** zonal wind stress (N m^{-2})

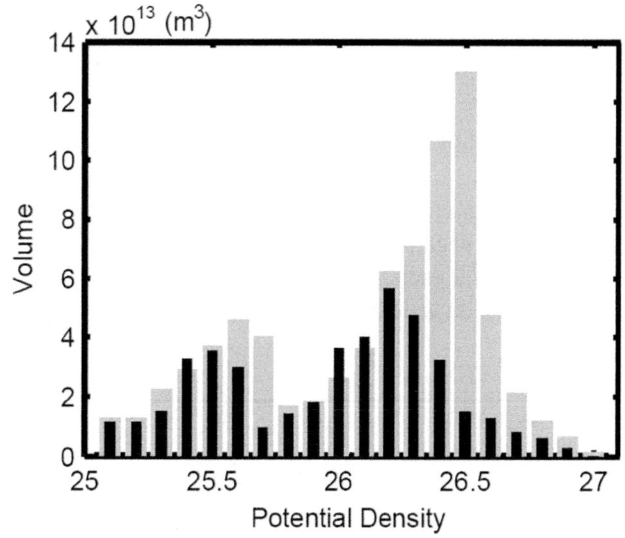

Fig. 4 Total volume (m^3) of the minimum PV layer over the North Pacific (140°E–140°W, 20–40°N) for each density class in 20C3M (*gray bars*) and SRESA1B (*black bars*)

$S_{mean} = \vec{v}_m \cdot \nabla h_m + \omega_e$, where ω_e is the Ekman pumping velocity, $\vec{v}_m \cdot \nabla h_m$ represents the lateral induction that occurs as water flows from a region of deep mixed layer to a region of shallow mixed layer, h_m is MLD, and $\vec{v}_m$ is the current velocity in the mixed layer. We calculated both the annual subduction rate (Qiu and Hwang 1995) and the March subduction rate, and found that the March subduction rate contributes most to the annual subduction rate in this model. In this study we use the March subduction rate for comparison. As the climate warms, the subduction rate decreases along the MLD front especially in 180°E–160°W, 33–38°N. The corresponding subduction rate averaged in the box in 20C3M is 7.1×10^{-6} m s^{-1}, and in SRESA1B decreases to 4.5×10^{-6} m s^{-1}. Further calculations show that the weakened MLD front is the major cause of reduced subduction, whereas the effects of the Ekman pumping velocity and the anomalous current on lateral induction are negligible (Fig. 7c).

As a result of the reduced subduction rate, less mode waters are formed (Fig. 4). Figure 4 compares density distributions of mode water cores between 20C3M and SRESA1B. The mode water is characterized by a vertical minimum in PV. PV is defined as $\frac{f}{\rho}\frac{\Delta\rho}{\Delta z}$, where f is the Coriolis parameter, ρ is sea water density, and $\frac{\Delta\rho}{\Delta z}$ is the vertical gradient of potential density. A low-PV core is defined as a vertical PV minimum ($<1.5 \times 10^{-10}$ m^{-1} s^{-1}) in the density range of 25.0–27.0 σ_θ. We chose this density range because it covers the surface density in the STMW formation area during winter and the base of CMW, which is closely related to the STCC (not shown here). We calculated the actual volume of such low-PV cores as a function of density. A PV minimum tends to occur in two distinct density ranges (i.e., 25.4–25.7 σ_θ and 26.2–26.6 σ_θ in 20C3M; 25.4–25.6 σ_θ and 26.0–26.4 σ_θ in SRESA1B) (Fig. 4). The peak at 25.4–25.7 σ_θ in 20C3M (25.4–25.6 σ_θ in SRESA1B) corresponds to the simulated STMW core, whereas the peak of 26.2–26.6 σ_θ in 20C3M (26.0–26.4 σ_θ at SRESA1B) corresponds to the simulated CMW core. Both peaks in low-PV occurrence are markedly diminished and each moves to a lighter density, similar to the results reported by Luo et al. (2009), i.e., the mode waters are produced on lighter isopycnal surfaces and are significantly weakened as the climate warms on the basis of a set of IPCC AR4 models. As a result of a larger decrease in subduction rate at the CMW formation region, the changes in CMW are much more prominent than those in the STMW in this GFDL model, which we will show lead to large changes in STCC.

Figure 5 plots the March low-PV ($<1.5 \times 10^{-10}$ m^{-1} s^{-1}) distributions on 26.5 $_\theta$ for 20C3M and 26.2 σ_θ for SRESA1B. These isopycnals represent the cores of CMW in the respective periods (Fig. 4). In the northwest subtropical gyre, the isopycnal PV minimum forms at the intersection between the MLD front and the outcrop line, and then flows southward by the gyre circulation. The low-PV tongue extends eastward and then turns southwestward. Compared to 20C3M, the low-PV tongue is much weaker in SRESA1B, with the area of PV less than 1.5×10^{-10} m^{-1} s^{-1} contracted. The subduction point for the CMW core also shifts from the southeast flank of the MLD front to the eastern edge, consistent with the similar shift in maximum subduction rate (Fig. 1a, b). We compared mode waters formed at a similar subduction point (e.g., 26.5 σ_θ in

20C3M and 26.0 σ_θ in A1B), and found a marked weakening in PV minimum. There are some changes in the path of low-PV water, consistent with circulation changes.

3.3 Relationship between mode waters and STCC

As mode waters form and circulate in the subtropical gyre, they produce negative PV gradients that support a density front above that anchors STCC. This subsection relates the weakened mode waters to STCC under global warming.

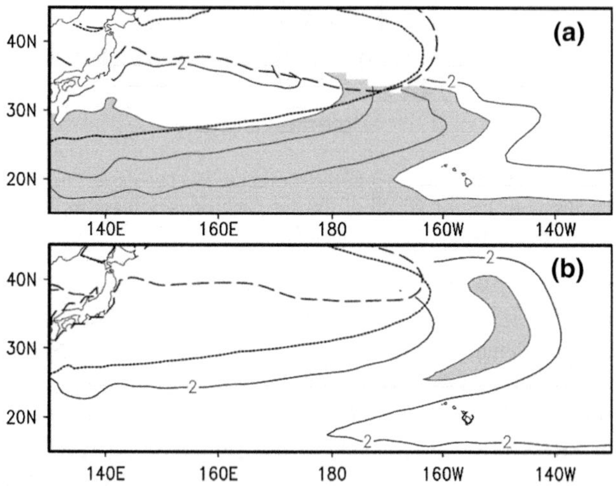

Fig. 5 March PV distribution (only PV < 2.0×10^{-10} m^{-1} s^{-1} is plotted in *solid line* at 0.5×10^{-10} m^{-1} s^{-} intervals) on **a** 26.5-σ_θ isopycnal surface in 20C3M and **b** 26.2-σ_θ isopycnal surface in SRESA1B in addition, PV less than 1.5×10^{-10} m^{-1} s^{-1} is *shaded.* The 100-m MLD contour is plotted as a *dotted line* to mark the MLD front, and the outcrop line is a *dashed line*

Figure 6 shows a latitude–depth section of zonal velocity and potential density along the international date line. For 20C3M, the STCC is a surface eastward current confined to the upper 250 m, in thermal wind balance with the northward shoaling of the upper pycnocline ($\sigma_\theta < 26.25$ at this longitude). In 20–25°N where the STCC is found, the lower pycnocline slopes downward as predicted by the Sverdrup theory. The 26.5-σ_θ isopycnal begins to shoal northward as a thick layer of low-PV water is found underneath. The 26.25-σ_θ isopycnal shows an ever steeper northward shoaling, pushed by low-PV water in the 26.25–26.5-σ_θ layer (26°N). Thus, the STCC is closely associated with mode waters underneath, which is in turn determined by upstream conditions in the ventilated thermocline. This is consistent with Kobashi (2006) who relates the eastward shear to negative PV gradients in the underneath mode-water layer. For SRESA1B, this overall vertical structure of STCC and its relationship to mode waters do not change, indicating that the STCC is still tied to mode waters in SRESA1B. The upper ocean has a lighter density field (~ 1.0 kg m^{-3} lighter), and is more stratified over the entire depth, similar to Luo et al.'s (2009) result based on a set of IPCC AR4 models. As mode waters move to lighter isopycnal surfaces, the upper pycnocline that starts to shoal northward changes to 25.5 σ_θ from 26.5 σ_θ in 20C3M. In addition, the northward shoaling is much gentler thanks to the weakened mode waters underneath. Consequently, compared to 20C3M, the mean zonal velocity of STCC reduces from 0.1 m s^{-1} in 20C3M to 0.04 m s^{-1} in SRESA1B and the corresponding STCC depth also shoals from 250 m to 150 m, indicating that the STCC decelerates as the climate warms in this coupled model.

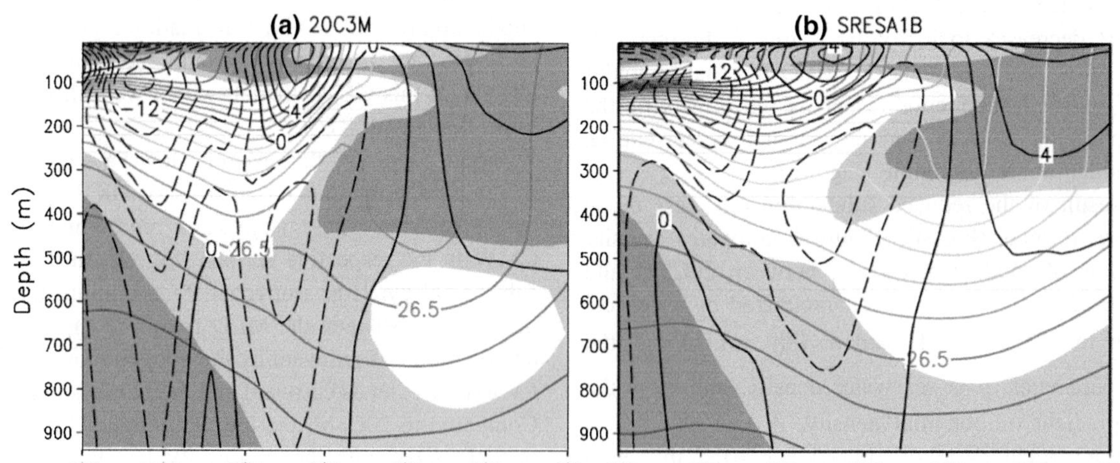

Fig. 6 March mean potential density (*color contours* at 0.25-σ_θ intervals) and zonal velocity (*black contours* at 0.02 m s^{-1}) at 180°E as a function of latitude and depth, along with PV < 1.5×10^{-10} m^{-1} s^{-1} in *gray shading* (PV < 1.0×10^{-10} m^{-1} s^{-1} in *dark gray shading*) in **a** 20C3M and **b** SRESA1B. The lower thermocline water shaded in the sections is not mode water but just low-PV waters below the thermocline

We further investigated the evolution of STCC and mode waters. Figure 7a shows time series of 50-m zonal current velocity, averaged in 170°E–140°W, 20–30°N with the zonal velocity larger than 0 m s^{-1}. The mean zonal velocity of STCC is ~0.1 m s^{-1}, with a decadal variability at a typical timescale of 50 years during the twentieth century. The STCC weakens sharply as the atmospheric CO_2 concentrations increase from 2001 to 2100 (Fig. 7a). After 2100 when the CO_2 concentrations stabilize in SRESA1B, the STCC speed also becomes stable at ~0.04 m s^{-1} without a distinct linear trend. In addition, we calculated the vertical PV minimum averaged in 170°E–140°W, 25–35°N in the density range of 25.4–26.6 σ_θ with the condition of PV less than 1.5×10^{-10} m^{-1} s^{-1}. The minimum PV is about 2.0×10^{-11} m^{-1} s^{-1} in the twentieth century but it increases to 4.0×10^{-11} m^{-1} s^{-1} as the climate warms, illustrating the weakened mode waters (Fig. 7b). The correlation between minimum PV (Fig. 7b) and the STCC speed (Fig. 7a) is approximately −0.9, suggestive of a close relationship between STCC variability and mode waters. This high correlation mainly emanates from the rising minimum PV trend and the decreasing STCC strength trend as the climate warms. Furthermore, the lateral induction and STCC speed are also closely related, with a correlation of ~0.72 when lateral induction leads STCC by 3 years.

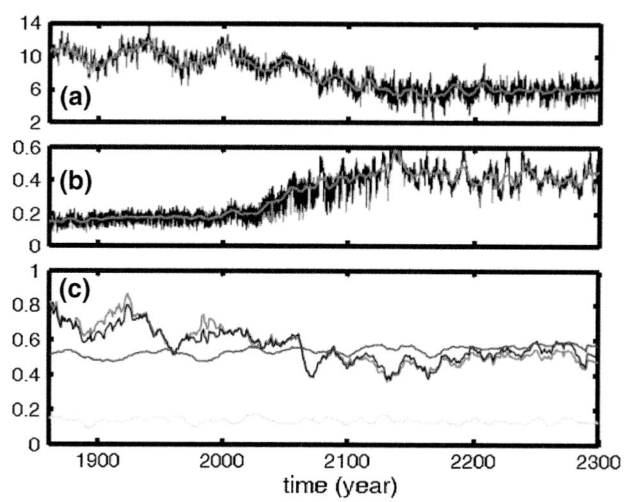

Fig. 7 Time series of **a** 50-m zonal velocity (cm s^{-1}) averaged in 170°E–140°W, 20–30°N, where the high SSH variance band of STCC is located; **b** vertical minimum PV (in 10^{-10} m^{-1} s^{-1}) averaged between 25.4 and 26.6 σ_θ in 170°E–140°W, 25–35°N; the *red lines* in **a** and **b** are the 5-year running mean for the original time series (*black lines*); and **c** area-averaged Ekman pumping velocity (*green line* in 10^{-5} m s^{-1}), and lateral induction $u_m \cdot \nabla h_m$ (*red line* in 10^{-5} m s^{-1}) averaged in 175°E–175°W, 30–40°N with the subduction rate larger than 0.4×10^{-5} m s^{-1}. The *black curve* in **c** represents lateral induction with current velocity in the mixed layer fixed at its climatological mean (in 10^{-5} m s^{-1}), and the *blue curve* with the MLD fixed at its climatological mean. The statistical period for climatological mean is from 1861 to 2300

3.4 STCC-induced SSH and SST response to global warming

The STCC is located in the central North Pacific (20–30°N) where the isohypses of the SSH veer northeastward. The horizontal gradient of SSH there indicates the strength of STCC. From 20C3M to SRESA1B, the overall horizontal structure of mean SSH does not change, but the STCC ridge in SSH decays significantly, indicative of a weakened STCC (Fig. 8). A band of negative SSH anomalies is collocated with the mean SSH ridge. The SSH anomalies represent a weakening of STCC (Xie et al. 2010). In the SST field, a northeast slanted dipole of surface warming is found in the central subtropical gyre. The SST warming dipole, however, is displaced west of the SSH anomalies band, suggestive of the importance of meridional thermal advection. According to Xie et al. (2011), the northeast-slanted bands of temperature and circulation in the central subtropical gyre are characteristic of a natural mode of STCC variability associated with changes in mode-water ventilation. It seems that this natural mode of STCC variability is excited by global warming, resulting in banded structures in sea surface warming.

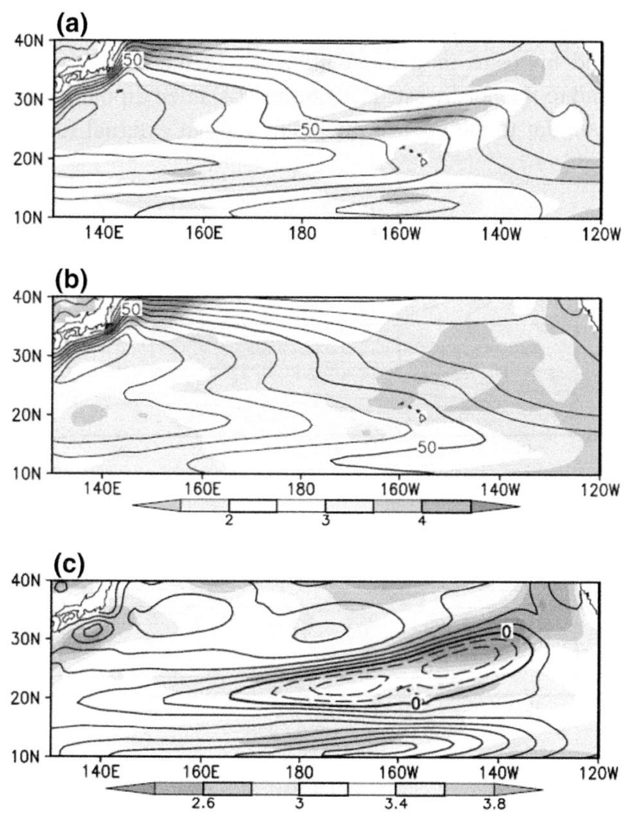

Fig. 8 Standard deviation of SSH (*shading* in cm) superimposed on the annual mean SSH (*contours* at 10-cm intervals) from **a** 20C3M, **b** SRESA1B, and **c** SRESA1B–20C3M SSH change (*contours* in 3 cm), superimposed on the SST change (*shading* in °C)

4 Response of STCC variability to global warming

STCC variability is the dominant mode of SSH in the central subtropical gyre on decadal timescales in CM2.1 (Xie et al. 2011). This section investigates how this STCC mode evolves as the climate warms, and explores the mechanisms. In order to focus on decadal variability, a band-pass filter (10–100 years) is applied.

4.1 STCC mode response to global warming

In the interior subtropical gyre away from KOE, a high-variance band of SSH is found in 20C3M (Fig. 8a), roughly coinciding with the STCC ridge in mean SSH and stretching from the western Pacific through the north of Hawaii. This high variance band of STCC almost disappears as climate warms (Fig. 8b). Following Xie et al. (2011), we perform an empirical orthogonal function (EOF) analysis for the band-pass filtered SSH from 1861 to 2300 in the central subtropical gyre domain (165°E–145°W, 20–35°N) where the STCC high-variance band is located. Figure 9 shows the spatial pattern and principal component (PC) of the leading mode—the STCC mode hereafter—which explains 43.7% of the total variance in the central gyre domain. The EOF mode features an east-westward dipole, with the SSH anomalies in the STCC band opposite in phase with those in the high-variance band to the northwest. This northeast-slanted dipole pattern is similar to the control run result without external forcing

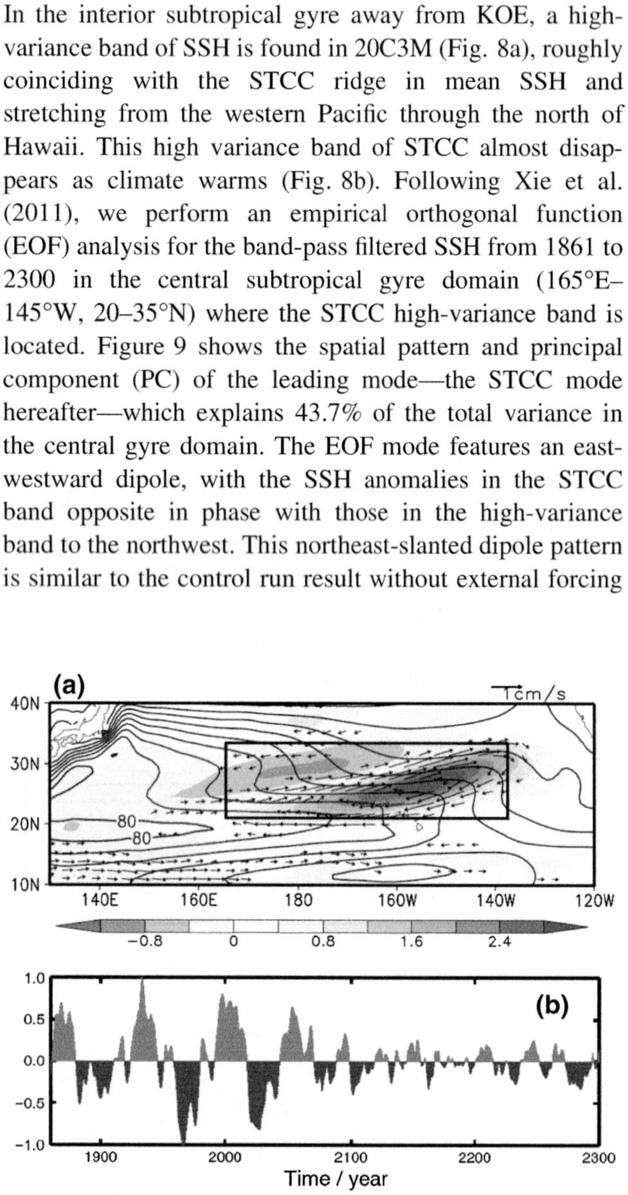

Fig. 9 **a** First EOF mode for SSH (*color shading* in cm) in the central subtropical gyre (*box*), and **b** the time series. The mean SSH in 20C3M (*black contours* at 10-cm intervals) and regression of 50-m current velocity upon the PC are superimposed in **a**

(Xie et al. 2011), representing a combination of the STCC strength and displacement.

The PC time series shows that the STCC mode decays substantially as climate warms from the late twenty-first century onward. By the end of 2100 when the CO_2 concentrations double their present-day level, the amplitude of the STCC mode is reduced about three times (Fig. 9b). The typical timescale of the STCC mode in the twentieth century is $\sim$50 years. The STCC decadal variability in the twentieth century seems to be associated with the model's Pacific Decadal Oscillation (PDO), with the latter affecting MLD in the KOE region and thus the subduction rate (Qu and Chen 2009). The correlation coefficient between the PDO index and the KOE subduction rate is $\sim$0.65 in this model. The timescale of the STCC mode then decreases quickly with the increase in CO_2 concentrations to $\sim$25 years by the end of 2100 (Fig. 9b). We performed separate EOF analysis for SSH for 2100–2300 in the same domain, but the STCC mode no longer appears as the leading mode (not shown here). The timescale of PDO also decreases under global warming, which may explain the apparent decrease in the timescale of the STCC mode.

4.2 Mechanisms

This subsection investigates why the STCC mode decays as the climate warms. Figure 10b shows the band-pass filtered time series of the area-averaged mode-water thickness in 25.4 σ_θ–26.6 σ_θ with the vertical PV minimum less than 1.5×10^{-10} m^{-1} s^{-1} in 170°E–140°W, 25–35°N north of the STCC, and Fig. 10c the lateral induction averaged in the subduction region to the northwest (175°E–175°W, 30–40°N) with the total subduction rate greater than 0.4×10^{-5} m s^{-1}, the same region as in Fig. 7c except for the band-pass filter here. As the climate warms, variability of the mode-water thickness and lateral induction weakens greatly, and the typical timescale decreases. The close correlation among the three time series in Fig. 10 indicates that decadal variability of STCC is closely related to changes in mode water and lateral induction, which is further caused by MLD variability as discussed in the last section. In the present-day climate, the correlation between STCC mode and mode water thickness is $\sim$0.82, and the maximum correlation between lateral induction and STCC mode is $\sim$0.79 when the former leads by 3 years. This close connection is weakened as the climate warms, to $\sim$0.40 between SSH-PC1 and mode water thickness, and $\sim$0.25 between SSH-PC1 and lateral induction.

Following Kobashi (2006), the shoaling of an upper pycnocline may be expressed as a function of PV (q) distribution,

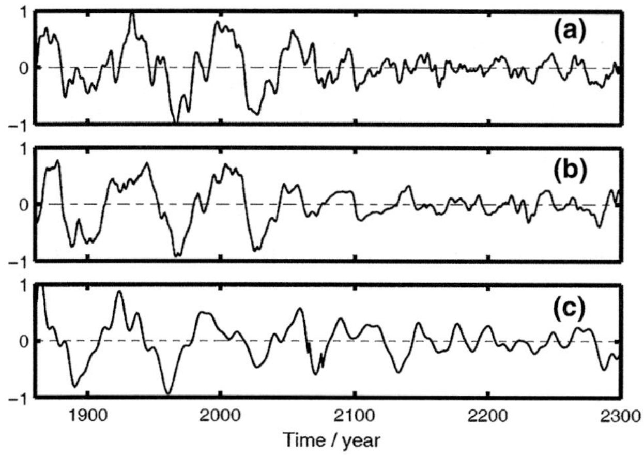

Fig. 10 **a** Time series of the first principle component of SSH-EOF; **b** mode water thickness (with the vertical PV minimum less than 1.5×10^{-10} m^{-1} s^{-1} in the 25.4–26.6-σ_θ range) north of STCC averaged in 170°E–140°W, 25–35°N; and **c** lateral induction averaged in 175°E–175°W, 30–40°N where with values larger than 0.4×10^{-5} m s^{-} . All time series are normalized and band-pass filtered

$$\left(\frac{\partial z(\rho)}{\partial y}\right)_\rho = -\frac{1}{\rho_0}\int_{\rho_b}^{\rho} \frac{1}{q(\rho)}\left(\beta - \frac{f}{q(\rho)}\left(\frac{\partial q(\rho)}{\partial y}\right)_\rho\right)\mathrm{d}\rho + \left(\frac{\partial z_0(\rho_b)}{\partial y}\right)_\rho, \qquad (1)$$

where z_0 is the depth of the reference isopycnal surface ρ_b ($\geq \rho$) and the subscript ρ denotes the partial derivative taken at a constant ρ surface. The left-hand side of Eq. (1) is the meridional slope of the isopycnal surface that is related to the STCC strength, and the first term on the right side is the vertical integration of the deviation of the meridional PV gradient from the ambient vorticity gradient β. Thus, the meridional slope of an isopycnal surface and the STCC strength are related to the PV (q) gradient below.

This PV gradient effect is further inversely proportional to PV value itself in Eq. (1). It is conceivable that changes in both the mean and variability of PV lead to changes in STCC variability. We qualify the effect of changes in the mean and variability of PV on STCC by examining the following four combinations for calculating the meridional slope of the 24.5-σ_θ isopycnal: (1) $q = \bar{q}_{20C} \pm q'_{20C}$; (2) $q = \bar{q}_{20C} \pm q'_{A1B}$; (3) $q = \bar{q}_{A1B} \pm q'_{20C}$; (4) $q = \bar{q}_{A1B} \pm q'_{A1B}$, where $\bar{q}$ is the mean PV, q' is the PV standard deviation, and the subscript 20C (A1B) denotes output from the 20C3M (SRESA1B) run. For example, the variability of $\left(\frac{\partial z(\rho)}{\partial y}\right)_\rho$ in case 2 is derived from the difference based on Eq. (1) using $q = \bar{q}_{20C} - q'_{A1B}$ and $q = \bar{q}_{20C} + q'_{A1B}$, namely by holding mean PV to the 20C3M values but using anomalies from the SRESA1B projection. The comparison of case 2 to cases 4 and 1 gives an estimate of the effects of the mean state and variability, respectively.

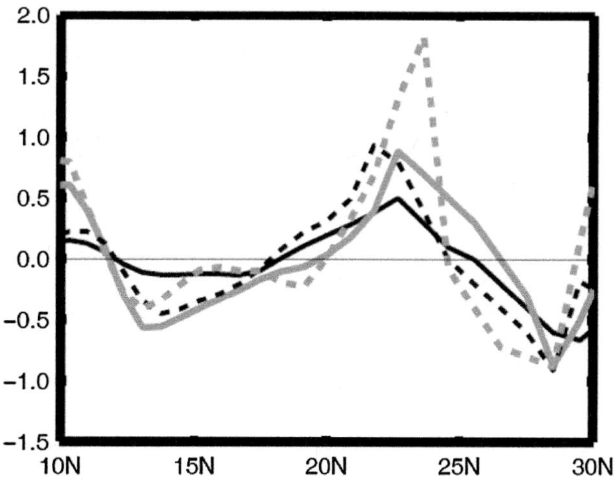

Fig. 11 Diagnosed anomalies of STCC (as represented by the 24.5-σ_θ isopycnal surface's meridional slope derived from Eq. 1 in four cases: (1) when the mean PV is from 20C3M and the PV variance is from 20C3M (*dash gray line*); (2) when the mean PV is from 20C3M and the PV variance is from SRESA1B (*dash black line*); (3) when the mean PV is from SRESA1B and the PV variance is from 20C3M (*solid gray line*); (4) when the mean PV is from SRESA1B and the PV variance is from SRESA1B (*solid black line*)

Figure 11 compares the meridional slope anomaly $\left(\frac{\partial z(\rho)}{\partial y}\right)'_\rho$ at 24.5 σ_θ along the international date line among four cases. We chose 24.5 σ_θ because it generally represents the maximum meridional slope of the upper pycnocline in both 20C3M and SRESA1B runs (Fig. 6). For a given PV variability, the STCC anomaly is larger for the 20C3M mean state with a strong mode-water ventilation than for the SRESA1B mean state. On the other hand, for a given mean state, the STCC anomaly is bigger with 20C3M mode-water variability (q'_{20C}). The comparison in Fig. 11 indicates that the $\bar{q}$ change (dashed black curve) is equally important as the decrease in q' (gray solid curve) for STCC variability. In global warming, the PV minimum in the mean weakens because of the shoaling MLD in KOE whereas PV variability decreases as a result of the decreased variance of MLD.

5 Summary

We have examined the response of the mean STCC and its variability to global warming in the GFDL coupled model CM2.1, which is forced by increased greenhouse gas concentrations. Because land warming is stronger than over the ocean as the climate warms, the offshore advection from Asia decreases the sea–air temperature difference over KOE. As a result, the ocean-to-atmosphere heat loss is reduced by 30 W m^{-2} in the KOE region. Such a change in the heat flux contributes to an ocean warming that is large near the surface and decreases with depth, causing the MLD to shoal. The maximum reduction in MLD appears in the KOE region where the mean MLD is the largest, weakening the MLD front and decreasing the subduction rate (mainly determined by lateral induction). Both the warming and the freshening at the sea surface contribute to decreasing surface density, causing mode waters to form on lighter isopycnals. As a result of both the decreased subduction rate and the decreased sea surface density, less mode waters are formed on lighter isopycnals. The southward advection of weakened mode waters causes the upper pycnocline to shoal less than in the present-day climatology and decelerate the STCC. The STCC mode of decadal variability decays in amplitude as the climate warms. In fact, it ceases to be the dominate mode of SSH in the central subtropical gyre for the twenty-second to twenty-third century. The decay of the STCC mode is due both to the weakened mode water in the mean state and to reduced variability in MLD and subduction over the KOE region.

Compared with observations by Kobashi (2006) the eastern STCC is too strong, a bias likely due to too strong mode waters. The same bias also appears in the control run of GFDL CM2.1, as documented by Xie et al. (2011). Non-eddy-resolving models simulate too strong mode waters and may exaggerate mode-water dynamics. There is evidence for the mode-water effect on STCC variability in observations (Kobashi 2006) and eddy-resolving models (Yamanaka et al. 2008; Nonaka et al. 2011; Sasaki et al. 2011; this issue). It is important to bridge the gap between coarse-resolving and eddy-resolving models and with observations, and quantify the role of mode-water ventilation in climate variability.

Acknowledgments We thank the anonymous reviewers for constructive and helpful comments. This work is supported by the Qianren project, Changjiang Scholar Program, Natural Science Foundation of China (40830106, 40921004), National Key Program for Developing Basic Science of China 2007CB411803 and 2010CB428904, the US National Science Foundation, and the Japan Agency for Marine Earth Science and Technology.

Appendix: Comparison with observations

Our recent study (Sect. 3a of Xie et al. 2011) compared the control run of GFDL CM2.1, the same model used here, with observations; here we just provide a brief comparison of key variables from 20C3M against previous observational studies.

Using hydrographic data, Suga et al. (2004) showed that the winter MLD exhibits two distinct maximum, one to the south of the KE and one to the north, which are the formation regions for the STMW and the CMW, respectively. Compared to observations, Fig. 1a exhibits a broad, zonally elongated region of deep mixed layer, resulting in too extensive formation of mode waters (Suga et al. 2004; Thompson and Cheng 2008). The zonal elongated region of deep MLD may have caused spatial biases in the maximum subduction rate in Fig. 1a displaced too east compared to observations (e.g., Qiu and Hwang 1995; Qu and Chen 2009). According to Suga et al. (2004), small cross-isopycnal flow in the mixed layer plays a key role in the CMW formation, whereas it is the lateral induction in this model.

In observations, the STMW (CMW) is found in a density range of 25.3–25.7 σ_θ (26.1–26.5 σ_θ) [Suga et al. 1989, 1997, 2004; Nakamura 1996; Hanawa and Talley 2001]. For 20C3M, the STMW (CMW) is clustered in 25.4–25.7 σ_θ (26.2–26.6 σ_θ). The model mode waters are slightly denser by 0.1 σ_θ, but the density range of the two mode waters is broadly consistent with observations. However, the typical minimum PV of mode waters is $\sim 1.5 \times 10^{-10}$ m^{-1} s^{-1} in the observed climatology, whereas it is $\sim 0.5 \times 10^{-10}$ m^{-1} s^{-1} in the model (Fig. 7b). Moreover, the low-PV tongue on the isopycnal is much more diffused in observations than in the model, with a broader spatial structure and larger PV values (Fig. 5b of Xie et al. 2011).

In the observed meridional transect (Fig. 3b of Xie et al. 2011), the maximum zonal velocity of STCC is ~ 3 m s^{-1} at 26.5°N, anchored by PV minima on 25.5–26.5 σ_θ to the north. The observed zonal velocity of STCC is much weaker than in the model, and these PV minima north of STCC are not as pronounced as in the model. The eastern STCC that we mainly focused on in this study seems to be too strong in 20C3M compared to observations, a bias likely due to too strong mode waters.

References

Aoki Y, Suga T, Hanawa K (2002) Subsurface subtropical fronts of the North Pacific as inherent boundaries in the ventilated thermocline. J Phys Oceanogr 32:2299–2311

Delworth TL et al (2006) GFDL's CM2 global coupled climate models. Part I: formulation and simulation characteristics. J Climate 19(5). doi:10.1175/JCLI3629.1

GFDL Global Atmospheric Model Development Team (2004) The new GFDL global atmosphere and land model AM2-LM2: evaluation with prescribed SST simulations. J Climate 17:4641–4673

Hanawa K, Talley LD (2001) Mode waters. In: Siedler G, Church J, Gould J (eds) Ocean circulation and climate. International Geophysics. Elsevier, New York, pp 373–386

Kobashi F, Mitsudera H, Xie S-P (2006) Three subtropical fronts in the North Pacific: observational evidence for mode water-induced subsurface frontogensis. J Geophys Res Oceans 111:C09033. doi:10.1029/2006JC003479

Kobashi F, Xie S-P, Iwasaka N, Sakamoto T (2008) Deep atmospheric response to the North Pacific oceanic subtropical front in spring. J Climate 21:5960–5975

Kubokawa A (1997) A two-level model of subtropical gyre and subtropical countercurrent. J Oceanogr 53:231–244
Kubokawa A (1999) Ventilated thermocline strongly affected by a deep mixed layer: a theory for subtropical countercurrent. J Phys Oceanogr 29:1314–1333
Kubokawa A, Inui T (1999) Subtropical countercurrent in an idealized ocean GCM. J Phys Oceanogr 29:1303–1313
Lee H-C (2009) Impact of atmospheric CO_2 doubling on the North Pacific subtropical mode water. Geophys Res Lett 36:L06602. doi:10.1029/2008GL037075
Lin S-J (2004) A "vertically Lagrangian" finite-volume dynamical core for global models. Mon Weather Rev 132:2293–2307
Luo Y, Liu Q, Rothstein LM (2009) Simulated response of North Pacific mode waters to global warming. Geophys Res Lett 36:L23609. doi:10.1029/2009GL040906
Nakamura H (1996) A pycnostad on the bottom of the ventilated portion in the central subtropical North Pacific: its distribution and formation. J Oceanogr 52:171–188
Nonaka M, Xie S-P, Sasaki H (2011) Interannual variations in low potential vorticity water and the subtropical countercurrent in an eddy-resolving OGCM. J Oceanogr (in press)
Qiu B, Hwang X (1995) Ventilation of the North Atlantic and North Pacific: subduction versus obduction. J Phys Oceanogr 25:2374–2390
Qu T, Chen J (2009) A North Pacific decadal variability in subduction rate. Geophys Res Lett 36:L22602. doi:10.1029/2009GL040914
Sakamoto T, Hasumi H, Ishii M, Emori S, Suzuki T, Nishimura T, Sumi A (2005) Responses of the Kuroshio and the Kuroshio extension to global warming in a high-resolution climate model. Geophys Res Lett 32:L14617. doi:10.1029/2005GL023384
Sasaki H, Xie S-P, Taguchi B, Nonaka M, Hosoda S, Masumoto Y (2011) Interannual variations of the Hawaiian Lee countercurrent induced by low potential vorticity water ventilation in the subsurface. J Oceanogr (submitted)
Sato Y, Yukimoto S, Tsujino H, Ishizaki H, Noda A (2006) Response of North Pacific ocean circulation in a Kuroshio-resolving ocean model to an Arctic oscillation (AO)-like change in Northern Hemisphere atmospheric circulation due to greenhouse-gas forcing. J Meteorol Soc Jpn 84:295309
Stouffer RJ et al (2006) GFDL's CM2 global coupled climate models. Part IV: idealized climate response. J Climate 19:723–740
Suga T, Hanawa K, Toba Y (1989) Subtropical mode water in the 137°E section. J Phys Oceanogr 19:1605–1618
Suga T, Takei Y, Hanawa K (1997) Thermostad distribution in the North Pacific subtropical gyre: the central mode water and the subtropical mode water. J Phys Oceanogr 27:140–152
Suga T, Motoki K, Aoki Y, McDonald AM (2004) The North Pacific climatology of winter mixed layer and mode waters. J Climate 34:3–22
Sutton RT, Dong B, Gregory JM (2007) Land/sea warming ratio in response to climate change: IPCC AR4 model results and comparison with observations. Geophys Res Lett 34:L02701. doi:10.2029/2006GL028164
Takeuchi K (1984) Numerical study of the subtropical front and the subtropical countercurrent. J Oceanogr Soc Jpn 40:371–381
Thompson L, Cheng W (2008) Water masses in the Pacific in CCSM3. J Climate 21:4514–4528
Uda M, Hasunuma K (1969) The eastward subtropical countercurrent in the western North Pacific Ocean. J Oceanogr Soc Jpn 25:201–210
White WB, Hasunuma K, Solomon H (1978) Large-scale seasonal, secular variability of the subtropical front in the western North Pacific from 1954 to 1974. J Geophys Res 83:4531–4544
Williams RG (1991) The role of the mixed layer in setting the potential vorticity of the main thermocline. J Phys Oceanogr 21:1803–1814
Xie S-P, Kunitani T, Kubokawa A, Nonaka M, Hosoda S (2000) Interdecadal thermocline variability in the North Pacific for 1958–1997: a GCM simulation. J Phys Oceanogr 30:2798–2813
Xie S-P, Deser C, Vecchi GA, Ma J, Teng H, Wittenberg AT (2010) Global warming pattern formation: Sea surface temperature and rainfall. J Climate 23:966–986
Xie S-P, Xu L, Liu Q, Kobashi F (2011) Dynamical role of mode water ventilation in decadal variability in the central subtropical gyre of the North Pacific. J Climate 24:1212–1225
Yamanaka G, Ishizaki H, Hirabara M, Ishikawa I (2008) Decadal variability of the subtropical front of the western North Pacific in an eddy-resolving ocean general circulation model. J Geophys Res 113:C12027. doi:10.1029/2008JC005002
Yoshida K, Kidokoro T (1967) A subtropical countercurrent in the North Pacific—an eastward flow near the subtropical convergence. J Oceanogr Soc Jpn 23:88–91

GPSR Compliance

The European Union's (EU) General Product Safety Regulation (GPSR) is a set of rules that requires consumer products to be safe and our obligations to ensure this.

If you have any concerns about our products, you can contact us on ProductSafety@springernature.com

In case Publisher is established outside the EU, the EU authorized representative is:

Springer Nature Customer Service Center GmbH
Europaplatz 3
69115 Heidelberg, Germany

Batch number: 10632828

Printed by Printforce, the Netherlands